T0122591

Lecture Notes in Electrical Engineering

Volume 897

The book series *Lecture Notes in Electrical Engineering* (LNEE) publishes the latest developments in Electrical Engineering - quickly, informally and in high quality. While original research reported in proceedings and monographs has traditionally formed the core of LNEE, we also encourage authors to submit books devoted to supporting student education and professional training in the various fields and applications areas of electrical engineering. The series cover classical and emerging topics concerning:

- Communication Engineering, Information Theory and Networks
- Electronics Engineering and Microelectronics
- Signal, Image and Speech Processing
- Wireless and Mobile Communication
- Circuits and Systems
- Energy Systems, Power Electronics and Electrical Machines
- Electro-optical Engineering
- Instrumentation Engineering
- Avionics Engineering
- Control Systems
- Internet-of-Things and Cybersecurity
- Biomedical Devices, MEMS and NEMS

For general information about this book series, comments or suggestions, please contact leontina.dicecco@springer.com.

To submit a proposal or request further information, please contact the Publishing Editor in your country:

China

Jasmine Dou, Editor (jasmine.dou@springer.com)

India, Japan, Rest of Asia

Swati Meherishi, Editorial Director (Swati.Meherishi@springer.com)

Southeast Asia, Australia, New Zealand

Ramesh Nath Premnath, Editor (ramesh.premnath@springernature.com)

USA, Canada:

Michael Luby, Senior Editor (michael.luby@springer.com)

All other Countries:

Leontina Di Cecco, Senior Editor (leontina.dicecco@springer.com)

**** This series is indexed by EI Compendex and Scopus databases. ****

More information about this series at https://link.springer.com/bookseries/7818

Santanu Saha Ray · H. Jafari · T. Raja Sekhar · Suchandan Kayal
Editors

Applied Analysis, Computation and Mathematical Modelling in Engineering

Select Proceedings of AACMME 2021

Editors
Santanu Saha Ray
Department of Mathematics
National Institute of Technology Rourkela
Rourkela, Odisha, India

H. Jafari
Department of Mathematical Sciences
University of South Africa
Florida, Gauteng, South Africa

T. Raja Sekhar
Department of Mathematics
Indian Institute of Technology Kharagpur
Kharagpur, West Bengal, India

Suchandan Kayal
Department of Mathematics
National Institute of Technology Rourkela
Rourkela, Odisha, India

ISSN 1876-1100 ISSN 1876-1119 (electronic)
Lecture Notes in Electrical Engineering
ISBN 978-981-19-1826-1 ISBN 978-981-19-1824-7 (eBook)
https://doi.org/10.1007/978-981-19-1824-7

This Springer imprint is published by the registered company Springer Nature Singapore Pte Ltd.
The registered company address is: 152 Beach Road, #21-01/04 Gateway East, Singapore 189721, Singapore

Preface

This edited book, "**Lecture Notes in Electrical Engineering**", is an outcome of the **International Conference on Applied Analysis, Computation and Mathematical Modelling in Engineering (AACMME-2021)**. The contents of this book are intended to present an overall idea about the recent advances in latest developments and researches in the field of Mathematical Science and its applications.

This book focuses on the comparative study of some wavelet based numerical methods to solve initial value problems. It also investigates the enhancement of natural convection heat transfer using hybrid nanofluids over a moving vehicle plate. This book addresses the provoked flow pattern due to an impulsive motion of porous wavy wall with no slip suction velocity under the influence of magnetic field.

In this book, a linear stability analysis is applied to study the onset of bio-convection in a suspension of negatively geotactic swimmers saturated with a non-Darcy porous fluid layer under the effect of high frequency and Small-amplitude vertical vibrations. This book also studies the impact of two temperatures on a generalized thermoelastic plate with thermal loading.

Baffle spacing has a decisive effect on heat transfer and pumping power. The development of baffle spacing significantly dominates the turbulence created inside the shell and tube heat exchanger and heat transfer. This book studies the impact of baffle spacing in both global and local thermohydraulic characteristics.

In this book, Kudryashov and modified Kudryashov techniques have been implemented to acquire new exact solutions of the time fractional (2+1)-dimensional CBS equation. This book also explores the impact of double dispersion effects on the nonlinear convective flow of power-law fluid along an inclined plate.

This book emphasizes the Soret and viscous dissipation effects on mixed convective flow of an incompressible micropolar fluid over a vertical frustum of a cone embedded in a non-Darcy porous medium subject to convective boundary condition. It proposes convergence and comparison theorems for three-step alternating iteration method for rectangular linear system. This book also studied thermal hydraulic performance of helical baffle shell and tube heat exchanger using RSM Method.

This book investigates a newly proposed dual-mode Kawahara equation. It finds out the soliton and periodic solutions of the Kawahara equation. In this book, the

Lie transformation method has been used to find out the group invariant solutions of (2+1)-dimensional modified Calogero-Bogoyavlenskii-Schiff (mCBS) equation.

This book addresses the estimation and classification of two logistic distributions with a common scale and different location parameters. Bayes estimates are computed using Metropolis-Hastings method using gamma and normal prior distributions. The Bayes estimates are compared with some of the existing estimates with respect to the bias and mean squared error. Utilizing these estimates some classification rules are proposed to classify a single observation into one of the two logistic populations under the same model.

The book considers the problem of testing of hypothesis for the quantile when independent random samples are drawn from two normal populations with a common mean and order restricted variances. Several test procedures are proposed and are evaluated through their sizes and powers using a simulation procedure.

In this book, various geometrical parameters of the planted roof are studied to optimize the dimensional parameters by means of independent and dependent variables using an exact mathematical model. Using experiment, the factors influencing the performance of the planted roof activity are identified to optimize the performance of the heat flow through planted roof.

This book deals with the modal analysis of a Jeffcott functionally graded (FG) rotor system, consisting of an FG shaft mounted on linear bearings at the ends. The material gradation is applied following the exponential gradation law, whereas the thermal gradients across the radius of the FG shaft are achieved through the exponential temperature distribution method 3D finite element modelling and the modal analysis of the FG rotor system are carried out using ANSYS software. The influence of the material gradation and temperature gradients on the rotor-bearing system's natural and whirl frequencies are studied.

This book presents five-point finite difference method to solve the two-dimensional Laplace and Poisson equations on regular and irregular regions. Dirichlet and Robin boundary conditions are considered for solving the system of equations in each iteration. The obtained numerical results are compared with analytical solutions.

This book also focuses on the selection of the best ultra-sound machine using ELECTRE method based on the user's criteria. This study considers six criteria to select best one from five alternatives.

This book examines the processes included for initiation along with expansion of a crack on the web of the rail weldment in order to anticipate the direction of fracture crack and secondary, the intervals of weld inspections. The finite element study for the expected cracking is performed to measure the brief history of stress intensity factors. Computational simulations and experimental findings made by RDSO on three-dimensional growth of fatigue crack are compared.

This book deals with a higher-order wave equation with delay term and variable exponents. Under suitable conditions, they prove the nonexistence of solutions in a finite time. There is no research related to higher-order wave equations with delay term and variable exponents.

In this book, the existence result of a solution to continuous nonlinear, initial value problem is studied. A special type of problem representing the time evolution of particle number density due to the coagulation, multi-fragmentation events among the particles present in a system has been considered. The proof of the main theorem is based on the contraction mapping principle. Initially the local existence of nonnegative solutions for these compactly supported kernels has been also proved in this book. The study is completed by examining the mass conservation law of the existing solution.

This book also introduces a new sequence of Szasz—Kantorovich type operators based on Boas - Buck type polynomials which include Brenke type polynomials, Sheffer polynomials and Appell polynomials. The error is estimated in the approximation by these operators in terms of the Lipschitz type maximal function, Peetre's K-functional and Ditzian–Totik modulus of smoothness. The order of convergence is also studied of these operators for unbounded functions by using the weighted modulus of continuity. This study also covers quantitative-Voronovoskaya-type theorem and Gruss Voronovskaya-type theorem.

A study on the numerical modeling and simulation of heat distribution inside the skin tissue for cancer treatment with external exponential heating is also presented in this book. The two-dimensional Pennes bio-heat model for thermal therapy based on Fourier's law of heat conduction is considered in this study. The mathematical model's numerical solution is obtained using Crank Nicolson finite difference approximation and radial basis function approximation for time and space. The effects of thermophysical properties of the skin on the temperature profile in the tissue are also explained.

Overall, the chapters create new avenues and present intriguing information to comprehend the difficulties and provide answers for various challenges, which would assist readers grasp and implement for the new development and mathematically analyse physical problems.

The editors would like to express their appreciation to Springer, the Springer Editor, for publishing these chapters in "Lecture Notes in Electrical Engineering." We are also grateful to the anonymous reviewers who provided worthwhile review reports that resulted in significant modifications and enhancements to these chapters.

Rourkela, India — Santanu Saha Ray
Pretoria, South Africa — H. Jafari
Kharagpur, India — T. Raja Sekhar
Rourkela, India — Suchandan Kayal

Contents

About the Editors

Santanu Saha Ray is a Professor and Head in the Department of Mathematics, National Institute of Technology Rourkela, India. He earned his Ph.D. from Jadavpur University, India, in 2008. His research interests include fractional calculus, differential equations, wavelet transforms, stochastic differential equations, integral equations, nuclear reactor kinetics with simulation, numerical analysis, operations research, mathematical modeling, mathematical physics, and computer applications. He has published over 70 research papers in numerous fields and various international journals of repute.

H. Jafari is a Professor in the Department of Mathematical Sciences, University of South Africa. He got his B.Sc. degree from the University of Mazandaran, Babolsar, Iran, and M.Sc. degree from Tarbiat Modares University, Tehran, Iran, in 1998 and 2001, respectively. He earned his Ph.D. from Pune University, Pune, India, in 2006. He has got 3 books, 138 papers in journals, and 25 papers in conference proceedings published to his credit. He was the supervisor of more than 50 M.Sc. students. His current research interests include fractional differential equations and their applications, symmetries and conservation laws, new iterative methods, q-calculus, local fractional differential equations.

T. Raja Sekhar is an Associate Professor in the Department of Mathematics, Indian Institute of Technology Kharagpur, India. He earned his Ph.D. from the Indian Institute of Technology Bombay, India, in 2008. Presently, he is working on a quasi-linear hyperbolic system of partial differential equations involving classical and non-classical nonlinear waves such as shock waves, rarefaction waves, contact discontinuities, delta shock waves, and nonlinear wave interactions. His research areas are theoretical and computational differential equations, groups of symmetries, analysis, and geometry. She has published several research papers in international journals and conference proceedings.

Suchandan Kayal is an Assistant Professor in the Department of Mathematics, National Institute of Technology Rourkela, India. He earned his Ph.D. from the Indian Institute of Technology Kharagpur, India, in 2011. His research interests include applied probability, statistical inference, statistical information theory, statistical decision theory, and order statistics. He has published several research papers in international journals and conference proceedings.

Comparative Study of Some Wavelet-Based Numerical Methods to Solve Initial Value Problems

Kshama Sagar Sahu and Mahendra Kumar Jena

Abstract Ordinary differential equations, in particular initial value problems, play a vital role in applied mathematics. There are many methods available to solve these initial value problems. The operational matrix method based on wavelet is a recent one. In this paper, we briefly review some operational matrix methods. The operational matrix method from the Haar wavelet, the frame, and the Legendre wavelet is considered. We give a comparison of the solution by providing several numerical examples.

Keywords Frame · Haar wavelet · Legendre wavelet · Operational matrices

1 Introduction

Many mathematical models in real-life problems involve ordinary differential equations (ODEs). The solution of these ODE plays a vital role in solving real-life problems. Sometimes, it is not easy to find an analytical solution to the ODE. In such cases, we depend upon the numerical solution. Several methods exist to solve ODE numerically, but the operational matrix method based on wavelet is a recent trend. Many researchers use the operational matrix in solving differential equations. The method using an operational matrix to solve ODE is called as operational matrix method. In this method, the given ODE is converted to an algebraic equation. Solving the algebraic equation, we get the solution of the given ODE.

Wavelet theory is a wide field in science and engineering. It constitutes a family of functions constructed from dilation and translation of a single function called the mother wavelet [14]. Wavelets are used in signal processing, image compression, and many more. In 1997, Chen and Hsiao [2] introduce the operational matrix of

K. S. Sahu (✉) · M. K. Jena
Department of Mathematics, Veer Surendra Sai University of Technology, Burla, Sambalpur, India
e-mail: sahu.kshama@yahoo.in

M. K. Jena
e-mail: mkjena_math@vssut.ac.in

S. S. Ray et al. (eds.), *Applied Analysis, Computation and Mathematical Modelling in Engineering*, Lecture Notes in Electrical Engineering 897,
https://doi.org/10.1007/978-981-19-1824-7_1

integration from Haar wavelet. Many researchers used this operational matrix to find the solution of differential equations numerically. This method got the popularity as it is simple and easy. Later, so many operational matrices based on wavelets have been introduced. Now, this operational matrix method is not limited to solve ODE only. It is widely used to solve fractional differential equations [8, 16], partial differential equations [18], integral equations [1], integro-differential equations [17].

Some well-known operational matrix methods are the Haar wavelet operational matrix (HWOM) method, Legendre's wavelet operational matrix (LWOM) method, and the frame operational matrix (FOM) method. All these methods have been derived using an operational matrix of integration.

Frames first appear in 1952 [3, 7]. These are considered some kinds of alternatives to wavelets. They are more useful when compactly supported and obtained from a single prototype function by dilation and translation. Like wavelets, a function in $L^2(\mathbb{R})$ can also be expressed as a linear combination of frame elements [3]. In this paper, we consider the frame constructed from the linear cardinal B-spline. First, we find out the operational matrices, and then with the help of these operational matrices, we find the approximate solutions of initial value problems (IVPs).

The remaining part of the paper is organized as follows. In Sect. 2, we review the Haar wavelet operational matrix method. In Sect. 3, we present the frame operational matrix method. Legendre wavelet operational matrix method is outlined in Sect. 4. Some numerical examples are given in Sect. 5. A conclusion is given in Sect. 6.

2 Haar Wavelet Operational Matrix Method

In this section, we first find out the operational matrices from the Haar wavelet for the different resolutions. These operational matrices are then used to solve the IVPs. The given IVP is transferred to an algebraic equation which involves the operational matrices. The algebraic equation is then solved, and as a result, we get an approximate solution of the IVP.

Definition 1 (*Haar Wavelet*) Let $m = 2^j$, $j = 0, 1, \ldots J$, $k = 0, 1, \ldots m - 1$, and $i = m + k + 1$. Here, i and j denote wavelet number and the level of wavelet respectively, whereas k is the translation parameter. The maximum level of resolution is J. The minimum value of $i = 2$, and the maximum value is $2M$. The Haar wavelet family for $t \in [A, B]$ is given as [10, 11]

$$h_i(t) = \begin{cases} 1 & t \in [\xi_1(i), \xi_2(i)] \\ -1 & t \in [\xi_2(i), \xi_3(i)] \\ 0 & \text{otherwise} \end{cases},$$

where

$$\xi_1(i) = A + 2k\mu\Delta x,$$
$$\xi_2(i) = A + (2k+1)\mu\Delta x,$$
$$\xi_3(i) = A + 2(k+1)\mu\Delta x,$$

$$\mu = \frac{M}{m}, \quad \text{and} \quad \Delta x = \frac{(B-A)}{2M}.$$

The scaling function is $h_1(t) = 1$ for $t \in [A, B]$ and 0 elsewhere. The Haar wavelet are orthogonal to each other:

$$\int_A^B h_i(t)\, h_l(t) = 2^{-j}\delta_{ij} = \begin{cases} 2^{-j}, & i = l = 2^j + k \\ 0, & i \neq l. \end{cases} \tag{1}$$

Haar wavelets form a good basis for this orthogonal property. Any function $y(t)$ which is square-integrable in the interval $[A, B]$ can be expanded into a Haar wavelet expansion

$$y(t) = \sum_{i=1}^{2M} a_i h_i(t),$$

where $a_i = 2^j \int_A^B y(t)\, h_i(t)\, dt$.

2.1 Operational Matrix of Integration

Let us define Haar wavelet matrix [2, 10, 11, 13] of order $2M \times 2M$ by

$$H_{2M\times 2M} = (h_i(t_l))_{i=1,l=1}^{2M\times 2M} = \begin{bmatrix} h_1(t_1) & h_1(t_2) & \cdots & h_1(t_{2M}) \\ h_2(t_1) & h_2(t_2) & \cdots & h_2(t_{2M}) \\ \vdots & \vdots & \vdots & \vdots \\ h_{2M}(t_1) & h_{2M}(t_2) & \cdots & h_{2M}(t_{2M}) \end{bmatrix}.$$

In general,

$$H_{2M\times 2M} = \left[h_{2M}\left(\tfrac{1}{4M}\right) h_{2M}\left(\tfrac{3}{4M}\right) \cdots h_{2M}\left(\tfrac{4M-1}{4M}\right) \right].$$

The operational matrices are defined as follows:

$$(PH)_{il} = \int\limits_{0}^{t_l} h_i(t)\,dt, \tag{2}$$

$$(QH)_{il} = \int\limits_{0}^{t_l} dt \int\limits_{0}^{t} h_i(t)\,\mathrm{d}t, \tag{3}$$

where t_l are collocation points and $t_l = \frac{l-0.5}{2M}$.
H, P, and Q are matrices of order $2M \times 2M$. Taking $2M = 2$ and $2M = 4$, we have

$$H_{2\times 2} = \begin{bmatrix} 1 & 1 \\ 1 & -1 \end{bmatrix}, (PH)_{2\times 2} = \frac{1}{4}\begin{bmatrix} 1 & 3 \\ 1 & 1 \end{bmatrix}, P_{2\times 2} = \frac{1}{4}\begin{bmatrix} 2 & -1 \\ 1 & 0 \end{bmatrix}.$$

$$H_{4\times 4} = \begin{bmatrix} 1 & 1 & 1 & 1 \\ 1 & 1 & -1 & -1 \\ 1 & -1 & 0 & 0 \\ 0 & 0 & 1 & -1 \end{bmatrix}, (PH)_{4\times 4} = \frac{1}{8}\begin{bmatrix} 1 & 3 & 5 & 7 \\ 1 & 3 & 3 & 1 \\ 1 & 1 & 0 & 0 \\ 0 & 0 & 1 & 1 \end{bmatrix},$$

$$P_{4\times 4} = \frac{1}{16}\begin{bmatrix} 8 & -4 & -2 & -2 \\ 4 & 0 & -2 & 2 \\ 1 & 1 & 0 & 0 \\ 1 & -1 & 0 & 0 \end{bmatrix}.$$

Similarly,

$$(QH)_{2\times 2} = \frac{1}{32}\begin{bmatrix} 1 & 9 \\ 1 & 15 \end{bmatrix}, \qquad Q_{2\times 2} = \frac{1}{32}\begin{bmatrix} 5 & -4 \\ 8 & -7 \end{bmatrix}, \text{ and}$$

$$(QH)_{4\times 4} = \frac{1}{128}\begin{bmatrix} 1 & 9 & 25 & 49 \\ 1 & 9 & 23 & 31 \\ 1 & 7 & 8 & 8 \\ 0 & 0 & 1 & 7 \end{bmatrix}, \qquad (Q)_{4\times 4} = \frac{1}{128}\begin{bmatrix} 21 & -16 & -4 & -12 \\ 16 & -11 & -4 & -4 \\ 6 & -2 & -3 & 0 \\ 2 & -2 & 0 & -3 \end{bmatrix}.$$

Chen and Hsiao [2] have derived the following formula

$$P_{2M\times 2M} = \frac{1}{4M}\begin{pmatrix} 4M\,P_{M\times M} & -H_{M\times M} \\ H^{-1}_{M\times M} & O \end{pmatrix}.$$

Notation: We have used the symbols:

$$H_{2M}^{(0)} := H_{2M\times 2M},\, H_{2M}^{(1)} := (PH)_{2M\times 2M}\,,\, P_{2M}^{(1)} := H_{2M}^{(1)}\left(H_{2M}^{(0)}\right)^{-1}.$$

2.2 Method for First-Order Linear IVP

Consider the first-order linear ordinary differential equation

$$U^{'} = a\,(t)\,U + b\,(t)\,, \quad t \in [0, T]\,, U\,(0) = U_0. \tag{4}$$

Let us divide the interval $[0, T]$ into n equal subinterval such that $t_{i+1} - t_i = d_i$. Let introduce the local coordinate $\tau = \frac{t-t_i}{d_i}$ in the interval $\left[t_i, t_{i+1}\right]$. Define the collocation points in the interval [0, 1] by

$$\tau_j = \frac{\left(j - \frac{1}{2}\right)}{2M}, \quad j = 1, 2, \ldots, 2M.$$

Now, define $u\,(\tau) = U\,(t)$ and the given IVP becomes

$$\frac{\mathrm{d}u}{\mathrm{d}\tau} = d_i\,[a\,(\tau)\,u\,(\tau) + b\,(\tau)]\,, \quad u\,(0) = U_i. \tag{5}$$

Introducing the row vector of order $1 \times 2M$

$$\mathbf{u} = \left[u\,(\tau_1)\; u\,(\tau_2)\; \cdots\; u\,(\tau_{2M})\right],$$

the equation (2.5), can be written as

$$\frac{\mathrm{d}\mathbf{u}}{\mathrm{d}\tau} = d_i\,[\mathbf{u}A\,(\tau) + B\,(\tau)] \tag{6}$$

where

$$A\,(\tau) = \begin{bmatrix} a\,(\tau_1 d_i + t_i) & 0 & 0 \cdots & 0 \\ 0 & a\,(\tau_2 d_i + t_i) & 0 \cdots & 0 \\ \vdots & \vdots & \vdots \cdots & \vdots \\ 0 & 0 & 0 \cdots & a\,(\tau_{2M} d_i + t_i) \end{bmatrix},$$

and

$$B\,(\tau) = (b\,(\tau_1 d_i + t_i)\,, b\,(\tau_2 d_i + t_i)\,, \ldots, b\,(\tau_{2M} d_i + t_i))\,,$$

Following [2, 11], we take

$$\frac{\mathbf{du}}{d\tau} = cH_{2M}^{(0)}, \tag{7}$$

where $c = [c(1), c(2), \ldots c(2M)]$.
Integrating (2.7) we have

$$\mathbf{u} = cH_{2M}^{(1)} + U_i E, \tag{8}$$

where $E = [1, 1, 1, \ldots, 1]$ and $U_i = U(t_i)$.
Comparing (2.6) & (2.7) and putting the value of $\mathbf{u}$ from (2.8) to get c.
Now,

$$c = d_i U_i Y S^{-1} + d_i B A^{-1} \left(H_{2M}^{(0)}\right)^{-1} S^{-1},$$

where

$$S = \left(H_{2M}^{(0)} A^{-1} \left(H_{2M}^{(0)}\right)^{-1} - d_i P_{2M}^{(1)}\right),$$

and

$$Y = E \left(H_{2M}^{(0)}\right)^{-1}.$$

Taking all $\tau_j = 1$, the approximation is

$$U_{i+1} = c(1) + U_i.$$

2.3 *Method for Second-Order Linear IVP*

Let us consider the second-order linear differential equation

$$\frac{\mathrm{d}^2 U}{\mathrm{d}t^2} = F\left(t, U, \frac{\mathrm{d}U}{\mathrm{d}t}\right), \quad t \in [0, T], U(t_0) = U_0, U'(t_0) = V_0. \tag{9}$$

We follow [10] to find the solution of the above equation. Let $V = \frac{\mathrm{d}U}{\mathrm{d}t}$. Then the given differential equation becomes the first-order linear system as follows:

$$\frac{\mathrm{d}V}{\mathrm{d}t} = \frac{\mathrm{d}^2 U}{\mathrm{d}t^2} = F(t, U, V).$$

Here also, we divide the interval $[0, T]$ into n equal subinterval of length d_i. Let us consider the interval $\left[t_i, t_{i+1}\right]$ and define the collocation points τ_j as in previous section. In this interval, define $u(\tau) = U(t)$ and $v(\tau) = V(t)$, where τ is the local coordinate in $\left[t_i, t_{i+1}\right]$. Let $U(t_i) = U_i$ and $V(t_i) = V_i$ are known approximations. Now, the converted system of differential equations can be written as

$$\frac{\mathrm{d}u}{\mathrm{d}\tau} = d_i v \quad \text{and} \quad \frac{\mathrm{d}v}{\mathrm{d}\tau} = d_i F(t_i + \tau d_i, u, v).$$

Let us introduce the row vectors **u** and **v** as given below:

$$\mathbf{u} = \left[u(\tau_1)\ u(\tau_2) \cdots u(\tau_{2M}) \right] \text{and} \quad \mathbf{v} = \left[v(\tau_1)\ v(\tau_2) \cdots v(\tau_{2M}) \right].$$

Following Chen & Hsiao [2] and Lepik [11],

$$\frac{\mathrm{d}\mathbf{u}}{\mathrm{d}\tau} = aH_{2M}^{(0)} \quad \text{and} \quad \frac{\mathrm{d}\mathbf{v}}{\mathrm{d}\tau} = d_i \left(bH_{2M}^{(1)} + V_i E \right),$$

where a and b are row matrix of order $1 \times 2M$. The converted system of ODE becomes

$$aH_{2M}^{(0)} = d_i \left(bH_{2M}^{(1)} + V_i E \right), \tag{10}$$

and

$$bH_{2M}^{(0)} = d_i F \left(t_i + \tau d_i, aH_{2M}^{(1)} + U_i E, bH_{2M}^{(1)} + V_i E \right). \tag{11}$$

Solving Eqs. (10) and (11) we get a and hence b. Taking all $\tau_j = 1$, the approximations are

$$U_{i+1} = a(1) + U_i, \quad V_{i+1} = b(1) + V_i,$$

where $a(1)$ and $b(1)$ are first elements of a and b.

3 Frame Operational Matrix Method

The frame of linear cardinal B-spline is considered to construct the operational matrix. It is an operational matrix of integration.

3.1 A Short Literature Review

Recently, the frame operational matrix method has been used to solve the initial value problems [15]. This operational matrix is obtained from a frame of linear cardinal B-spline. Frames are considered as some kinds of alternatives to wavelets. They are useful when they have compact supports and are obtained from a refinable function.

Definition 2 (*Refinable Function*) [5] A function $\phi \in L^2(\mathbb{R})$ is called a refinable function if there exists scalars $p_k \in \mathbb{R}, \quad k \in Z$ such that

$$\phi(x) = \frac{1}{2} \sum_{k \in Z} p_k \phi(2x - k).$$

Definition 3 (*Multiresolution Analysis*) [5] Let $\phi \in L^2(\mathbb{R})$ is a refinable function and $V_j = closure\left\{\phi_{j,k} : k \in Z\right\}$. The collection of subspaces $\left\{V_j\right\}_{j \in Z}$ of $L^2(\mathbb{R})$ generates an multiresolution analysis (MRA) of $L^2(\mathbb{R})$ if they have the following properties:

1. $\cdots \subset V_{-1} \subset V_0 \subset V_1 \subset \cdots$
2. $span\left(\bigcup_{j \in Z} V_j\right) = L^2(\mathbb{R})$
3. $\bigcap_{j \in Z} V_j = \{0\}$
4. $V_{j+1} = V_j + W_j, \quad j \in Z$
5. $f(x) \in V_j \Leftrightarrow f(2x) \in V_{j+1}, \quad j \in Z$

Definition 4 (*Tight Frame*) [5] A family $\Psi = \{\psi_1, \psi_2, \ldots, \psi_N\} \subset L^2(\mathbb{R})$ is called tight frame of $L^2(\mathbb{R})$ if it satisfies

$$\sum_{i=1}^{N} \sum_{j,k \in Z} \left|\left\langle f, \psi_{i;j,k}\right\rangle\right|^2 = \|f\|^2, \quad \text{all} \quad f \in L^2(\mathbb{R}),$$

where $\psi_{i;j,k} = 2^{j/2}\psi_i\left(2^j \cdot -k\right)$.

Definition 5 (*Linear Cardinal B-spline*) [5] Let us define the linear cardinal B-spline by

$$\phi(x) = \begin{cases} x, x \in [0, 1] \\ 2 - x, x \in [1, 2] \end{cases}.$$

It is refinable with $p_0 = \frac{1}{2}$, $p_1 = 1$, $p_2 = \frac{1}{2}$ and $p_k = 0$ for $k \neq 0, 1, 2$.

Definition 6 (*MRA Tight(wavelet)Frame*) [5] A family $\Psi = \{\psi_1, \psi_2, \ldots, \psi_N\} \subset L^2(\mathbb{R})$ is called an MRA tight(wavelet) frame if it is a tight frame and is associated with a refinable function that generates an MRA and $\Psi \subset V_1$.

We now consider linear cardinal B-spline ϕ to construct an operational matrix method. Define

$$\psi_0(x) = \phi(2x),$$

$$\psi_1(x) = \frac{1}{\sqrt{2}}\left(\psi_0(2x) - \psi_0(2x - 1)\right),$$

$$\text{and} \quad \psi_2(x) = \frac{1}{2}\left(\psi_0(2x) - 2\psi_0\left(2x - \frac{1}{2}\right) + \psi_0(2x - 1)\right).$$

Note that ψ_0 generates an MRA. Moreover, $\Psi \subset V_1$ and is a tight frame [5] and also minimum energy(tight)frame [5]. All functions ψ_0, ψ_1 and ψ_2 have support in [0, 1].

3.2 Frame Operational Matrices

The collection $\Delta = \{\psi_0, \psi_{l,j,k} : l = 1, 2 \text{ and } j, k \in Z\}$, where $\psi_{l,j,k} = 2^{j/2}\psi_l(2^j x - k)$ forms a minimum energy (tight)frame for $L^2(\mathbb{R})$ [5]. The parameter $j \geq 0$ in Δ is called the resolution level. Let J denotes the maximal resolution. Let $M = 1 + 2(1 + 2 + \cdots + 2^J)$. Suppose the grid points are $t_i = (i-1)/M,\ i = 1, 2, \ldots, M+1$ and the collocation points are

$$\tau_n = \frac{t_n + t_{n+1}}{2},\quad n = 1, 2, \ldots, M.$$

Following Sahu and Jena [15], we have the frame matrix and the frame operational matrix as given below:

For fixed J, the frame matrix F_0 is a matrix of order $M \times M$, defined by

$$F_0 := \begin{pmatrix} \Psi_0 \\ \Psi_{1,0} \\ \Psi_{1,1} \\ \vdots \\ \Psi_{1,J} \\ \Psi_{2,0} \\ \Psi_{2,1} \\ \vdots \\ \Psi_{2,J} \end{pmatrix}$$

where $\Psi_0 = (\psi_0(\tau_1), \psi_0(\tau_2), \ldots, \psi_0(\tau_M))$ and for $l = 1, 2$ and $j = 0, 1, \ldots, J$,

$$\Psi_{l,j} = \begin{pmatrix} \psi_{l,j,0}(\tau_1) & \cdots & \psi_{l,j,0}(\tau_M) \\ \vdots & & \vdots \\ \psi_{l,j,2^j-1}(\tau_1) & \cdots & \psi_{l,j,2^j-1}(\tau_M) \end{pmatrix}.$$

Let us define α-th order integrations ψ_0^α and $\psi_{l;j,k}^\alpha$, $\alpha \geq 1$ by [11]

$$\psi_0^\alpha(x) = \int_0^x \int_0^x \cdots \int_0^x \psi_0(t)\,\mathrm{d}t^\alpha = \frac{1}{(\alpha-1)!}\int_0^x (x-t)^{\alpha-1}\,\psi_0(t)\,\mathrm{d}t,$$

$$\psi_{l;j,k}^\alpha(x) = \int_0^x \int_0^x \cdots \int_0^x \psi_{l;j,k}(t)\,\mathrm{d}t^\alpha = \frac{1}{(\alpha-1)!}\int_0^x (x-t)^{\alpha-1}\,\psi_{l;j,k}(t)\,\mathrm{d}t.$$

The higher-order frame matrices $F_\alpha, \alpha \geq 1$ are matrices of order $M \times M$, defined by

$$F_\alpha := \begin{pmatrix} \Psi_0^\alpha \\ \Psi_{1,0}^\alpha \\ \Psi_{1,1}^\alpha \\ \vdots \\ \Psi_{1,J}^\alpha \\ \Psi_{2,0}^\alpha \\ \Psi_{2,1}^\alpha \\ \vdots \\ \Psi_{2,J}^\alpha \end{pmatrix}$$

where $\Psi_0^\alpha = \left(\psi_0^\alpha(\tau_1), \psi_0^\alpha(\tau_2), \ldots, \psi_0^\alpha(\tau_M)\right)$ and for $l = 1, 2$ and $j = 0, 1, \ldots, J$,

$$\Psi_{l,j}^\alpha = \begin{pmatrix} \psi_{l;j,0}^\alpha(\tau_1) & \cdots & \psi_{l;j,0}^\alpha(\tau_M) \\ \vdots & & \vdots \\ \psi_{l;j,2^j-1}^\alpha(\tau_1) & \cdots & \psi_{l;j,2^j-1}^\alpha(\tau_M) \end{pmatrix}.$$

The α-th-order frame operational matrix P_α is now defined by

$$P_\alpha = F_\alpha F_0^{-1}.$$

In particular, frame matrices F_0, F_1 and frame operational matrix P_1 are given below.

Frame Matrix for $J = 1$ ($M = 7$)

$$F_0 = \begin{pmatrix} \frac{2}{7} & \frac{6}{7} & \frac{10}{7} & 2 & \frac{10}{7} & \frac{6}{7} & \frac{2}{7} \\ -\frac{398}{985} & -\frac{1194}{985} & -\frac{796}{985} & 0 & \frac{796}{985} & \frac{1194}{985} & \frac{398}{598} \\ -\frac{796}{985} & -\frac{398}{985} & \frac{1194}{985} & 0 & 0 & 0 & 0 \\ 0 & 0 & 0 & 0 & -\frac{1194}{985} & \frac{398}{985} & \frac{796}{985} \\ -\frac{2}{7} & -\frac{6}{7} & \frac{2}{7} & 2 & \frac{2}{7} & -\frac{6}{7} & -\frac{2}{7} \\ -\frac{4}{7} & \frac{8}{7} & -\frac{4}{7} & 0 & 0 & 0 & 0 \\ 0 & 0 & 0 & 0 & -\frac{4}{7} & \frac{8}{7} & -\frac{4}{7} \end{pmatrix}.$$

$$F_1 = \begin{pmatrix} \frac{1}{98} & \frac{9}{98} & \frac{25}{98} & \frac{1}{2} & \frac{73}{98} & \frac{89}{98} & \frac{97}{98} \\ -\frac{226}{15661} & -\frac{253}{1948} & -\frac{447}{1511} & -\frac{1189}{3363} & -\frac{447}{1511} & -\frac{253}{1948} & -\frac{226}{15661} \\ -\frac{253}{4383} & -\frac{371}{1094} & -\frac{393}{1757} & 0 & 0 & 0 & 0 \\ 0 & 0 & 0 & 0 & -\frac{393}{1757} & -\frac{371}{1094} & -\frac{253}{4383} \\ -\frac{1}{98} & -\frac{9}{98} & -\frac{8}{49} & 0 & \frac{8}{49} & \frac{9}{98} & \frac{1}{98} \\ -\frac{2}{49} & -\frac{11}{98} & \frac{15}{98} & 0 & 0 & 0 & 0 \\ 0 & 0 & 0 & 0 & -\frac{15}{98} & \frac{11}{98} & \frac{2}{49} \end{pmatrix}.$$

Operational Matrix for $J = 1$ $(M = 7)$

$$P_1 = \begin{pmatrix} \frac{1}{2} & \frac{1121}{2378} & \frac{66}{19601} & \frac{65}{152} & -\frac{1}{4} & \frac{2}{105} & -\frac{44}{105} \\ -\frac{580}{3361} & 0 & -\frac{17}{420} & \frac{17}{420} & -\frac{165}{39202} & -\frac{33}{19601} & -\frac{33}{19601} \\ -\frac{336}{3713} & \frac{43}{336} & -\frac{13}{420} & 0 & \frac{336}{3713} & -\frac{83}{2293} & 0 \\ -\frac{336}{3713} & -\frac{43}{336} & 0 & \frac{13}{420} & \frac{336}{3713} & 0 & -\frac{83}{2293} \\ 0 & \frac{529}{4834} & -\frac{130}{2413} & -\frac{130}{2413} & 0 & \frac{1}{60} & -\frac{1}{60} \\ 0 & 0 & \frac{195}{2032} & 0 & 0 & -\frac{9}{140} & 0 \\ 0 & 0 & 0 & \frac{195}{2032} & 0 & 0 & \frac{9}{40} \end{pmatrix}.$$

3.3 Method for First-Order Linear IVP

This method is very much similar to the HWOM method. Let us consider the ODE

$$U^{'}(t) = A(t)U(t) + B(t), \quad U(t_0) = U_0, \quad t \in [t_0, T]. \tag{12}$$

We divide the whole interval of discretization into n equal segments with $h_i = t_{i+1} - t_i, \quad i = 0, 1, \ldots, n-1$. Let us consider the interval $\left[t_i, t_{i+1}\right]$. Assume that U_i is a known approximation to $U(t_i)$. The local coordinate $\tau = \frac{t-t_i}{h_i}$ to the interval $[t_i, t_{i+1}]$. This is now belongs to [0, 1]. The given Eq. (12) becomes a new IVP with local coordinate,

$$\dot{u}(\tau) := \frac{\mathrm{d}u}{\mathrm{d}\tau} = h_i(a(\tau)u(\tau) + b(\tau)), \quad u(0) = U_i. \tag{13}$$

Introduce

$$\mathbf{u} = [u(\tau_1), u(\tau_2), \ldots, u(\tau_M)],$$

From (13), we get

$$\frac{\mathrm{d}\mathbf{u}}{\mathrm{d}\tau} = h_i\left[\mathbf{u}A + \mathbf{b}\right], \tag{14}$$

where

$$\mathbf{A} = \begin{pmatrix} a(t_1^*) & 0 & \cdots & 0 \\ 0 & a(t_2^*) & \cdots & 0 \\ & & \vdots & \\ 0 & 0 & \cdots & a(t_M^*) \end{pmatrix},$$

$t_j^* = t_i + h_i \tau_j$
and

$$\mathbf{b} = \left[b(t_1^*), b(t_2^*), \ldots, b(t_M^*)\right].$$

Following Sahu and Jena [15], and simplifying,

$$\mathbf{c} = h_i\left(dY + \mathbf{b}\mathbf{A}^{-1}F_0^{-1}\right)(F_0\mathbf{A}^{-1}F_0^{-1} - h_i P_1)^{-1}, \tag{15}$$

where $Y = EF_0^{-1}$ and $P_1 = F_1 F_0^{-1}$.

The approximation is $u(1) = c(1) + u(0) = c(1) + U_i$, where $c(1)$ is the first entry of $\mathbf{c}$.

3.4 Method for Second Order Linear IVP

Let us consider the differential equation

$$U'' + p^*U' + q^*U = f(t), \quad U(t_0) = U_0, \quad U'(t_0) = V_0, \tag{16}$$

where p^*, q^*, and $f(t)$ are function of t. The equation (16) is reduced to a system of first-order ODE by taking

$$\frac{dU}{dt} = V, \qquad \frac{dV}{dt} = -p^*V - q^*U + f(t). \tag{17}$$

Let us consider the interval $[t_i, t_{i+1}]$. Assume that the known U_i and V_i are approximation to $U\ (t_i)$ and $V\ (t_i)$=$U^{'}\ (t_i)$, respectively. The local coordinate in the interval $[t_i, t_{i+1}]$ is $\tau = (t - t_i)/h_i$, where $h_i = t_{i+1} - t_i$. In terms of this local coordinate, we have $u(\tau) = U(t), \quad v(\tau) = V(t)$. Introduce

$$\mathbf{u} = [u(\tau_1), u(\tau_2), \ldots, u(\tau_M)],$$
$$\mathbf{v} = [v(\tau_1), v(\tau_2), \ldots, v(\tau_M)].$$

We have the following relation from (17)

$$\dot{\mathbf{u}} = h_i\mathbf{v}, \tag{18}$$
$$\dot{\mathbf{v}} = -\mathbf{vp} - \mathbf{uq} + \Phi, \tag{19}$$

where

$$\mathbf{p} = h_i \begin{bmatrix} p^*(t_1^*) & 0 & 0 \cdots & 0 \\ 0 & p^*(t_2^*) & 0 \cdots & 0 \\ \vdots & \vdots & \vdots \ \vdots & \vdots \\ 0 & 0 & 0 \cdots & p^*(t_M^*) \end{bmatrix},$$

$$\mathbf{q} = h_i \begin{bmatrix} q^*(t_1^*) & 0 & 0 \cdots & 0 \\ 0 & q^*(t_2^*) & 0 \cdots & 0 \\ \vdots & \vdots & \vdots \ \vdots & \vdots \\ 0 & 0 & 0 \cdots & q^*(t_M^*) \end{bmatrix},$$

and

$$\Phi = h_i \left[f(t_1^*)\ f(t_2^*)\ f(t_3^*) \cdots f(t_M^*) \right].$$

Here $t_j^* = \tau_j h_i + t_i, \quad j = 1, 2, \ldots, M$. Following Chen [2], Lepik [11], and simplifying as [15] we have

$$a = h_i b P_1 + h_i V_i Y \tag{20}$$

and

$$b = -V_i E\mathbf{p}F_0^{-1}S^{-1} - h_i V_i Y F_1\mathbf{q}F_0^{-1}S^{-1} - U_i E\mathbf{q}F_0^{-1}S^{-1} + \Phi F_0^{-1}S^{-1}, \quad (21)$$

where $S = I + F_1\mathbf{p}F_0^{-1} + h_i P_1 F_1\mathbf{q}F_0^{-1}$. Here $Y = EF_0^{-1}$ and $P_1 = F_1F_0^{-1}$. The approximation becomes

$$U_{i+1} = a\,(1) + U_i$$
$$V_{i+1} = b\,(1) + V_i.$$

4 Legendre's Wavelet Operational Matrix Method

The "Legendre wavelets" $\psi_{n,m}(t) = \psi(k, n, m.t)$ is defined as follows: [9, 12, 14]

Definition 7 Legendre Polynomial

$$\psi_{n,m}(t) = \begin{cases} \sqrt{m + \frac{1}{2}}2^{\frac{k}{2}} P_m\left(2^k t - 2n + 1\right), & t \in [\xi_1, \xi_2] \\ 0 & \text{otherwise} \end{cases},$$

where $\xi_1 = \frac{2n-2}{2^k}$, $\xi_2 = \frac{2n}{2^k}$, $m = 0, 1, \ldots, M-1$, $n = 1, 2, \ldots, 2^{k-1}$, and $P_m(t)$ are Legendre polynomial of degree m. In particular, $P_0(t) = 1$ and $P_1(t) = t$.

Any function $f(t)$ can be represented in Legendre wavelet series in [0, 1) by Razzaghi and Yousefi [14]

$$f(t) = \sum_{n=1}^{\infty}\sum_{m=1}^{\infty} c_{n,m}\psi_{n,m}(t),$$

where $c_{n,m} = \langle f(t), \psi_{n,m}(t)\rangle$, in which $\langle\cdots\rangle$ is the inner product.

4.1 Function Approximation

Consider the Legendre wavelet as [9, 14]

$$\Psi = \left(\psi_{1,0} \cdots \psi_{2^{k-1},0}\ \psi_{1,1} \cdots \psi_{2^{k-1},1} \cdots \psi_{1,M-1} \cdots \psi_{2^{k-1},M-1}\right)^T$$

Let f be an arbitrary function in $L^2[0, 1]$ then there exist unique coefficients $c_{n,m}$ such that

$$f(t) \simeq \sum_{n=1}^{2^{k-1}}\sum_{m=0}^{M-1} c_{n,m}\psi_{n,m}(t) = C^T\Psi(t),$$

where $c_{n,m} = \int_0^1 f(t)\,\psi_{n,m}(t)\,dt$, and

$$C = \left(c_{1,0} \cdots c_{2^{k-1},0}\; c_{1,1} \cdots c_{2^{k-1},1} \cdots\; c_{1,M-1} \cdots c_{2^{k-1},M-1}\right)^T,$$

Collocation points are given by [9]

$$t_i = \frac{2i-1}{2^k M}, \quad i = 1, 2, \ldots, 2^{k-1}M$$

Following [9, 14] we get the operational matrices as below:

$$\int_0^t \Psi(t)\,\mathrm{d}t = P\Psi(t),$$

where P is the operational matrix of order $2^{k-1}M \times 2^{k-1}M$.

In general, the operational matrix P is given by Razzaghi and Yousefi [14]

$$P = \frac{1}{2^k}\begin{bmatrix} L & F & F & \cdots & F \\ 0 & L & F & \cdots & F \\ 0 & 0 & L & \cdots & F \\ \vdots & \vdots & \vdots & \vdots & \vdots \\ 0 & 0 & \cdots & 0 & L \end{bmatrix},$$

where F and L are square matrix of order M as follows:

$$F = \begin{bmatrix} 2 & 0 & \cdots & 0 \\ 0 & 0 & \cdots & 0 \\ \vdots & \vdots & \vdots & \vdots \\ 0 & 0 & \cdots & 0 \end{bmatrix},$$

and

$$F = \begin{bmatrix} 1 & \frac{1}{\sqrt{3}} & 0 & 0 & \cdots & 0 & 0 & 0 \\ -\frac{\sqrt{3}}{3} & 0 & \frac{\sqrt{3}}{3\sqrt{5}} & 0 & \cdots & 0 & 0 & 0 \\ 0 & -\frac{\sqrt{5}}{5\sqrt{3}} & 0 & \frac{\sqrt{5}}{5\sqrt{7}} & \cdots & 0 & 0 & 0 \\ 0 & 0 & -\frac{\sqrt{7}}{7\sqrt{5}} & 0 & \cdots & 0 & 0 & 0 \\ \vdots & \vdots & \vdots & \vdots & \cdots & \vdots & \vdots & \vdots \\ 0 & 0 & 0 & 0 & \cdots & -\frac{\sqrt{2M-3}}{(2M-3)\sqrt{2M-5}} & 0 & \frac{\sqrt{2M-3}}{(2M-3)\sqrt{2M-1}} \\ 0 & 0 & 0 & 0 & \cdots & 0 & -\frac{\sqrt{2M-1}}{(2M-1)\sqrt{2M-3}} & 0 \end{bmatrix}.$$

In particular, the matrices Ψ and P for $M = 3$ and $k = 2$ are given in the following:

$$\Psi_{6\times6} = \begin{pmatrix} \sqrt{2} & \frac{-2}{3}\sqrt{6} & \frac{\sqrt{10}}{6} & 0 & 0 & 0 \\ \sqrt{2} & 0 & \frac{\sqrt{10}}{2} & 0 & 0 & 0 \\ \sqrt{2} & \frac{2}{3}\sqrt{6} & \frac{\sqrt{10}}{6} & 0 & 0 & 0 \\ 0 & 0 & 0 & \sqrt{2} & \frac{-2}{3}\sqrt{6} & \frac{\sqrt{10}}{6} \\ 0 & 0 & 0 & \sqrt{2} & 0 & \frac{\sqrt{10}}{2} \\ 0 & 0 & 0 & \sqrt{2} & \frac{2}{3}\sqrt{6} & \frac{\sqrt{10}}{6} \end{pmatrix}$$

$$P_{6\times6} = \frac{1}{4}\begin{pmatrix} 1 & \frac{\sqrt{2}}{\sqrt{6}} & 0 & 2 & 0 & 0 \\ -\frac{\sqrt{3}}{3} & 0 & \frac{\sqrt{3}}{3\sqrt{5}} & 0 & 0 & 0 \\ 0 & -\frac{\sqrt{5}}{5\sqrt{3}} & 0 & 0 & 0 & 0 \\ 0 & 0 & 0 & 1 & \frac{\sqrt{2}}{\sqrt{6}} & 0 \\ 0 & 0 & 0 & -\frac{\sqrt{3}}{3} & 0 & \frac{\sqrt{3}}{3\sqrt{5}} \\ 0 & 0 & 0 & 0 & -\frac{\sqrt{5}}{5\sqrt{3}} & 0 \end{pmatrix}$$

5 Results and Discussions

In this section, we consider some numerical examples to compare the solutions obtain from different operational matrix methods described above.

Example 1 Consider the singular initial value problem [9]

$$U'' + \frac{4}{t}U' + \left(\frac{2}{t^2} + t\right)U = 20t + t^4, \quad U\,(0) = U'\,(0) = 0.$$

The exact solution of the given IVP is $U\,(t) = t^3$. Numerical comparisons of solutions obtain from LWOM, HWOM, and FOM are presented in Table 1. Comparison of the solution from HWOM, FOM, and exact solution presented graphically in Fig. 1.

Example 2 Consider the singular initial value problem [9]

$$U'' + \frac{1}{t}U' = \left(\frac{8}{8 - t^2}\right)^2, \quad U\,(0) = 0, U'\,(0) = 0.$$

The exact solution of the given IVP is $U\,(t) = 2\log\left(\frac{7}{8-t^2}\right)$. Numerical comparison of solution from LWOM, HWOM, and FOM is presented in the Table 2. Comparison of the solution from HWOM, FOM, and exact solution is presented graphically in Fig. 2.

Table 1 L_2−norm Comparison of solution of Example 1

t	LWOM [9]	HWOM	FOM	Exact
0.1	0.001000	0.0010	0.0010	0.001000
0.2	0.008003	0.0079	0.0080	0.008000
0.3	0.027008	0.0269	0.0269	0.027000
0.4	0.064045	0.0639	0.0639	0.064000
0.5	0.125131	0.1249	0.1249	0.125000
0.6	0.216534	0.2158	0.2158	0.216000
0.7	0.345017	0.3428	0.3428	0.343000
0.8	0.513002	0.5118	0.5118	0.512000
0.9	0.730000	0.7288	0.7288	0.729000

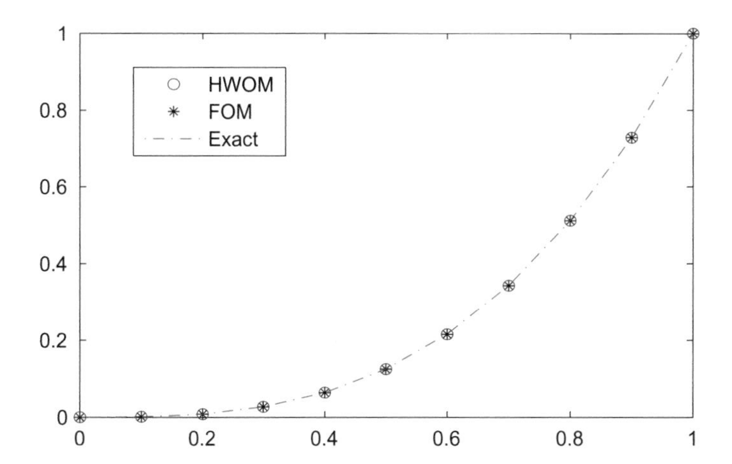

Fig. 1 Comparison of solution Example 1

Table 2 L_2−norm comparison of solution of Example 2

t	LWOM [9]	HWOM	FOM	Exact
0.1	−0.26456123	−0.2646	−0.2646	−0.26456122
0.2	−0.25703772	−0.2570	−0.2571	−0.25703770
0.3	−0.24443526	−0.2444	−0.2446	−0.24443526
0.4	−0.22665738	−0.2267	−0.2268	−0.22665737
0.5	−0.20356540	−0.2036	−0.2037	−0.20356538
0.6	−0.17497491	−0.1750	−0.1751	−0.17497490

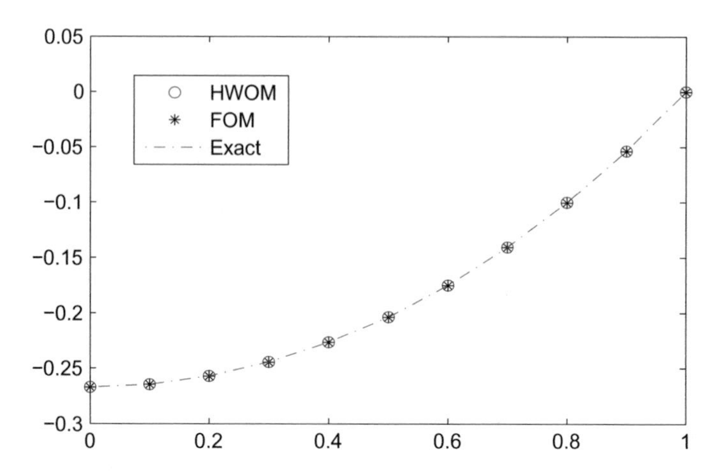

Fig. 2 Comparison of solution Example 2

6 Conclusion

In this paper, numerical comparison of the solutions from different operational matrix methods is computed. It is observed from the results that the HWOM and FOM are the same rate of convergence. LWOM solutions are more close to the exact solution than the other two methods presented with the same step size and in the same interval.

References

1. Babolian E, Masouri Z (2008) Direct method to solve Volterra integral equation of the first kind using operational matrix with block-pulse functions. J Comput Appl Math 220:51–57
2. Chen C, Hsiao C (1997) Haar wavelet method for solving lumped and distributed parameter systems. IEEE Proc Cont Theory Appl 144:87–94
3. Christensen O (2003) An introduction to frames and riesz basis. Birkhauser, Berlin
4. Chui CK (1992) An introduction to wavelets. Academic Press, San Diego, California
5. Chui CK, Wenjie H (2000) Compactly supported tight frames associated with refinable functions. Appl Comput Harmon Anal 8:293–319
6. Daubechies: ten lectures on wavelet. SIAM Philadelphia (1992)
7. Duffin RJ, Schaeffer AC (1952) A class of nonharmonic fourier series. Trans Am Math Soc 72:341–366
8. Ganji RM, Jafari H, Kgarose M, Mohammadi A (2021) Numerical solutions of time-fractional Klein-Gordon equations by clique polynomials. Alex Eng J 60(5):4563–4571
9. Iqbal J, Abass R, Kumar P (2018) Solution of linear and nonlinear singular boundary value problems using legendre wavelet method. Ital J Pure Appl Math 40:311–328
10. Jena MK, Sahu KS (2017) Haar wavelet operational matrix method to solve initial value problems: a short survey. Int J Appl Comput Math 3(4):3961–3975
11. Lepik U (2005) Numerical solution of differential equations using Haar wavelets. Math Comput Simul 68:127–143

12. Mirzaee F, Samadyar N (2018) Convergence of Legendre wavelet collocation method for solving nonlinear Stratonovich Volterra integral equations. Comput Methods Differ Equ 6(1):80–97
13. Patra A, Ray SS (2014) Numerical simulation based on Haar wavelet operational method to solve neutron point kinetics equation involving sinusoidal and pulse reactivity. Ann Nucl Energy 73:408–412
14. Razzaghi M, Yousefi S (2001) The Legendre wavelets operational matrix of integration. Int J Syst Sci 32(4):495–502
15. Sahu KS, Jena MK (2020) Solution of initial value problems using an operational matrix. Int J Appl Comput Math 6:61. https://doi.org/10.1007/s40819-020-00810-9
16. Tuan NH, Ganji RM, Jafari H (2020) A numerical study of fractional rheological models and fractional Newell-Whitehead-Segel equation with non-local and non-singular kernel. Chin J Phys 68:308–320
17. Tuan NH, Nemati S, Ganji RM, Jafari H (2020) Numerical solution of multi-variable order fractional integro-differential equations using the Bernstein polynomials. Eng Comput. https://doi.org/10.1007/s00366-020-01142-4
18. Zaky MA, Baleanu D, Alzaidy JF, Hashemizadeh E (2018) Operational matrix approach for solving the variable-order nonlinear Galilei invariant advection-diffusion equation. Adv Differ Equ 102. https://doi.org/10.1186/s13662-018-1561-7

Effects of MHD and Thermal Radiation on Unsteady Free Convective Flow of a Hybrid Nanofluid Past a Vertical Plate

V. Rajesh, M. Kavitha, and M. P. Mallesh

Abstract This paper investigates the enhancement of natural convection heat transfer using hybrid nanofluids over a moving vertical plate. Two different kinds of fluids, namely $Cu - Al_2O_3$/water and Cu/water, are selected for investigation. The principal equations modelling the flow are solved using a robust finite difference numerical method. Impacts of main parameters, such as magnetic parameter, radiation parameter, Grashof number, and nanoparticle volume fraction, are analysed on skin friction coefficient, velocity profiles, Nusselt number, and temperature profiles. The numerical results are validated by comparing them with analytical correlations. The present work finds application to the cooling of nuclear reactors, microelectronics, and solar energy conservation systems.

Keywords Unsteady flow · Free convection · Hybrid nanofluid · Vertical plate · Finite difference numerical method · Magnetohydrodynamics (MHD) · Thermal radiation

1 Introduction

Magnetohydrodynamic nanofluid flow problems play a prominent role in industries such as oil exploration, geothermal energy extractions, and boundary layer control due to the interaction of electrically conducting fluid and magnetic field in aerodynamics, MHD power generators, pumps, bearings, and electromagnetic accelerators. Sparrow and Cess [1] studied the magnetic field's effect on the natural convection

V. Rajesh (✉) · M. Kavitha (✉)
Department of Mathematics, GITAM (Deemed to be University), Hyderabad Campus, Rudraram Village, Medak, Patancheru (M), Telangana, India
e-mail: v.rajesh.30@gmail.com

M. Kavitha
e-mail: kavitha.itikala@gmail.com

M. P. Mallesh
Department of Mathematics, Koneru Lakshmaiah Education Foundation, Hyderabad Campus, Aziz Nagar Village, R R Dist, Moinabad (M), Telangana, India

S. S. Ray et al. (eds.), *Applied Analysis, Computation and Mathematical Modelling in Engineering*, Lecture Notes in Electrical Engineering 897,
https://doi.org/10.1007/978-981-19-1824-7_2

heat transfer. Vajravelu [2] analysed the hydromagnetic convective flow in detail for a continuous moving surface. Other pertinent studies subject to MHD were inspected by Rajesh et al. [3, 4]. MHD effects on the flow with several cases were presented by Sarma [5] and Sambath [6]. Other admissible investigations to nanofluid were scrutinized by Rajesh et al. [7–10]. The next-generation nanofluids are hybrid nanofluids obtained by dispersing a homogenous (or non-homogenous) mixture of composite nanopowder of several nanoparticles in one or more base fluids. Hybridization of nanoparticles yields new chemical and physical bonds, making hybrid nanofluids offer better heat transfer performance and thermophysical properties than conventional heat transfer fluids and nanofluids with single nanoparticles. Hybrid nanofluids have many applications in almost all heat transfer fields, such as nuclear cooling, microelectronics, microfluidics, transportation, manufacturing, medical, defence, acoustics, naval structures propulsion, solar energy conservation system, pasteurization, medical, and drug decrease. Numerical evaluation of hybrid nanofluids based on Al_2O_3, TiO_2, and SiO2 nanoparticles with different approaches was presented by Minea [11]. Numerically, Sahoo et al. [12] reported that the hybrid nanofluids were used as radiator coolant and Al_2O_3 + Ag/water-based hybrid nanofluid had high efficiency. Anjali Devi and Surya Uma Devi [13] numerically investigated hydromagnetic hybrid Cu-Al_2O_3/water nanofluid flow over a permeable stretching sheet with suction. Moldoveanu et al. [14] prepared the $Al_2O_3 - SiO_2$ nanofluids and their hybrid. They measured the hybrid nanofluid viscosity variation with temperature. They found that increasing the temperature reduced the viscosity of hybrid nanofluid. Moldoveanu et al. [15] presented an experimental study on the viscosity of stabilized Al_2O_3, TiO_2 nanofluids, and their hybrid. Ghadikolaeia and Gholiniab [16] investigated 3D mixed convection MHD flow of GO–MoS_2 hybrid nanoparticles in H_2O–$(CH_2OH)_2$ hybrid base fluid under the effect of H_2 bond. Few other imperative scrutinies on nano and hybrid nanofluid flow problems are prospected [17–23].

This paper highlights unsteady natural convection 2D hydromagnetic flow and heat transfer of hybrid nanofluid past a moving vertical plate. Hybrid nanofluid is considered by suspending two different nanoparticles Cu and Al_2O_3 in water. This study may be viewed as an extension of Rajesh et al. [8]. The novelties compared to Rajesh et al. [8] are (i) advanced type of nanofluid called hybrid nanofluid (Cu-Al2O3) is considered and (ii) effects of thermal radiation is introduced. The controlling equations with boundary conditions are solved by the implicit finite difference method. Numerical study of $Cu - Al_2O_3$/water hybrid nanofluid flow comparing with single nanoparticle Cu-water nanofluid is provided in terms of magnetic parameter (M), radiation parameter (N), Grashof number (Gr), and nanoparticle volume fraction (δ_2), on the coefficient of skin friction, velocity, Nusselt number, and temperature profiles.

2 Formulation of the Problem

In this article, a Cartesian coordinate system (x^*, y^*) is taken such that x^*—coordinate is along with the plate, and y^*—coordinate is normal to the plate. In the beginning, both the plate and fluid are at rest at free stream temperature θ_∞^* for all $t^* \leq 0$. Subsequently, at a time $t^* > 0$, the plate moves with invariable velocity $u_{1_0}^*$. Plate temperature is elevated to $\theta^* = \theta_w^*$. A magnetic field of uniform strength B_0 is considered to act in the y^*−axis direction, and the consequence of an induced magnetic field is negligible, which is reasonable when the magnetic Reynolds number is small. The viscous dissipation, Ohmic heating, ionslip, and Hall effects are assumed to be negligible. We have premeditated copper (Cu) and aluminium oxide (Al_2O_3) nano-size particles with a base fluid as water in the present study. In this model, a copper nanoparticle is initially scattered into the base fluid with 0.1 vol. solid volume fraction (i.e. $\delta_1 = 0.1$, which is fixed throughout the problem) to make nanofluid $Cu - H_2O$. To develop the targeted hybrid nanofluid $Cu - Al_2O_3$/water, aluminium oxide nanoparticle with different volume fraction is dispersed in nanofluid Cu/water. Under the above assumptions, we adopt Tiwari and Das [24] nanofluid model and Boussinesq approximation (Schlichting and Gersten, [25]) in this flow. The constitutive controlling boundary layer equations of the hybrid nanofluid flow and heat transfer are as follows (Fig. 1):

$$\frac{\partial u_1^*}{\partial x^*} + \frac{\partial u_2^*}{\partial y^*} = 0. \tag{1}$$

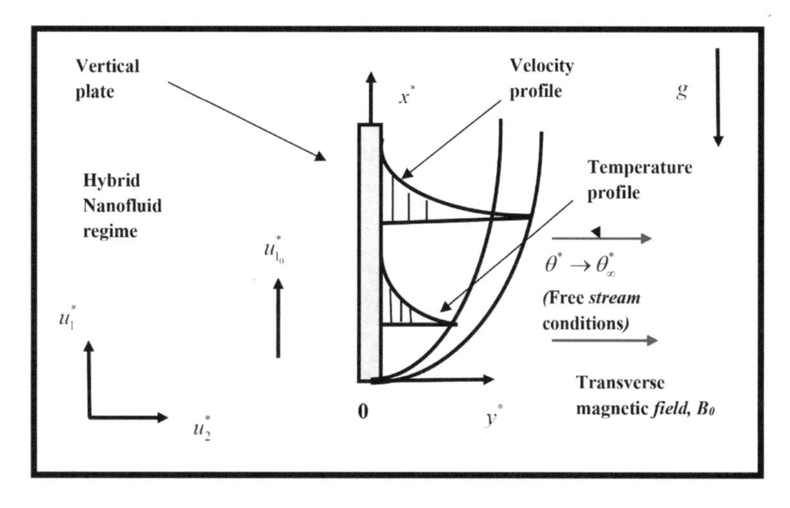

Fig. 1 Physical model and coordinate system

$$\rho_{\text{hnf}}\left[\frac{\partial u_1^*}{\partial t^*}+u_1^*\frac{\partial u_1^*}{\partial x^*}+u_2^*\frac{\partial u_1^*}{\partial y^*}\right]$$
$$=\mu_{\text{hnf}}\frac{\partial^2 u_1^*}{\partial y^{*^2}}+(\rho\beta)_{\text{hnf}}g\left(\theta^*-\theta_\infty^*\right)-\sigma_{\text{hnf}}B_0^2u_1^*. \tag{2}$$

$$\left(\rho C_p\right)_{\text{hnf}}\left[\frac{\partial \theta^*}{\partial t^*}+u_1^*\frac{\partial \theta^*}{\partial x^*}+u_2^*\frac{\partial \theta^*}{\partial y^*}\right]=\kappa_{\text{hnf}}\frac{\partial^2\theta^*}{\partial y^{*^2}}-\frac{\partial q_R^*}{\partial y^*}. \tag{3}$$

where q_R^* *is r*adiative heat flux and it is defined as

$$q_R^*=-\frac{4\sigma_s}{3k_e}\frac{\partial\theta^{*^4}}{\partial y^*}. \tag{4}$$

Here, σ_s and k_e denote the coefficient of absorption and Stefan–Boltzmann constant, respectively. By applying Rosseland approximation, we confined our analysis to optically thick nanofluids. If differences of temperature within the flow are small, such that θ^{*^4} may be expressed as a linear function of the temperature, applying Taylor's series for θ^{*^4} about θ_∞^*, after neglecting higher-order terms as:

$$\theta^{*^4}\cong 4\theta_\infty^{*^3}\theta^*-3\theta_\infty^{*^4}. \tag{5}$$

Then, we obtain

$$\frac{\partial q_R^*}{\partial y^*}=-\frac{16\sigma_s\left(\theta_\infty^*\right)^3}{3k_e}\frac{\partial^2\theta^*}{\partial y^{*^2}}. \tag{6}$$

On considering Eqs. (4–6), Eq. (3) becomes

$$\left(\rho C_p\right)_{\text{hnf}}\left[\frac{\partial \theta^*}{\partial t^*}+u_1^*\frac{\partial \theta^*}{\partial x^*}+u_2^*\frac{\partial \theta^*}{\partial y^*}\right]$$
$$=\kappa_{\text{hnf}}\frac{\partial^2\theta^*}{\partial y^{*^2}}+\frac{16\sigma_s\left(\theta_\infty^*\right)^3}{3k_e}\frac{\partial^2\theta^*}{\partial y^{*^2}}. \tag{7}$$

The relevant conditions are as follows:

$$\begin{aligned}
t^*\le 0: u_1^*=0,\quad u_2^*=0,\ \theta^*=\theta_\infty^* \text{ for all } x^* \text{and } y^*.\\
t^*>0: u_1^*=u_{1_0}^*,\ u_2^*=0,\ \theta^*=\theta_w^* \text{ at } y^*=0.\\
u_1^*=0,\qquad\qquad \theta^*=\theta_\infty^* \text{ at } x^*=0.\\
u_1^*\to 0,\qquad\qquad \theta^*=\theta_\infty^* \text{ as } y^*\to\infty.
\end{aligned} \tag{8}$$

The fundamental thermal parameters of hybrid nanoliquid, namely density ρ_{hnf}, dynamic viscosity μ_{hnf}, heat capacity $(\rho C_p)_{\text{hnf}}$, heat expansion coefficient $(\rho\beta)_{\text{hnf}}$, heat conductivity κ_{hnf}, and electrical conductivity σ_{hnf} are stated in Table 1. Thermal

Table 1 Thermophysical properties of $Cu - Al_2O_3$/water

Property	Hybrid nanofluid ($Cu - Al_2O_3$/water)
Density	$\rho_{\text{hnf}} = [(1-\delta_2)\{(1-\delta_1)\rho_f + \delta_1\rho_{s1}\}] + \delta_2\rho_{s2}$
Dynamic viscosity	$\mu_{\text{hnf}} = \frac{\mu_f}{(1-\delta_1)^{2.5}(1-\delta_2)^{2.5}}$
Heat capacity	$(\rho C_P)_{\text{hnf}} =$ $[(1-\delta_2)\{(1-\delta_1)(\rho C_P)_f + \delta_1(\rho C_P)_{s1}\}] + \delta_2(\rho C_P)_{s2}$
Thermal expansion coefficient	$(\rho\beta)_{\text{hnf}} = [(1-\delta_2)\{(1-\delta_1)(\rho\beta)_f + \delta_1(\rho\beta)_{s1}\}] + \delta_2(\rho\beta)_{s2}$
Thermal conductivity	$\kappa_{\text{hnf}} = \kappa_{\text{bf}}\frac{\kappa_{s2}+(n_1-1)\kappa_{\text{bf}}-(n_1-1)\delta_2(\kappa_{\text{bf}}-\kappa_{s2})}{\kappa_{s2}+(n_1-1)\kappa_{\text{bf}}+\delta_2(\kappa_{\text{bf}}-\kappa_{s2})}$, where $\kappa_{\text{bf}} = \kappa_f\frac{\kappa_{s1}+(n_1-1)\kappa_f-(n_1-1)\delta_1(\kappa_f-\kappa_{s1})}{\kappa_{s1}+(n_1-1)\kappa_f+\delta_1(\kappa_f-\kappa_{s1})}$
Electrical conductivity	$\sigma_{\text{hnf}} = \sigma_{\text{bf}}\left[\frac{\sigma_{s2}(1+2\delta_2)+2\sigma_{\text{bf}}(1-\delta_2)}{\sigma_{s2}(1-\delta_2)+\sigma_{\text{bf}}(2+\delta_2)}\right]$, where $\sigma_{\text{bf}} = \sigma_f\left[\frac{\sigma_{s1}(1+2\delta_1)+2\sigma_f(1-\delta_1)}{\sigma_{s1}(1-\delta_1)+\sigma_f(2+\delta_1)}\right]$

Table 2 Thermophysical properties of water and nanoparticles

	$\rho(\text{kg/m}^3)$	C_p(J/kgK)	κ(W/mK)	σ(s/m)	β(1/K)
H_2O (f)	997.1	4179	0.613	5.5×10^{-6}	21×10^{-5}
Cu ($s1$)	8933	385	401	59.6×10^{6}	1.67×10^{-5}
Al_2O_3 ($s2$)	3970	765	40	35×10^{6}	0.85×10^{-5}

characteristics of water and nanoparticles are presented in Table 2.

The relevant dimensionless variables and parameters are as follows:

$$x = \frac{x^* u_{1_0}^*}{\nu_f} = \frac{x^*}{L_{\text{ref}}},\ y = \frac{y^* u_{1_0}^*}{\nu_f} = \frac{y^*}{L_{\text{ref}}},\ t = \frac{t^* u_{1_0}^{*^2}}{\nu_f},\ u_1 = \frac{u_1^*}{u_{1_0}^*},$$
$$u_2 = \frac{u_2^*}{u_{1_0}^*},\ \theta = \frac{\theta^* - \theta_\infty^*}{\theta_w^* - \theta_\infty^*},\ \text{Gr} = \frac{\nu_f g\beta_f(\theta_w^* - \theta_\infty^*)}{u_{1_0}^{*^3}},\ \text{Pr} = \frac{\nu_f}{\alpha_f},$$
$$N = \frac{k_f k_e}{4\sigma_s\theta_\infty^{*^3}},\quad M = \frac{\sigma_f B_0^2\nu_f}{\rho_f u_{1_0}^2}. \tag{9}$$

After non-dimensionlization, Eqs (1), (2), and (7) become

$$\frac{\partial u_1}{\partial x} + \frac{\partial u_2}{\partial y} = 0. \tag{10}$$

$$\frac{\partial u_1}{\partial t} + u_1\frac{\partial u_1}{\partial x} + u_2\frac{\partial u_1}{\partial y} = \frac{A_2}{A_1}\frac{\partial^2 u_1}{\partial y^2} + \frac{A_3}{A_1}\text{Gr}\theta - \frac{A_4}{A_1}\text{Mu}_1. \tag{11}$$

$$\frac{\partial\theta}{\partial t} + u_1\frac{\partial\theta}{\partial x} + u_2\frac{\partial\theta}{\partial y} = \frac{1}{\text{Pr}}\frac{A_6}{A_5}\frac{\partial^2\theta}{\partial y^2} + \frac{4}{3N\,\text{Pr}\,A_5}\frac{\partial^2\theta}{\partial y^2}. \tag{12}$$

With initial and boundary conditions:

$$
\begin{aligned}
t \le 0 &: u_1 = 0, \quad u_2 = 0, \ \theta = 0 \ \text{ for all } x \text{ and } y. \\
t > 0 &: u_1 = 1, \quad u_2 = 0, \ \theta = 1 \ \text{ at } y = 0. \\
& \quad u_1 = 0, \qquad\qquad\quad \theta = 0 \ \text{ at } x = 0. \\
& \quad u_1 \to 0, \qquad\qquad\quad \theta \to 0 \text{ as } y \to \infty.
\end{aligned}
\tag{13}
$$

where

$$A_1 = \left[(1-\delta_2)\left\{(1-\delta_1) + \delta_1 \frac{\rho_{s_1}}{\rho_f}\right\}\right] + \delta_2 \frac{\rho_{s_2}}{\rho_f}.$$

$$A_2 = \frac{1}{(1-\delta_1)^{2.5}(1-\delta_2)^{2.5}}.$$

$$A_3 = \left[(1-\delta_2)\left\{(1-\delta_1) + \delta_1 \frac{(\rho\beta)_{s_1}}{(\rho\beta)_f}\right\}\right] + \delta_2 \frac{(\rho\beta)_{s_2}}{(\rho\beta)_f}.$$

$$A_4 = \frac{\sigma_{bf}}{\sigma_f}\left[\frac{\sigma_{s_2}(1+2\delta_2) + 2\sigma_{bf}(1-\delta_2)}{\sigma_{s_2}(1-\delta_2) + \sigma_{bf}(2+\delta_2)}\right].$$

$$A_5 = \left[(1-\delta_2)\left\{(1-\delta_1) + \delta_1 \frac{(\rho C_p)_{s_1}}{(\rho C_p)_f}\right\}\right] + \delta_2 \frac{(\rho C_p)_{s_2}}{(\rho C_p)_f}.$$

$$A_6 = \frac{\kappa_{bf}}{\kappa_f}\frac{\left[\kappa_{s_2} + (n_1-1)\kappa_{bf} - (n_1-1)\delta_2(\kappa_{bf} - \kappa_{s_2})\right]}{\left[\kappa_{s_2} + (n_1-1)\kappa_{bf} + \delta_2(\kappa_{bf} - \kappa_{s_2})\right]}.$$

The analytical solution of Eq. (12) in the absence of inertial terms and radiation parameter, subject to the boundary conditions (13) by using the Laplace transform method, is given by

$$\theta = \operatorname{erfc}\left[\frac{y\sqrt{\Pr}}{2\sqrt{\frac{A_6}{A_5}t}}\right]. \tag{14}$$

The most important physical quantities which are defined on the wall are C_f and Nu as mentioned below:

$$C_f = \frac{\tau_w}{\rho_f u_{1_0}^{*2}},\ \mathrm{Nu} = \frac{q_w L_{\text{ref}}}{\kappa_f(\theta_w^* - \theta_\infty^*)}. \tag{15}$$

Here, τ_w denotes skin friction (shear stress) and q_w means heat flux (rate of heat transfer) as:

$$\tau_w = \mu_{\text{hnf}}\left(\frac{\partial u_1^*}{\partial y^*}\right)_{y^*=0}, q_w = -\kappa_{\text{hnf}}\left(\frac{\partial \theta^*}{\partial y^*}\right)_{y^*=0}. \tag{16}$$

Applying dimensionless variables defined in (9), we get the dimensionless form of

$$C_f = \frac{1}{(1-\delta_1)^{2.5}(1-\delta_2)^{2.5}}\left(\frac{\partial u_1}{\partial y}\right)_{y=0}. \tag{17}$$

and the dimensionless form of

$$\text{Nu} = -\frac{\kappa_{\text{hnf}}}{\kappa_f}\left(\frac{\partial \theta}{\partial y}\right)_{y=0}. \tag{18}$$

3 Numerical Technique

The prevailing set of partial differential Eqs. (10)–(12), together with the boundary conditions (13), are untangled by Crank–Nicolson type finite difference numerical procedure. The analogous finite difference equations are as follows:

$$\left[\frac{(u_1)_{i,j}^{n+1} - (u_1)_{i-1,j}^{n+1} + (u_1)_{i,j}^{n} - (u_1)_{i-1,j}^{n} + (u_1)_{i,j-1}^{n+1} - (u_1)_{i-1,j-1}^{n+1} + (u_1)_{i,j-1}^{n} - (u_1)_{i-1,j-1}^{n}}{4\Delta x}\right] + \left[\frac{\left[(u_2)_{i,j}^{n+1} - (u_2)_{i,j-1}^{n+1} + (u_2)_{i,j}^{n} - (u_2)_{i,j-1}^{n}\right]}{2\Delta y}\right] = 0. \tag{19}$$

$$\begin{aligned}
&\left[\frac{(u_1)_{i,j}^{n+1} - (u_1)_{i,j}^{n}}{\Delta t}\right] + (u_1)_{i,j}^{n}\left[\frac{(u_1)_{i,j}^{n+1} - (u_1)_{i-1,j}^{n+1} + (u_1)_{i,j}^{n} - (u_1)_{i-1,j}^{n}}{2\Delta x}\right] \\
&+ (u_2)_{i,j}^{n}\left[\frac{(u_1)_{i,j+1}^{n+1} - (u_1)_{i,j-1}^{n+1} + (u_1)_{i,j+1}^{n} - (u_1)_{i,j-1}^{n}}{4\Delta y}\right] \\
&= \frac{A_3}{A_1}\frac{Gr}{2}\left[\theta_{i,j}^{n+1} + \theta_{i,j}^{n}\right] \\
&+ \frac{A_2}{A_1}\left[\frac{(u_1)_{i,j-1}^{n+1} - 2(u_1)_{i,j}^{n+1} + (u_1)_{i,j+1}^{n+1} + (u_1)_{i,j-1}^{n} - 2(u_1)_{i,j}^{n} + (u_1)_{i,j+1}^{n}}{2(\Delta y)^2}\right] \\
&- \frac{A_4}{A_1}\frac{M}{2}\left[(u_1)_{i,j}^{n+1} + (u_1)_{i,j}^{n}\right].
\end{aligned} \tag{20}$$

$$\left[\frac{\theta_{i,j}^{n+1} - \theta_{i,j}^{n}}{\Delta t}\right] + (u_1)_{i,j}^{n}\left[\frac{\left(\theta_{i,j}^{n+1} - \theta_{i-1,j}^{n+1} + \theta_{i,j}^{n} - \theta_{i-1,j}^{n}\right)}{2\Delta x}\right]$$

$$+ (u_2)^n_{i,j}\left[\frac{\theta^{n+1}_{i,j+1} - \theta^{n+1}_{i,j-1} + \theta^n_{i,j+1} - \theta^n_{i,j-1}}{4\Delta y}\right]$$
$$= \left[\frac{1}{\Pr}\frac{A_6}{A_5} + \frac{4}{3N\Pr A_5}\right]\left[\frac{\theta^{n+1}_{i,j-1} - 2\theta^{n+1}_{i,j} + \theta^{n+1}_{i,j+1} + \theta^n_{i,j-1} - 2\theta^n_{i,j} + \theta^n_{i,j+1}}{2(\Delta y)^2}\right]. \tag{21}$$

The above equations are solved for the numerical solutions of temperature and velocity portraits via the Thomas algorithm [26]. Details of the method of solving difference equations, including stability and convergence, have been presented by Soundalgekar and Ganesan [27], Muthukumaraswamy and Ganesan [28], Ganesan and Rani [29], and Ramachandra Prasad et al. [30]. A grid-independent test has been conducted to obtain an economical and reliable grid system for the computations, which is presented in Fig. 2. The step sizes in x and y directions $\Delta x = 0.05$ and $\Delta y = 0.25$ were noted to give accurate results. Also, the time step size dependency has been carried out, from which $\Delta t = 0.01$ was found to give a reliable result. To ensure the correctness of the computational data, the temperature patterns of this research when $\delta_2 = 0$ and in the absence of radiation parameter are compared with the analytical solution presented by Eq. (14) in Fig. 3a and established to be in excellent agreement. Further, the velocity patterns of this research when Gr = 2, Pr = 6.2, $\delta_1 = 0$, $\delta_2 = 0$, and in the absence of magnetic and radiation parameters are compared with Soundalgekar [31] in Fig. 3b and proved to be in excellent agreement. This authenticates that the present computational technique is fit for this type of problem.

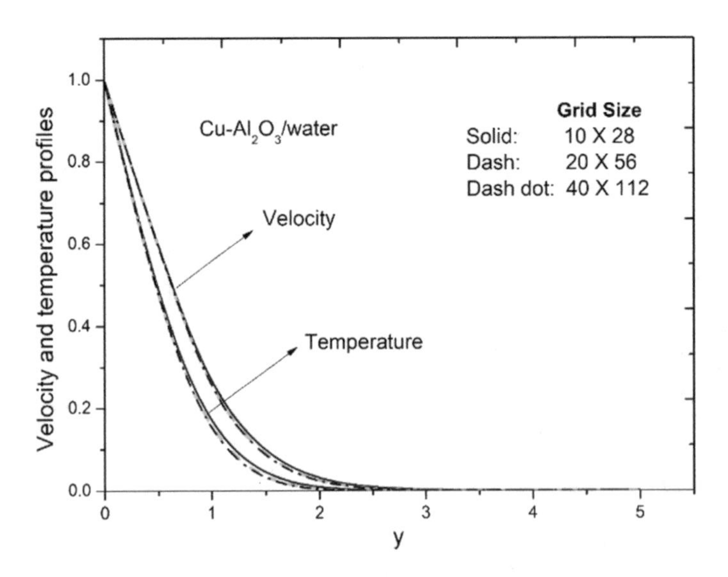

Fig. 2 The mesh independence analysis for velocity and temperature patterns

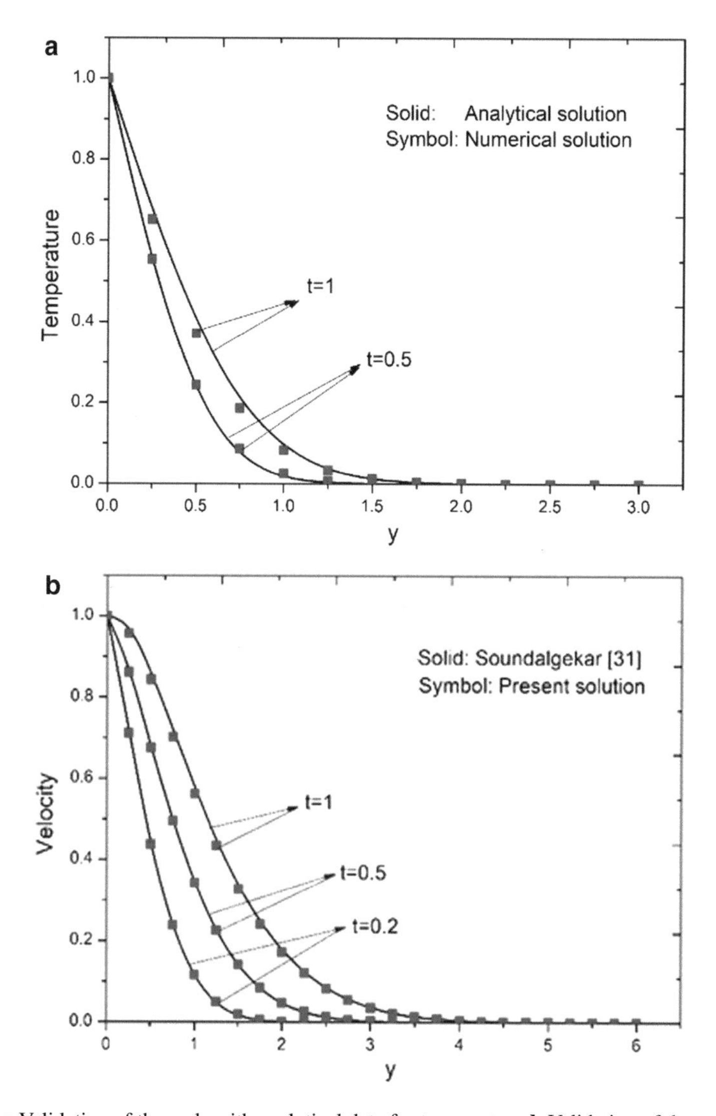

Fig. 3 **a** Validation of the code with analytical data for temperature **b** Validation of the code with Soundalgekar [31]

4 Results and Discussion

To comprehend the physics of the current problem, the impact of various relevant parameters on the velocity, temperature, Nusselt number, and skin friction are analysed through graphs 4–23. In this scrutiny, the default values of the parameters are set as $M = 2$, $\Pr = 6.2$, $\mathrm{Gr} = 5$, $\delta_1 = 0.1$, $\delta_2 = 0.04$, $X = 1$, $N = 3$, $t = 1$ unless otherwise defined.

The corollary of M over the velocity and temperature distributions is enlightened in Figs. 4 and 5 for both nanofluid and hybrid nanofluid. As the magnetic field strength increases, the retarding force increases. Consequently, the velocity decreases along with the momentum boundary layer thickness, but this is contrary in the temperature profile case. The presence of a transverse magnetic field creates Lorentz force, which arises from the magnetic field and electric field interaction

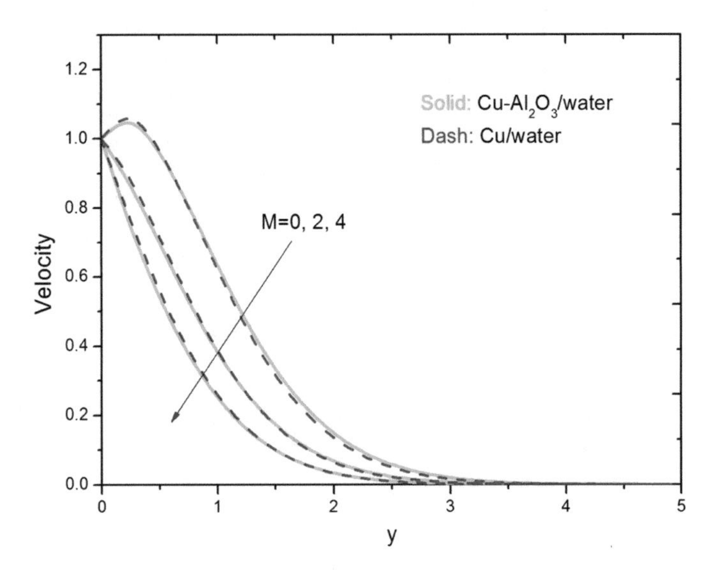

Fig. 4 The analogy of M on velocity field

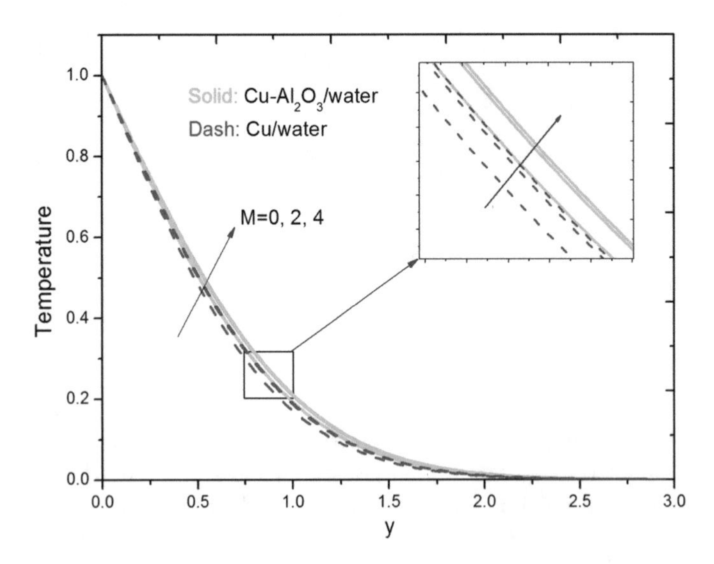

Fig. 5 The analogy of M on temperature field

during the motion of an electrically conducting fluid, which tends to suppress the velocity and enhance the temperature field. Figures 6 and 7 are drawn to witness the consequence of M on the skin friction coefficient and Nusselt number. Figures show that both skin friction coefficient and Nusselt number shrink with increasing M for both nanofluid and hybrid nanofluid cases. The consequence of Gr on the

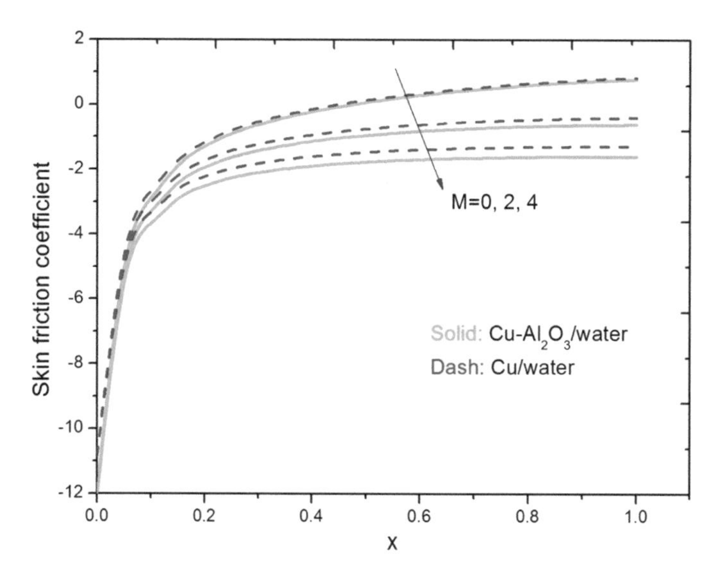

Fig. 6 The analogy of M on skin friction coefficient

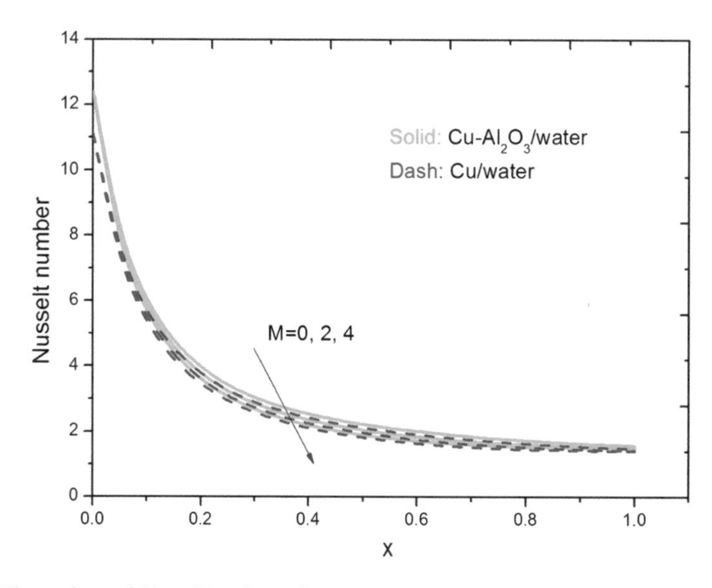

Fig. 7 The analogy of M on Nusselt number

velocity and temperature profiles is elucidated in Figs. 8 and 9. It is observed that as the thermal Grashof number increases, the velocity profiles increase. However, the reverse trend is observed in the temperature profile in Fig. 9 for both the fluids, because Gr is the ratio of thermal buoyancy force to viscous hydrodynamic force in the boundary layer. The skin friction coefficient and Nusselt number are raising functions of *Gr* for both the nano and hybrid nanofluids, presented in Figs. 10 and

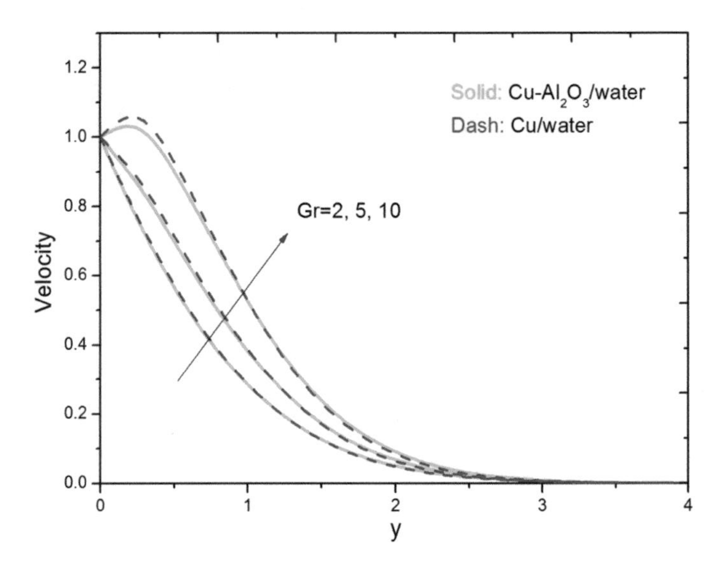

Fig. 8 The analogy of Gr on velocity field

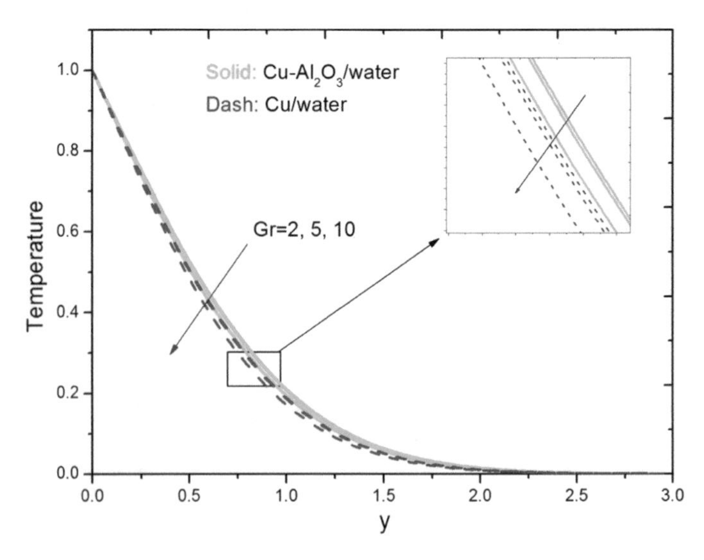

Fig. 9 The analogy of Gr on temperature field

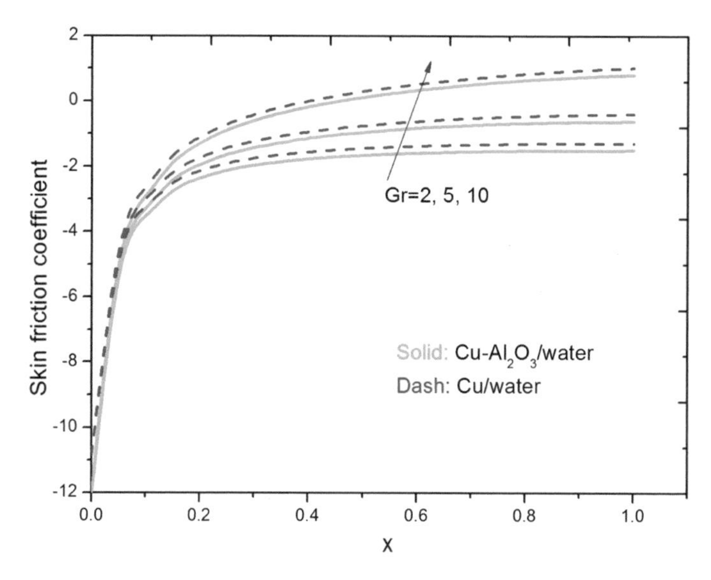

Fig. 10 The analogy of Gr on skin friction coefficient

11.

Figure 12 explicates that velocity is decreasing with increasing values of radiation parameter N for both the fluids. The radiation parameter versus temperature profile graph is given in Fig. 13. The nature of its curves shows that there is a decrease in temperature profiles with an increase in radiation parameter. Figure 14 is depicted to

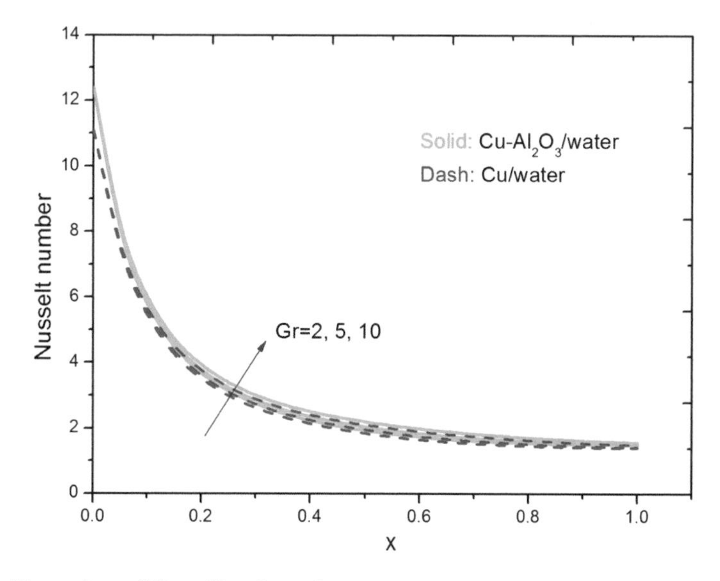

Fig. 11 The analogy of Gr on Nusselt number

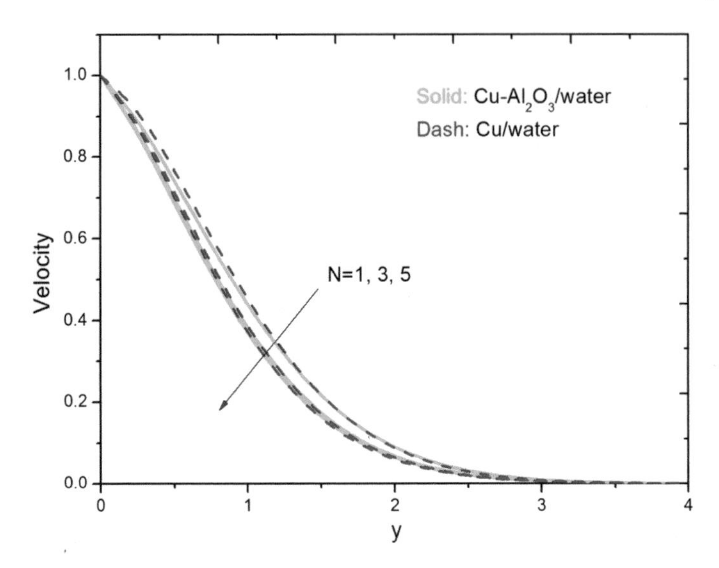

Fig. 12 The analogy of N on velocity field

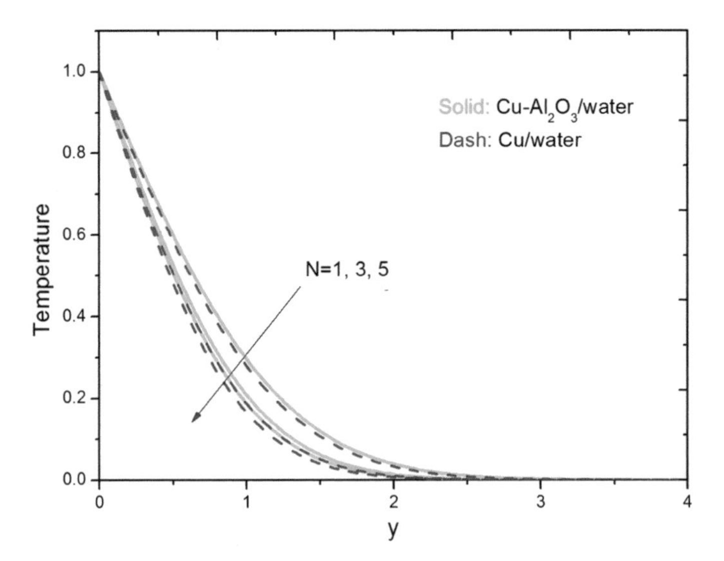

Fig. 13 The analogy of N on temperature field

explore the effect of the radiation parameter N on the skin friction profile for both the fluids; the skin friction coefficient is a declining radiation parameter function. Figure 15 explicates that Nusselt number is an increasing function of the radiation parameter for both the fluids. We can make out from Fig. 16 that velocity is a decreasing function of nanoparticle volume fraction near the plate. This effect is reversed away from the plate for both the nanofluids. The temperature distribution for

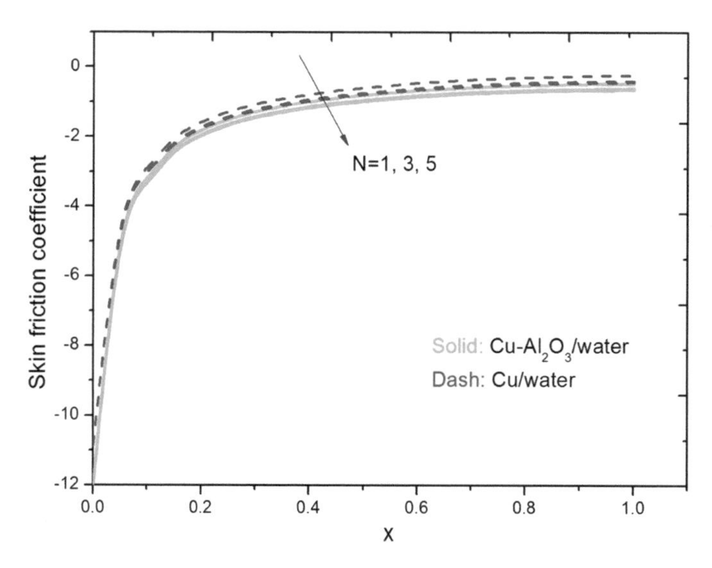

Fig. 14 The analogy of N on skin friction coefficient

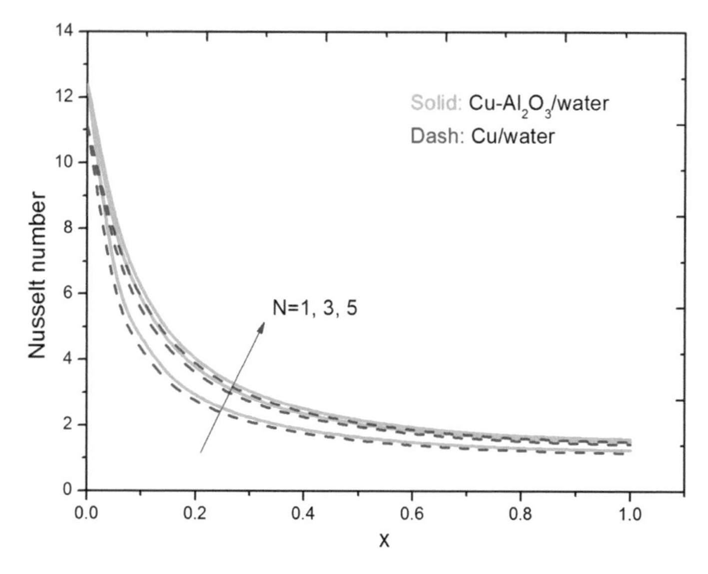

Fig. 15 The analogy of N on Nusselt number

nanoparticle volume fraction for both the fluids is shown in Fig. 17. Physically, the nanoparticles dissipate energy in the form of heat. Simultaneously adding up more nanoparticles may decelerate the flow field and exert more energy, enhancing the temperature and thickening the thermal boundary layer. An increase in solid volume fraction decelerates the flow velocity near the plate, reducing the skin friction for

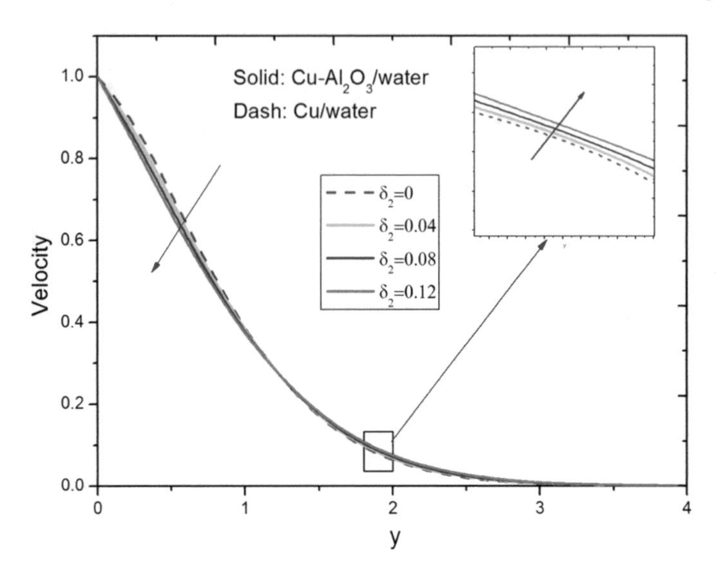

Fig. 16 The analogy of δ_2 on velocity field

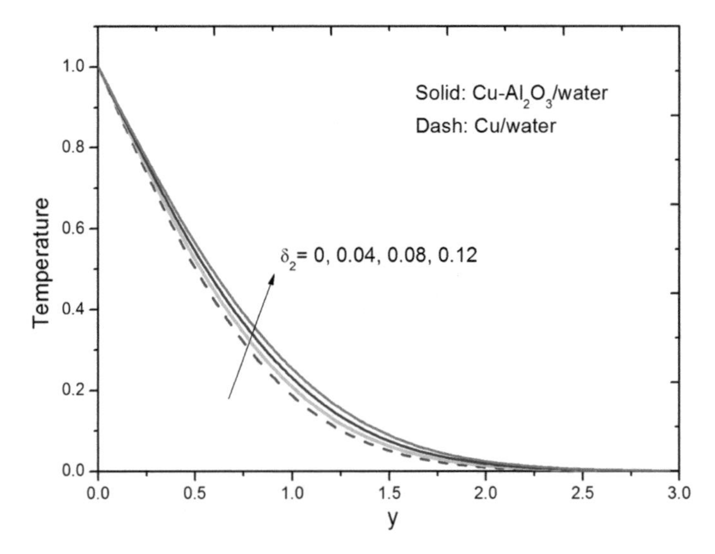

Fig. 17 The analogy of δ_2 on temperature field

both nanofluid and hybrid nanofluid, depicted in Fig. 18. It is explored in Fig. 19 that Nusselt number is increasing with nanoparticle volume fraction for both the fluids. The non-dimensional heat transfer rate through $Cu-Al_2O_3-H_2O$ is higher than that of $Cu-H_2O$. It is expounded in Figs. 20, 21, 22, and 23 that both the velocity and temperature profiles and their boundary thicknesses increase with time t for both

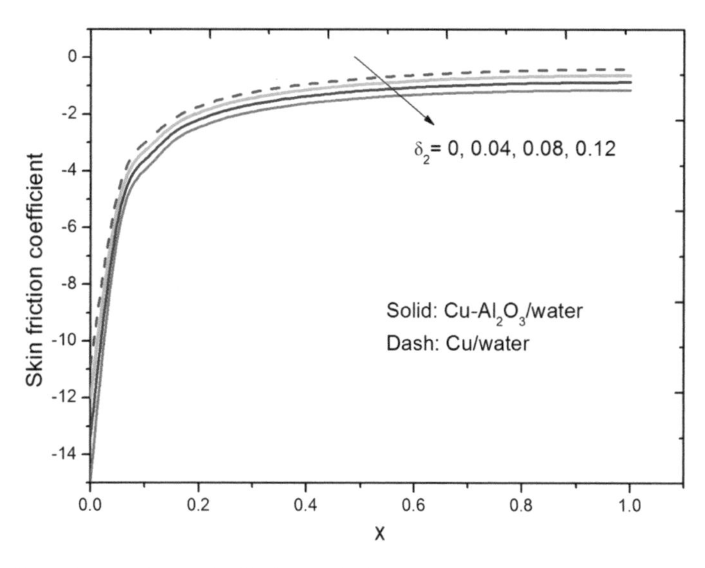

Fig. 18 The analogy of δ_2 on skin friction coefficient

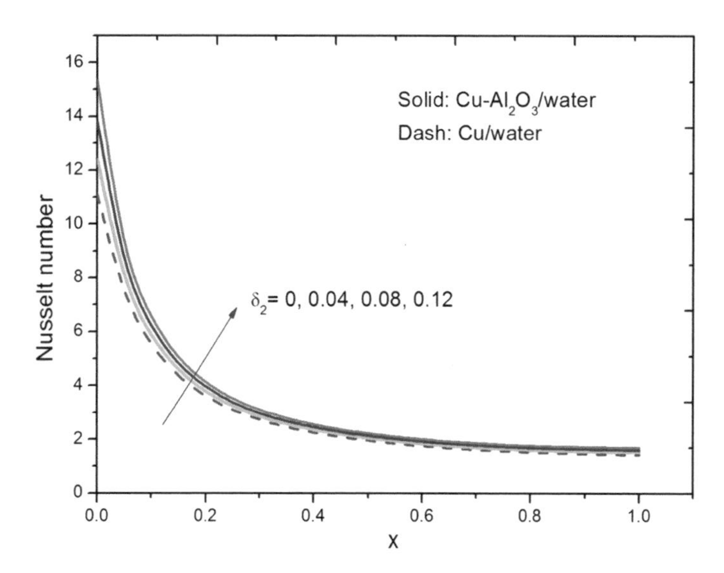

Fig. 19 The analogy of δ_2 on Nusselt number

the fluids. Further, the skin friction coefficient increases, and the Nusselt number declines with time t for both the fluids.

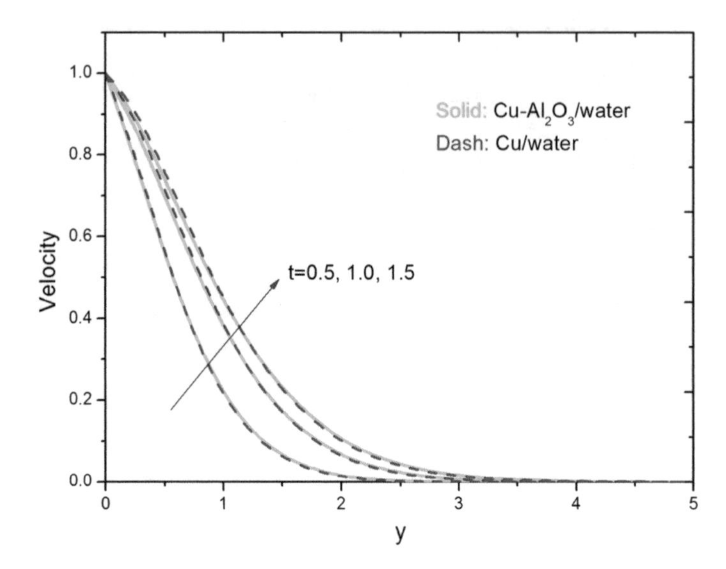

Fig. 20 The analogy of t on velocity field

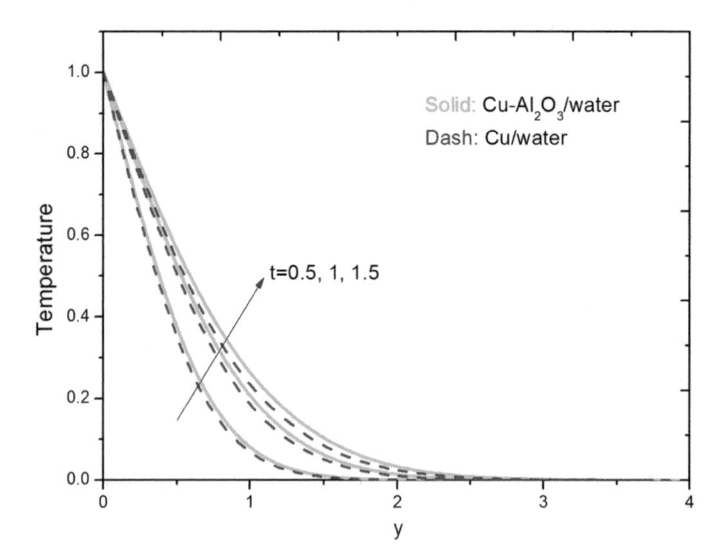

Fig. 21 The analogy of t on temperature field

5 Conclusions

In this study, impacts of vital parameters, namely magnetic parameter, radiation parameter, Grashof number, and nanoparticle volume fraction, are analysed on the unsteady free convective flow of a hybrid nanofluid past a moving vertical plate.

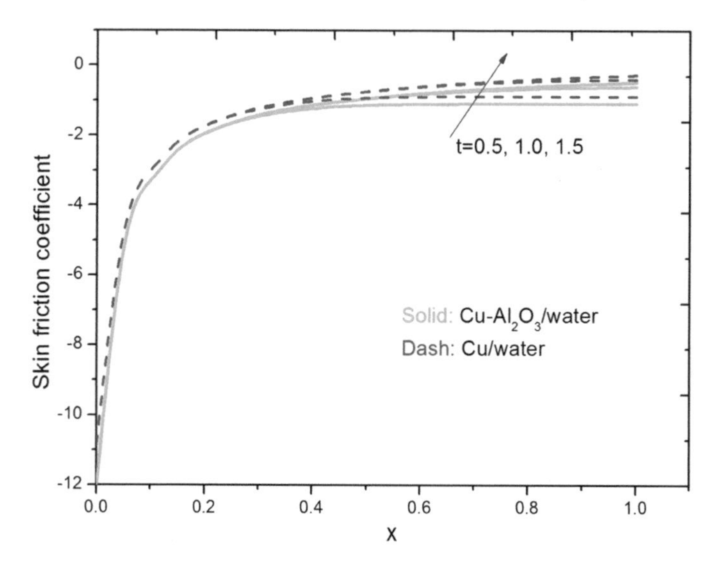

Fig. 22 The analogy of t on skin friction coefficient

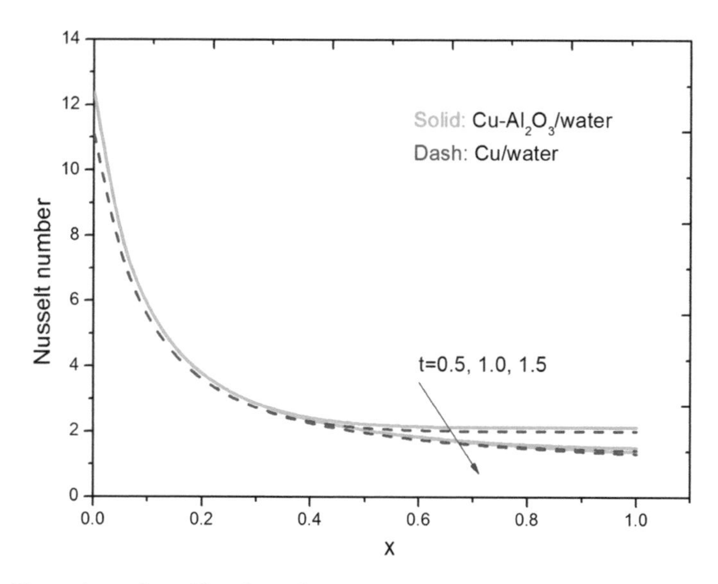

Fig. 23 The analogy of t on Nusselt number

The PDEs modelling the flow are solved using a robust finite difference numerical method. The outcomes of the study are precised as underneath.

1. With an increase in M, velocity profiles, skin friction coefficient, and Nusselt number decrease while the temperature rises with M for both nano and hybrid nanofluids.

2. As Gr increases, velocity, skin friction, and Nusselt number are escalating for both the fluids. However, the temperature decreases with Gr.
3. Velocity, temperature, and skin friction fields are declining with an increase in N for both the fluid cases. However, the Nusselt number shoots up with an increase in N.
4. With a rise in nanoparticle volume fraction, the temperature and Nusselt number enhance for both the fluids, while the skin friction coefficient diminishes.

References

1. Sparrow EM, Cess RD (1961) The effect of a magnetic field on free convection heat transfer. Int J Heat Mass Transf 3(4):267–274
2. Vajravelu K (1988) Hydromagnetic convection at a continuous moving surface. Acta Mech 72:223
3. Rajesh V, Varma SVK (2009) Radiation and mass transfer effects on MHD free convection flow past an exponentially accelerated vertical plate with variable temperature. ARPN J Eng Appl Sci 4(6):20–26
4. Rajesh V, Varma SVK (2010) Heat source effects on MHD flow past an exponentially accelerated vertical plate with variable temperature through a porous medium. Int J Appl Math Mech 6(12):68–78
5. Sarma D, Pandit KK (2018) Effects of Hall current, rotation and Soret effects on MHD free convection heat and mass transfer flow past an accelerated vertical plate through a porous medium. Ain Shams Eng J 9:631–646
6. Sambath P, Pullepu B, Kannan RM (2019) Unsteady free convective MHD radiative mass transfer flow past from an inclined vertical plate with heat source and sink. AIP Conf Proc 2112:020005
7. Rajesh V, Anwar BO, Mallesh MP (2016) Transient nanofluid flow and heat transfer from a moving vertical cylinder in the presence of thermal radiation: Numerical study. J Nanomater Nanoeng Nano Syst 230(1):3–16
8. Rajesh V, Chamkha AJ, Mallesh MP (2016) Transient MHD free convection flow and heat transfer of nanofluid past an impulsively started semi-infinite vertical plate. J Appl Fluid Mech 9(5):2457–2467
9. Rajesh V, Chamkha AJ, Sridevi C, Al-Mudhaf AF (2016) A numerical investigation of transient MHD free convective flow of a nanofluid over a moving semi-infinite vertical cylinder. Eng Comput 34(5):1393–1412
10. Rajesh V, Chamkha AJ, Mallesh MP (2016) Nanofluid flow past an impulsively started vertical plate with variable surface temperature. Int J Numer Meth Heat Fluid Flow 26(1):328–347
11. Minea AA (2017) Hybrid nanofluids based on, Al_2O_3, Tio_2 and SiO_2 Numerical evaluation of different approaches. Int J Heat Mass Transfer 104:852–860
12. Sahoo RR, Ghosh P, Sarkar J (2017) Performance analysis of a louvered fin automotive radiator using hybridnanofluid as coolant. Heat Transfer-Asian Res 46(7):978–995
13. Anjali Devi SP, Suriya Uma Devi S (2016) Numerical investigation of hydromagnetic hybrid Cu–Al_2O_3/water nanofluid flow over a permeable stretching sheet with suction. Int J Nonlinear Sci Numer Simul 17(5):249–257
14. Moldoveanu GM, Ibanescu C, Danu M, Minea AA (2018) Viscosity estimation of, Al_2O_3, SiO_2 nanofluids and their hybrid: an experimental study. J Mol Liq 253:188–196
15. Moldoveanu GM, Minea AA, Iacob M, Ibanescu C, Danu V (2018) Experimental study on viscosity of stabilized Al_2O_3, TiO_2 nanofluids and their hybrid. Thermo chimica Acta 659:203–212

16. Ghadikolaeia SS, Gholiniab M (2020) 3D mixed convection MHD flow of GO-MoS_2 hybrid nanoparticles in H_2O–$(CH_2OH)_2$ hybrid base fluid under effect of H_2 bond. Int Commun Heat Mass Transfer 110:104371
17. Loganathan P, Nirmal Chand P, Ganesan P (2013) Radiation effects on an unsteady natural convective flow of a nanofluid past an infinite vertical plate. NANO: Brief Rep Rev 8(1):1350001, 1–10
18. Suresh S, Venkitaraj KP, Selvakumar P, Chandrasekar M (2011) Synthesis of Al_2O_3–Cu/water hybrid nanofluids using two step method and its thermo physical properties. Colloids Surf A: Physicochem Eng Aspects 388:41–48
19. Suriya Uma Devi S, Anjali Devi SP (2017) Heat transfer enhancement of Cu–Al_2O_3/water hybrid nanofluid flow over a stretching sheet. J Niger Math Soc 36(2):419–433
20. Hayat T, Nadeem S (2018) An improvement in heat transfer for rotating flow of hybrid nanofluid: a numerical study. Can J Phys 96:1420–1430
21. Minea A, Moldoveanu MG (2018) Overview of hybrid nanofluids development and benefits. J Eng Thermophys 27(4):507–514
22. Waini I, Ishak A, Pop I (2019) Unsteady flow and heat transfer past a stretching/shrinking sheet in a hybrid nanofluid. Int J Heat Mass Transf 136:288–297
23. Chamkha J, Dogonchi AS, Ganji DD (2019) Magneto-hydrodynamic flow and heat transfer of a hybrid nanofluid in a rotating system among two surfaces in the presence of thermalradiation and Joule heating. AIP Advances 9(2):025103–1–025103–14
24. Tiwari RK, Das MK (2007) Heat transfer augmentation in a two-sided lid-driven differentially-heated square cavity utilizing nanofluids. Int J Heat Mass Transf 50(9–10):2002–2018
25. Schlichting H, Gersten K (2001) Boundary layer theory, 8th edn. Springer-Verlag, New York, USA
26. Carnahan B, Luther HA, Wilkes JO (1969) Applied numerical methods. John Wiley & Sons, New York
27. Soundalgekar VM, Ganesan P (1981) Finite difference analysis of transient free convection with mass transfer on an isothermal vertical flat plate. Int J Eng Sci 19:757–770
28. Muthukumaraswamy R, Ganesan P (1998) Unsteady flow past an impulsively started vertical plate with heat and mass transfer. Heat Mass Transf 34:187–193
29. Ganesan P, Rani HP (2000) Unsteady free convection MHD flow past a vertical cylinder with heat and mass transfer. Int J Therm Sci 39:265–272
30. Ramachandra Prasad V, Bhaskar Reddy N, Muthucumaraswamy R (2007) Radiation and mass transfer effects on two-dimensional flow past an impulsively started infinite vertical plate. Int J Thermal Sci 46:1251–1258
31. Soundalgekar VM (1977) Free convection effects on the stokes problem for an infinite vertical plate. J Heat Transfer 99:499–501

Rayleigh Streaming Past a Wavy Wall with No Slip Suction Under a Transverse Magnetic Field

Fathimunnisa and Neetu Srivastava

Abstract This paper investigates the provoked flow pattern due to an impulsive motion of porous wavy wall with no slip suction velocity under the influence of magnetic field. It is assumed that the amplitude of the wall is much smaller than the developed boundary-layer thickness. This engendered flow pattern has two significant flow regimes: Regime-I near to the wall boundary where flow is affected due to waviness and viscosity and Regime-II is adjacent to the boundary-layer region, i.e., in the core region. Flows in these regimes are governed by boundary-layer equations, and their solution is determined by considering the expansion in terms of smallness of the amplitude oscillations. Results obtained are depicted graphically. It is shown that the magnetic field and suction parameter tend to pull down the effect of Reynolds stress.

Keywords Boundary layer · Rayleigh streaming · Wavy wall · Magnetic field

Nomenclature

v_x, v_y	Velocity Component
a^*	Dimensionless amplitude of the wall
B_0	Applied magnetic field

Fathimunnisa
Research Scholar, Department of Mathematics, Amrita School of Engineering. Amrita Vishwa Vidyapeetham, Bengaluru 560035, India

N. Srivastava (✉)
Associate Professor, Department of Mathematics, Amrita School of Engineering, Amrita Vishwa Vidyapeetham, Bengaluru 560035, India
e-mail: s_neetu@blr.amrita.edu

S. S. Ray et al. (eds.), *Applied Analysis, Computation and Mathematical Modelling in Engineering*, Lecture Notes in Electrical Engineering 897,
https://doi.org/10.1007/978-981-19-1824-7_3

Greek symbols

ν	Dynamic viscosity
Ψ	Stream function
ρ	Fluid density
σ	Electrical conductivity
ε	Small amplitude of oscillation
λ	Wave length

Non-dimensional Number

Ω_ω	Magnetic interaction parameter
S	Suction parameter

1 Introduction

The Rayleigh problem is the subject of determining the flow caused by a sudden movement of an infinitely long plate from rest, which leads to the study of viscous boundary layer. Rayleigh [1] initiated the theoretical explanation on acoustic streaming. The properties of acoustic streaming are typically seen when the size of the flow region is very small compared with the wavelength (λ), but much greater than the boundary-layer thickness δ. Cuevas and Ramos [2] investigated the effects of a uniform, transverse magnetic field in the steady streaming associated with the oscillatory boundary-layer flow of an electrically conducting fluid. Fathimunnisa et al. [3] have analyzed the flow pattern generated due to an interaction of standing wave in the presence of transverse magnetic field with a fluid, which is slowly discharged from the porous wall. In a viscous incompressible fluid, Shankar and Sinha [4] investigated the fluid motion caused by the impulsive motion of a wavy wall. The steady streaming caused by an oscillating viscous flow over a wavy wall is investigated by Lyne [5] using the conformal transformation approach. Schlichting [6] originally treated the steady streaming due to an oscillating incompressible flow over a curved boundary. Assuming the amplitude of oscillation $(U_\infty/\omega * \lambda)$ and amplitude of the wall (A/δ) to be small, Kaneko and Honji [7] investigated the double structures of steady streaming in an oscillatory viscous flow over a wavy wall. Vittori and Verzicco [8] analyzed the viscous oscillatory flow over a wavy wall of small amplitude, taking into account nonlinear effects and considering amplitude of fluid displacement (a^*) to be small, equal, and greater than the wavelength (L^*) of the wall perturbation.

The problem analyzed in this paper focuses on the flow about an infinite porous wavy wall executing vibrations about the know solution of the flat wall [Rayleigh]. These drag reduction models depict the effect of transverse magnetic field on the eddies generated due to Reynolds stresses in an oscillatory electrically conducting fluid is commonly used in practical situation such as in the field of biomechanics, civil engineering, in the study of the interaction between gravity water waves and sea bottom in the near shore region, transpiration cooling of re-entry vehicles, and rocket boosters. In this paper, we studied the effect of transverse magnetic field on the fluid flow generated by the uniform impulsive motion of the porous wavy wall with no slip suction velocity. The equation is then solved by the perturbation method with the main objective to investigate the flow pattern. The influence of magnetic field, suction, and waviness of the boundaries on the flow has been shown analytically through the analysis of the velocity and skin friction. To achieve this goal, the paper is organized as follows: In Sect. 2, we have come up with the mathematical formulation of the problem. In Sect. 3, perturbation method is applied to obtain results. In Sect. 4, the results are discussed and conclusions are drawn.

2 Formulation of the Problem

Consider an incompressible electrically conducting fluid in a semi-infinite region bounded by a wavy wall $y = A\cos\left(\frac{2\pi x}{\lambda}\right)$ in the presence of a uniform transverse magnetic field with strength B_0. In our analysis, the assumptions made are as follows (Fig. 1):

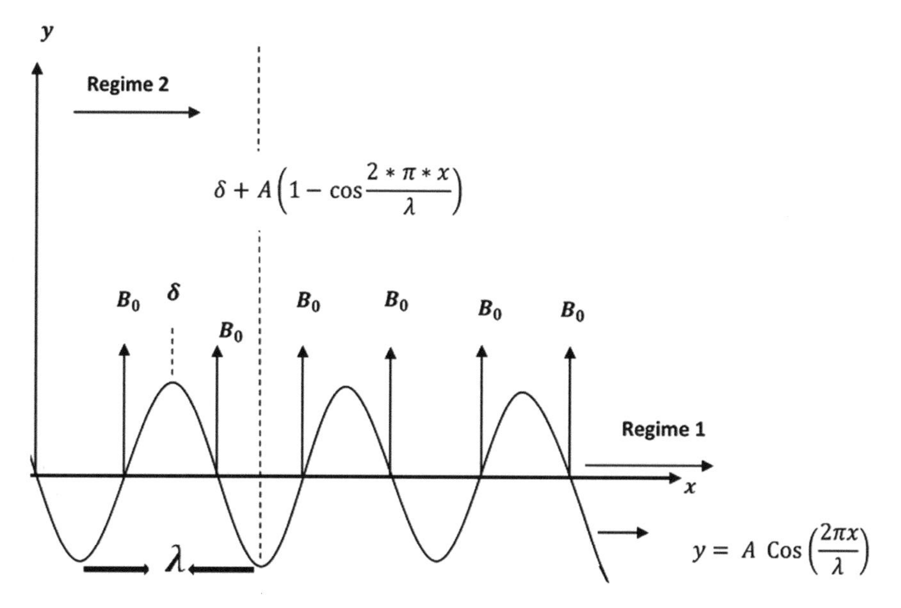

Fig. 1 Schematic of the problem

(i) We assumed that the amplitude of the wall is much smaller than the boundary-layer thickness.
(ii) We assumed that the quantity of fluid removed from the stream is so small that only fluid particles in the immediate neighborhood of wall are sucked away. This ensures the presence of no slip with suction.
(iii) We assumed that the induced magnetic field is much smaller than the applied magnetic field. This will be true, if the magnetic Reynolds number $R_m = \mu_e \sigma \lambda U_0$ is much less than unity.

For this two-dimensional situation, the equation and boundary conditions describing the motion of the incompressible fluid flow are as follows:

$$\frac{\partial v_x}{\partial x} + \frac{\partial v_y}{\partial y} = 0, \tag{1}$$

$$\frac{\partial v_x}{\partial t} + v_x \frac{\partial v_x}{\partial x} + v_y \frac{\partial v_x}{\partial y} = -\frac{1}{\rho}\frac{\partial P}{\partial x} + \nu\left(\frac{\partial^2 v_x}{\partial x^2} + \frac{\partial^2 v_x}{\partial y^2}\right) - \frac{\sigma B_0}{\rho}(E_z + v_x B_0), \tag{2}$$

$$\frac{\partial v_y}{\partial t} + v_x \frac{\partial v_y}{\partial x} + v_y \frac{\partial v_y}{\partial y} = -\frac{1}{\rho}\frac{\partial P}{\partial y} + \nu\left(\frac{\partial^2 v_y}{\partial x^2} + \frac{\partial^2 v_y}{\partial y^2}\right), \tag{3}$$

$$\left.\begin{array}{ll} v_x = 0 \quad v_y = V_0 & @ y = A\cos\left(\frac{2\pi x}{\lambda}\right) \\ v_x = e^{i\omega t} & \text{as} y \to \infty \end{array}\right\}, \tag{4}$$

where v_x, v_y is the velocity component in the x, y direction and a is the dimensionless amplitude of the wall. Since the boundary layer is thin, it is clear that the flow in it takes place parallel to the surface (i.e., y-component flow velocity v is very small as compared to x-component flow velocity). Since the velocity varies slowly along the x-axis, $\frac{\partial^2 v_x}{\partial x^2}$ may be neglected in comparison with $\frac{\partial^2 v_x}{\partial y^2}$ and comparing (2) with (3) we see that the derivative $\frac{\partial P}{\partial y}$ is small in comparison with $\frac{\partial P}{\partial x}$, i.e., $\frac{\partial P}{\partial y} \approx 0$.

Hence, the governing equations are given by

$$\frac{\partial v_x}{\partial x} + \frac{\partial v_y}{\partial y} = 0, \tag{5}$$

$$\frac{\partial v_x}{\partial t} + v_x \frac{\partial v_x}{\partial x} + v_y \frac{\partial v_x}{\partial y} = U\frac{\partial U}{\partial x} + \frac{\partial U}{\partial t} + \nu\left(\frac{\partial^2 v_x}{\partial y^2}\right) - \frac{\sigma B_0^2}{\rho}(v_x - U), \tag{6}$$

Because of the ponderomotive forces $\vec{J} \times \vec{B}$ per unit volume along x-axis which is equal to $\sigma B_0(E_z + v_x B_0)$, the condition for the pressure gradient along x-axis is as follows:

$$-\frac{1}{\rho}\frac{\partial p}{\partial x} - \frac{\sigma B_0}{\rho}(E_z + v_x B_0) = U\frac{\partial U}{\partial x} + \frac{\partial U}{\partial t}, \tag{7}$$

Equation (7) confirms the assumption that the velocity field generated in boundary layer will not be affected by the E_z, but it will modify the pressure gradient.

Let us define the dimensionless variables;

$$\tau = \omega t, \quad X = \frac{x}{\lambda}, \quad Y = \frac{y}{\delta}, \quad \delta = \sqrt{\frac{2\nu}{\omega}}, \quad v_x^* = \frac{v_x}{U_\infty},$$
$$v_y^* = \frac{v_y}{U_\infty * \delta/\lambda}, \quad U^* = \frac{U}{U_\infty}, \quad V_0^* = \frac{V_0}{U_\infty * \delta/\lambda}, \tag{7a}$$

After non-dimensionalization of Eqs. (5) and (6) using Eq. (7a) and on neglecting the asterisk, the equation of motion and the corresponding boundary conditions can be rewritten as follows:

$$\frac{\partial v_x}{\partial X} + \frac{\partial v_y}{\partial Y} = 0, \tag{8}$$

$$2\frac{\partial v_x}{\partial \tau} + \varepsilon\left(v_x * \frac{\partial v_x}{\partial X} + v_y * \frac{\partial v_x}{\partial Y}\right)$$
$$= 2 * \frac{\partial U}{\partial \tau} + \varepsilon * U\frac{\partial U}{\partial X} + \left(\frac{\partial^2 v_x}{\partial y^2}\right) - \Omega_\omega(v_x - U), \tag{9}$$

$$\left.\begin{matrix} v_x = 0 \quad v_y = S \ as \ Y = Y_w = a^* \cos(2 * \pi * X) \\ v_x = e^{i\tau} \qquad as \ Y \to \infty \end{matrix}\right\}, \tag{10}$$

where $\varepsilon = \frac{2*U_\infty}{\omega*\lambda}$, $\Omega_\omega = \frac{2*\sigma B_0^2}{\rho*\omega}$ and $S = \frac{V_0}{U_\infty*\delta/\lambda}$ are dimensionless parameters, respectively, and Ω_ω is the magnetic interaction parameter, which is defined as the ratio of the amplitude of the oscillation to the wave length λ and the ratio of the magnetic to the inertial forces, respectively. S is the suction parameter. In the present problem, we assume that there is small amplitude of oscillation, i.e., $\varepsilon \ll 1$ is considered; this assure that boundary-layer separation will not arise.

Let us define a new co-ordinate system in terms of old co-ordinates as follows:

$$X = \chi, \eta = Y - Y_w(X), \tag{11}$$

and therefore $$\frac{\partial}{\partial X} = \frac{\partial}{\partial \chi} + 2 * \pi * a^* \sin(2 * \pi * \chi)\frac{\partial}{\partial \eta}, \quad \frac{\partial}{\partial Y} = \frac{\partial}{\partial \eta}, \tag{12}$$

Equation (12) represents the change to the reference frame from the flat to wavy wall. Since waviness of the wall affects the volumetric flow conservation, it necessitates defining the flow defect equation.$U_\infty\delta$ is the volume flow defect at the crest of the wavy wall due to the presence of the boundary layer. The incompressibility condition signifies that this quantity must be preserved in any transversal section.

Hence, the volume defect balance equation is given as follows:

$$U_{\infty} * \delta = u_0 \left[\delta + A \left(1 - \cos \frac{2 * \pi * x}{\lambda} \right) \right],$$

where u_0 is the modified velocity due to the wavy wall. In dimensionless form, the above expression can be written as follows:

$$U_0(\chi) = \frac{u_0}{U_{\infty}} = \left[1 + a^* (1 - \cos 2 * \pi * \chi) \right]^{-1}, \tag{13}$$

Thus, the dimensionless outer flow is as follows:

$$U(\chi, \tau) = U_0(\chi) * e^{i\tau}.$$

Using the transformation (11), Eqs. (8)–(10) can be rewritten as follows:

$$\frac{\partial v_x}{\partial \chi} + 2 * \pi * a^* \sin(2 * \pi * \chi) * \frac{\partial v_x}{\partial \chi} + \frac{\partial v_y}{\partial \eta} = 0, \tag{14}$$

$$\begin{aligned} 2\frac{\partial v_x}{\partial \tau} + \varepsilon \left(v_x * \left[\frac{\partial v_x}{\partial \chi} + 2 * \pi * a^* \sin(2 * \pi * \chi) * \frac{\partial v_x}{\partial \eta} \right] + v_y * \frac{\partial v_x}{\partial \eta} \right) \\ = 2 * \frac{\partial U}{\partial \tau} + \varepsilon * U \frac{\partial U}{\partial \chi} + \left(\frac{\partial^2 v_x}{\partial \eta^2} \right) - \Omega_{\omega} (v_x - U), \end{aligned} \tag{15}$$

$$\left. \begin{array}{lll} v_x = 0 & v_y = S \; as & \eta = 0 \\ v_x = U_0(\chi) * e^{i\tau} & as & \eta \to \infty \end{array} \right\}, \tag{16}$$

To solve the above system of equations by the method of perturbation [Schlichting[6]], let us express the velocity in the form

$$\begin{aligned} v_x(\chi, \eta, \tau, \varepsilon, \Omega_{\omega}) = & v_x^{(1)}(\chi, \eta, \tau, \varepsilon, \Omega_{\omega}) \\ & + \varepsilon * v_x^{(2)}(\chi, \eta, \tau, \varepsilon, \Omega_{\omega}) + O(\varepsilon^2), \end{aligned} \tag{17}$$

where the superscripts 1 and 2 represent the first approximation and second approximation, respectively.

3 Solution of the Problem

3.1 *First Approximation* ($\varepsilon^{(0)}$)

Substituting Eq. (17) in Eq. (15) and equating the terms independent of ε, we get at $O(\varepsilon^0)$

$$2\frac{\partial v_x^{(1)}}{\partial \tau} - \frac{\partial^2 v_x^{(1)}}{\partial \eta^2} + \Omega_\omega\left(v_x^{(1)} - U\right) = 2 * \frac{\partial U}{\partial \tau}, \tag{18}$$

Let us substitute,
$v_x^{(1)} = F(\eta) * e^{i\tau}$ in Eq. (18), we obtain the solution as follows:

$$v_x^{(1)} = \mathrm{Re}\left\{\left[1 + a^*(1 - \cos 2 * \pi * \chi)\right]^{-1} * \left(1 - e^{J\eta}\right) * e^{i\tau}\right\}, \tag{19}$$

where $J = \sqrt{\Omega_\omega + 2 * I}$.

The corresponding stream function can be written as follows:

$$\Psi^{(1)} = -S\chi + \mathrm{Re}\left\{\left[1 + a^*(1 - \cos(2 * \pi * \chi))\right]^{-1} * \xi^{(1)}(\eta) * e^{i\tau}\right\}. \tag{20}$$

where $\xi^{(1)}(\eta) = \eta + \frac{e^{-J*\eta}}{J} - \frac{1}{J}$,

3.2 Second Approximation ($\varepsilon^{(1)}$)

The second approximation in terms of stream function can be written as follows:

$$2\frac{\partial^2 \Psi}{\partial \tau * \partial \eta} - \frac{\partial^3 \Psi^{(2)}}{\partial \eta^3} + \Omega_\omega \frac{\partial \Psi^{(2)}}{\partial \eta} = \varepsilon\left(U\frac{\partial U}{\partial \chi} - \frac{\partial \Psi^{(1)}}{\partial \eta} * \frac{\partial^2 \Psi^{(1)}}{\partial \chi * \partial \eta} + \frac{\partial \Psi^{(1)}}{\partial \chi} * \frac{\partial^3 \Psi^{(1)}}{\partial \eta^2}\right), \tag{21}$$

The right-hand side of Eq. (21) contains terms proportional to $\cos^2(\tau)$. This generates time-independent terms, which leads to steady streaming flow. Considering only this part of the velocity, the stream function $\Psi^{(2)}$ is written as follows:

$$\Psi^{(2)} = -S\chi - 2\pi a^* \sin(2 * \pi * \chi)\left[\xi^{(1)}(\eta) - \xi^{(2)}(\eta)\right]. \tag{22}$$

Once Eqs. (20) and (22) is introduced in Eq. (21), we arrive the equation which is as follows:

$$\xi^{(2)'''}(\eta) - \Omega_\omega \xi^{(2)'}(\eta) = -2 * e^{-a\eta}\sin(b\eta) - \frac{2 * \varepsilon}{[1 + a^*(1 - \cos(2 * \pi * \chi))]^3} - \frac{\varepsilon * e^{-2a\eta}}{[1 + a^*(1 - \cos(2 * \pi * \chi))]^3}\{1 + \cos(b\eta)\} + \frac{3 * \varepsilon * e^{-a\eta} * \cos(b\eta)}{[1 + a^*(1 - \cos(2 * \pi * \chi))]^3} - \frac{\varepsilon * e^{-a\eta} * \eta * a * \cos(b\eta)}{[1 + a^*(1 - \cos(2 * \pi * \chi))]^3}$$

$$- \frac{\varepsilon * e^{-a\eta} * \eta * b * \sin(b\eta)}{[1 + a^*(1 - \cos(2 * \pi * \chi))]^3}, \tag{23}$$

On solving Eq. (23) and eliminating the arbitrary constant with the boundary conditions, the equation obtained is as follows:

$$\begin{aligned} \xi^{(2)}(\eta) = & \frac{-[\gamma_{1p} + \gamma_{3p} + \gamma_{4p} + \gamma_{6p}]}{\sqrt{\Omega_\omega}} - D_{1p} + D_{4p} + D_{6p} \\ & + \frac{(\gamma_{1p} + \gamma_{3p} + \gamma_{4p} + \gamma_{6p})}{\sqrt{\Omega_\omega}} * e^{-\sqrt{\Omega_\omega}\eta} \\ & + e^{-a\eta}[D_{1p} * \cos(b\eta) + D_{2p} * \sin(b\eta)] + D_{3p} * \eta - D_{4p} * e^{-2a\eta} \\ & - e^{-2a\eta}[D_{5p} * \sin(b\eta) + D_{6p} * \cos(b\eta)] \\ & + e^{-a\eta} * \eta[D_{7p} \cos(b\eta) + D_{8p} * \sin(b\eta)], \end{aligned} \tag{24}$$

On differentiating Eq. (24), we get the equation as follows:

$$\begin{aligned} \xi^{(2)\prime}(\eta) = & -(\gamma_{1p} + \gamma_{3p} + \gamma_{4p} + \gamma_{6p}) * e^{-\sqrt{\Omega_\omega}\eta} \\ & + e^{-a\eta}[\gamma_{1p} * \cos(b\eta) + \gamma_{2p} * \sin(b\eta)] \\ & + \gamma_{3p} + \gamma_{4p} * e^{-2a\eta} \\ & + e^{-2a\eta}[\gamma_{5p} * \sin(b\eta) + \gamma_{6p} * \cos(b\eta)] \\ & + e^{-a\eta} * \eta[\gamma_{7p} \cos(b\eta) + \gamma_{8p} * \sin(b\eta)]. \end{aligned} \tag{25}$$

The above constants are defined in Appendix.

3.3 Skin Friction

The skin friction is defined as the shearing stress exerted by the fluid on the surface over which it flows, and is given by

$$\tau_{xy} = \mu * \frac{\partial v_x^{(1)}}{\partial \eta}, \tag{26}$$

On differentiating Eq. (19) and substituting in Eq. (26), we get the equation as follows:

$$\tau_{xy} = \frac{\mu}{[1 + a^*(1 - \cos(2 * \pi * \chi))]} * \{a * e^{-a\eta} \cos(\tau - b\eta) + b * e^{-a\eta} \sin(b\eta - \tau)\} \tag{27}$$

The skin friction coefficient at the wall is as follows:$c_f = \frac{2*\tau}{\rho * U^2}$,

Substituting Eq. (27) in above expression, the equation we get is as follows:

$$c_f = 2 * \nu * \left[1 + a^*(1 - \cos(2 * \pi * \chi))\right] \left\{a * e^{-a\eta} \cos(\tau - b\eta) + b * e^{-a\eta} \sin(b\eta - \tau)\right\}. \tag{28}$$

The above constants are defined in Appendix.

4 Result and Discussion

Intricacy of the considered problem is analyzed by the figures plotted using **MATHEMATICA 11.1**. These figures show the significant variation in the flow pattern due to the presence of the magnetic field and waviness. Fathimunnisa et.al [3] have presented the model of porous flat plate with the velocity distribution expression as follows:

For Flat Porous Plate

$$\begin{aligned}\zeta^{(2)} = & \delta\left\{-\frac{(\beta_{2p} + \beta_{3p} + \beta_{5p})}{\sqrt{2}\Gamma_B} - \alpha_{2p} + \alpha_{3p} - \alpha_{5p} + \frac{(\beta_{2p} + \beta_{3p} + \beta_{5p})}{\sqrt{2}\Omega_\omega}\right. \\ & * e^{-\sqrt{2}\Gamma_B\eta} + e^{-a\eta}\left[\alpha_{1p}\sin(b\eta) + \alpha_{2p}\cos(b\eta)\right] \\ & + \eta e^{-a\eta}\left[\alpha_{6p}\sin(b\eta) + \alpha_{7p}\cos(b\eta)\right] - \alpha_{3p}e^{-2a\eta} \\ & \left. + e^{-2a\eta}\left[\alpha_{4p}\sin(2b\eta) + \alpha_{5p}\cos(2b\eta)\right]\right\}.\end{aligned}$$

In this manuscript, the velocity expression will take a form.

For Porous Wavy Plate

$$\begin{aligned}\xi^{(2)}(\eta) = & \frac{-\left[\gamma_{1p} + \gamma_{3p} + \gamma_{4p} + \gamma_{6p}\right]}{\sqrt{\Omega_\omega}} - \beta_{1p} + \beta_{4p} + \beta_{6p} \\ & + \frac{(\gamma_{1p} + \gamma_{3p} + \gamma_{4p} + \gamma_{6p})}{\sqrt{\Omega_\omega}} * e^{-\sqrt{\Omega_\omega}\eta} \\ & + e^{-a\eta}\left[D_{1p} * \cos(b\eta) + D_{2p} * \sin(b\eta)\right] \\ & + D_{3p} * \eta - D_{4p} * e^{-2a\eta} - e^{-2a\eta}\left[D_{5p} * \sin(b\eta) + D_{6p} * \cos(b\eta)\right] \\ & + e^{-a\eta} * \eta\left[D_{7p}\cos(b\eta) + D_{8p} * \sin(b\eta)\right].\end{aligned}$$

where all the notations in the above expressions are presented in [3] appendix as well as in current appendix. The velocity distribution of the flow pattern developed in the current manuscript is compared with the velocity distribution established in [3].

Figure 2b represents the variation of stream function with magnetic interaction parameter. This graph depicts that the flow is influenced by the magnetic interaction

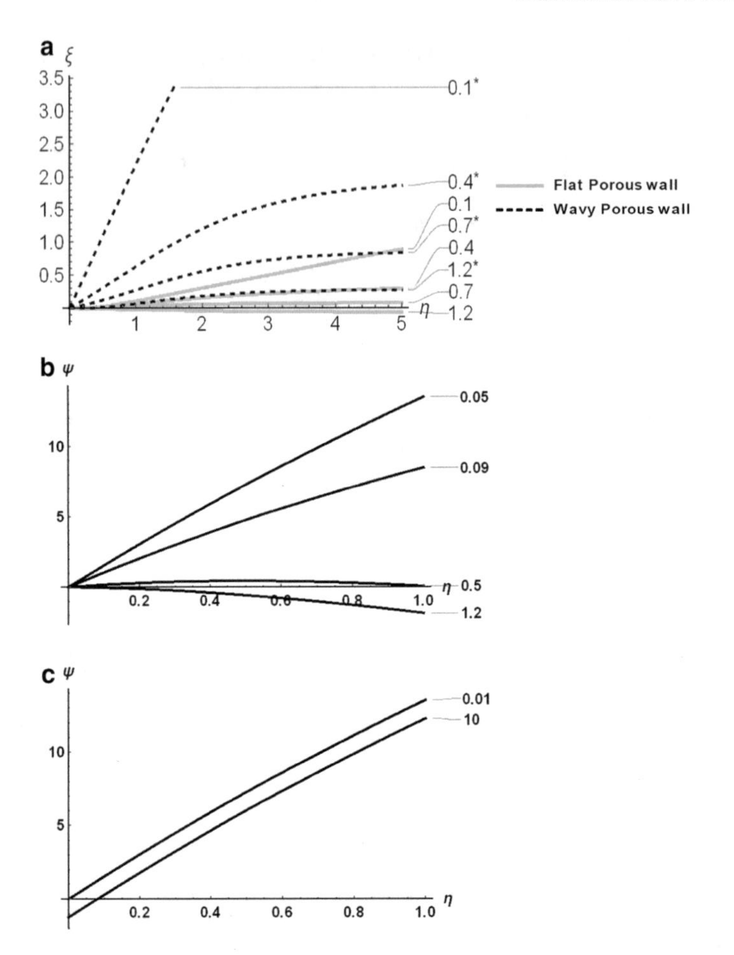

Fig. 2 **a** Velocity distribution perpendicular to porous wavy and flat wall for different Ω_ω. **b** Stream function plot for $S = 0.01$ and $\Omega_\omega = 0.05, 0.09, 0.5, 1.2$. **c** Stream function plot for $S = 0.01, 10$ and $\Omega_\omega = 0.05$

parameter, whereas Fig. 2c shows that for the fixed magnetic interaction parameter, the streamlines are damped as the suction parameter is increased.

Figure 3 represents the skin friction plot for different magnetic interaction parameters. It describes the flow Regime-I. It shows that skin friction is increasing with increase in magnetic field whereas it is decreasing from (0.4652, 0.0811) onwards. This figure depicts that the magnetic field will tend to retard the flow after $\eta = 0.4652$. From $\eta = 0$ to 0.46, flow will behave in a reverse way.

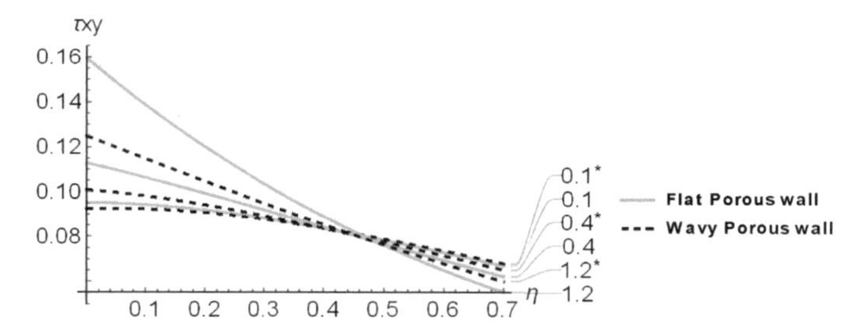

Fig. 3 Skin friction for Flat and Porous wavy wall for $\Omega_{\omega} = 0.1, 0.4, 1.2$

5 Conclusions

1. Damping phenomenon confirms that the increase in magnetic field tends to pull down the Reynold stress in both cases, i.e., wavy wall and flat plate. Hence, due to this phenomenon, a retarded zone is produced in the flow pattern.
2. Noteworthy increase in velocity for flow phenomena in case of wavy wall is seen due to the roughness of the wall.
3. In the considered flow field, the electromagnetic force is more dominant over the oscillatory motion present in the fluid. This helps in dampening the oscillatory flow. Also, this depicts that the flow changes due to the suction parameter.
4. Increase in magnetic parameter results the flow to have a retard zone due to the large shearing stress.
5. This figure also depicts that flat porous wall has large shearing stress near the boundary as compared to the wavy porous wall. This conveys that rough surface will help to reduce the large shearing stress as compared to the flat surface.

Our studies make it easier to develop powerful acoustofluidic devices for a variety of chemical and biomedical applications.

Acknowledgements The authors thank the university authorities for providing the necessary facilities. The authors express their sincere gratitude to Shaik Abdul Sayeed for his support.

Conflict of Interest Authors have no conflict of interest.

Appendix

$$a = \left(4 + \Omega_{\omega}^{2}\right)^{\frac{1}{4}} * \cos\left(\frac{1}{2}\text{Arg}[2 * i + \Omega_{\omega}]\right), \tag{29}$$

$$b = \left(4 + \Omega_{\omega}^{2}\right)^{\frac{1}{4}} * \sin\left(\frac{1}{2}\text{Arg}[2 * i + \Omega_{\omega}]\right), \tag{30}$$

$$A_1 = \frac{3b^2a - a^3 + \Omega_\omega * a}{3a^2 - \Omega_\omega - b^2}, \tag{31}$$

$$B_1 = \frac{6b^2a - 8a^3 + 2 * \Omega_\omega * a}{12a^2 - \Omega_\omega - b^2}, \tag{32}$$

$$E = \frac{A_1^2 - b^2}{2A_1}, \tag{33}$$

$$M_{1p} = \frac{2}{\left(3 * a^2 - \Omega_\omega - b^2\right) * \left(b^2 + A_1^2\right)}, \tag{34}$$

$$M_{2p} = \frac{2 * \varepsilon}{\Omega_\omega[1 + a^*(1 - \cos(2 * \pi * \chi))]^3}, \tag{35}$$

$$M_{3p} = \frac{\varepsilon}{[1 + a^*(1 - \cos(2 * \pi * \chi))]^3 * \left(2 * a * \Omega_\omega - 8 * a^2\right)}, \tag{36}$$

$$M_{4p} = \frac{\varepsilon}{[1 + a^*(1 - \cos(2 * \pi * \chi))]^3 * \left(12 * a^2 - \Omega_\omega - b^2\right) * \left(b^2 + B_1^2\right)}, \tag{37}$$

$$M_{5p} = \frac{3 * \varepsilon}{[1 + a^*(1 - \cos(2 * \pi * \chi))]^3 * \left(3 * a^2 - \Omega_\omega - b^2\right) * \left(b^2 + A_1^2\right)}, \tag{38}$$

$$M_{6p} = \frac{\varepsilon}{[1 + a^*(1 - \cos(2 * \pi * \chi))]^3 * \left(3 * a^2 - \Omega_\omega - b^2\right) * 2A_1\left(b^2 + E^2\right)}, \tag{39}$$

$$D_{1p} = M_{1p} * b + M_{5p} * A_1 + M_{6p} * a * E - M_{6p} * b^2, \tag{40}$$

$$D_{2p} = -M_{1p} * A_1 + M_{5p} * b + M_{6p} * a * b - M_{6p} * b * E, \tag{41}$$

$$D_{3p} = M_{2p}, \tag{42}$$

$$D_{4p} = M_{3p}, \tag{43}$$

$$D_{5p} = M_{4p} * b, \tag{44}$$

$$D_{6p} = M_{4p} * B_1, \tag{45}$$

$$D_{7p} = -M_{5p} * A_1 * a + M_{5p} * b^2, \tag{46}$$

$$D_{8p} = -M_{5p} * b * a - M_{5p} * A_1 * b, \tag{47}$$

$$\gamma_{1p} = -a * D_{1p} + b * D_{2p} + D_{7p}, \tag{48}$$

$$\gamma_{2p} = -b * D_{1p} - a * D_{2p} + D_{8p}, \tag{49}$$

$$\gamma_{3p} = D_{3p}, \tag{50}$$

$$\gamma_{4p} = 2 * a * D_{4p}, \tag{51}$$

$$\gamma_{5p} = 2 * a * D_{5p} + b * D_{6p} \tag{52}$$

$$\gamma_{6p} = -b * D_{5p} + 2 * a * D_{6p}, \tag{53}$$

$$\gamma_{7p} = -a * D_{7p} + b * D_{8p}, \tag{54}$$

$$\gamma_{8p} = -b * D_{7p} - a * D_{8p}. \tag{55}$$

References

1. Rayleigh L (1884) On the circulation of air observed in Kundt's tubes, and on some allied acoustical problems. Philos Trans R Soc Lond 175:1–21. https://doi.org/10.1098/rstl.1884.0002
2. Cuevas S, Ramos E (1997) Steady streaming in oscillatory viscous flow under a transverse magnetic field. Phys Fluids 9(5):1430–1434. https://doi.org/10.1063/1.869255
3. Fathimunnisa, Rakesh SG, Srivastava N (2019) Magnetohydrodynamic flow past a porous plate in presence of Rayleigh type streaming effect. J Porous Media 22(9). https://doi.org/10.1615/JPorMedia.2019026438
4. Shankar PN, Sinha UN (1976) The Rayleigh problem for a wavy wall. J fluid Mech 77(2):243–256. https://doi.org/10.1017/S0022112076002097
5. Lyne WH (1971) Unsteady viscous flow over a wavy wall. J Fluid Mech 50(1):33–48. https://doi.org/10.1017/S0022112071002441
6. Schlichting H (1955) Boundary layer theory. Pergamon, Oxford UK
7. Kaneko A, Honji H (1979) Double structures of steady streaming in the oscillatory viscous flow over a wavy wall. J Fluid Mech 93(4):727–736. https://doi.org/10.1017/S0022112079001993
8. Vittori G, Verzicco R (1998) Direct simulation of transition in an oscillatory boundary layer. J Fluid Mech 371:207–232. https://doi.org/10.1017/S002211209800216X

Non-Darcian Gravitactic Bioconvection with a Porous Saturated Vertical Vibration

K. Srikanth and Virendra Kumar

Abstract A linear stability analysis is applied to study the onset of bioconvection in a suspension of negatively geotactic (gravitactic) swimmers saturated with a non-Darcy porous fluid layer under the effect of high-frequency and small-amplitude vertical vibrations. The time-averaged formulation is used to write the closed system of equations for average quantities and amplitudes of pulsation quantities in the fluid, porous layer. The eigenvalue problem is solved using the Galerkin method. An analytical expression for the modified critical bioconvection Rayleigh–Darcy number dependence on parameters like vibrational Rayleigh–Darcy number, wave number, modified Darcy number, and Péclet number has been obtained for both rigid–rigid and rigid-free cases. The presence of a non-Darcy porous medium lessens the magnitude of critical bioconvection Rayleigh–Darcy number compared to its absence. Numerical results and discussions, along with their graphical comparisons, are explored.

Keywords Gravitactic swimmers · Rigid-free boundary · Non-Darcy porous medium · The time-averaged method · Vertical vibration

1 Introduction

In cultures of various protozoa, many of the microorganisms such as Paramecium, flagellate Euglena gracilis, Tetrahymena pyriformis, and ciliated protozoan freely swim preferentially upwards and gather near the top of the culture medium [1]. These kinds of microorganisms are named negative geotaxis (gravitaxis). Due to the

K. Srikanth (✉) · V. Kumar
Department of Mathematics, School of Mathematics and Computer Sciences, Central University of Tamil Nadu, Thiruvarur 610005, India
e-mail: srikanthehow@gmail.com

V. Kumar
e-mail: virendrakumar@cutn.ac.in

S. S. Ray et al. (eds.), *Applied Analysis, Computation and Mathematical Modelling in Engineering*, Lecture Notes in Electrical Engineering 897,
https://doi.org/10.1007/978-981-19-1824-7_4

density difference between fluid and cells, a layer with a higher density than the fluid below occurs, which can cause a Rayleigh Bénard type instability [1, 2].

Theoretical models on bioconvection and stability theory were developed by Pedly and coworkers [2, 3]. The effect of vibration can be utilized to control the stability of fluid systems and may be necessary for specific pharmaceutical and bioengineering processes [4]. Kuznetsov and Jiang first studied a model of bioconvection of negatively geotactic particles in a porous medium, which accounts for cell deposition and declogging [5], and observed the permeability, the rate of cell deposition are the important factors that affect the development of bioconvection. Nield and coworkers investigated the onset of instability of suspension of gyrotaxis in a horizontal porous layer [6]. The thermal behavior of the system is analyzed for the onset of convection in a vertically vibrated porous saturated fluid layer [7]. Nguyen and coworkers studied the suspension of gravitactic swimmers in a layer of finite depth in the absence of porous media [8]. The effect of high-frequency, low-amplitude vertical vibration in a suspension of different motile microorganisms confined in a shallow horizontal fluid layer was studied [9, 10]. They reported that the strength of vibration stabilizes the suspension. Bilgen and coworkers investigated the suspension of gravitactic swimmers in a horizontal thermally stratified fluid layer [11]. Gravitactic bioconvection with double diffusion in a thermally stratified porous layer was investigated and revealed that over-stability may take place when the diffusivity of the stabilizing quantity is weaker than that of the destabilizing quantity [12]. Many thin films have been designed by the 2D system of bio-fluid mechanics through the Hele-Shaw apparatus. Nguyen-Quang and coworkers investigated the 2D gravitactic bioconvection in a system of Hele-Shaw cells [13].

The other study, such as nanofluid bioconvection in a porous saturated layer, was addressed in [14]. Vertically vibrated suspension of active swimmers in a fluid layer and porous saturated fluid layer was analyzed by [15–17]. Saini and Sharma investigated the effect of vertical flow on the onset of nanofluid thermo-gravitactic bioconvection in porous media. They disclosed that vertical through-flow disturbs the formation of bioconvection patterns necessary for the growth of bioconvection [18]. Further, they studied linear and nonlinear stability analysis of thermal convection in a suspension of gravitactic swimmers in a fluid layer by energy method [19].

In recent times, researchers have published their work on the study of bioconvection formation due to suspension of motile microorganisms in porous media and utilized the Darcy–Brinkman model [14, 20]. The onset of instability of vertically vibrated suspension of gyrotactic swimmers in a thermally stratified fluid layer was analyzed by Kumar and Srikanth [21]. In this paper, the study by Kuznetsov [9] is continued. The mathematical model for this study has been based on the deterministic formulation of a suspension of gravitactic swimmers. Here the non-Darcy model is utilized to investigate the onset of bioconvection in a high-porosity porous layer subjected to vertical vibration. The porosity of the porous medium is presumed to be big enough so that the microorganism can freely swim. Weakly nonlinear analysis of the present study and the other studies, such as the phenomena of oscillatory instability of the bio-thermally vertically vibrated suspension of negatively geotactic microorganisms, would also be of great interest.

2 Model and Governing Equations

We consider a shallow horizontal sparsely packed high-porosity porous layer saturated with the suspension of gravitactic swimmers of depth l is confined between two parallel plates ($z = 0$ and $z = l$) with vertically downward gravity g acting on it as shown in the Fig. 1. It is assumed that the porous matrix does not absorb the microorganisms, imposed vertical vibration does not affect the behavior of gravitactic swimmers, and the fluid is incompressible, then mass conservation equation:

$$\nabla.\mathbf{v} = 0 \tag{1}$$

Here $\mathbf{v}$ is the fluid filtration velocity vector.

By volume averaging the equation and adding Brinkman term which account for the inertia effects [22], then the momentum equation:

$$\rho_0 c_a\left[\partial\mathbf{v}/\partial t\right] = -\nabla p + \tilde{\mu}\nabla^2\mathbf{v} - \left(\mu/K\right)\mathbf{v} + n\theta\Delta\rho\left(\mathbf{g} + \hat{b}\omega^2\cos\omega t\mathbf{k}\right) \tag{2}$$

Here ρ_0 and c_a are the density of fluid and the acceleration coefficient; p and $\tilde{\mu}$ are excess pressure and effective viscosity; μ and K are the dynamic viscosity of suspension and the permeability of the non-Darcy porous medium; n and $\mathbf{g}$ are number density of microorganisms and gravity vector; θ and $\Delta\rho$ are average volume of microorganism, density difference($\rho_c - \rho_0$); $\hat{b}$ and ω are vibration amplitude and vibration angular frequency; t and $\mathbf{k}$ are time and vertically upward unit vector.

And cell conservation equation:

$$\phi\left[\partial n/\partial t\right] = -\nabla.\left(n\mathbf{v} + nq_c\mathbf{k} - D_c\nabla n\right) \tag{3}$$

where q_c and $\mathbf{k}$ are the average upswimming velocity of microorganisms in the porous medium and upward unit vector in the z-direction; D_c and ϕ are the effective diffusivity of microorganisms in the porous medium and porosity.

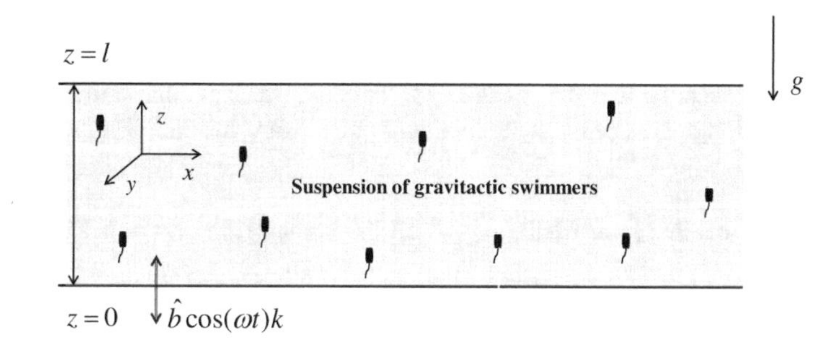

Fig. 1 Schematic diagram of the problem

To formulate the governing equation for the present study, the time-averaged method has been adopted [7]. This method provides the time-averaged system of equations which is valid if the following constraints are utilized [17]. Firstly, frequency of vibration is sufficiently high which makes the vibration period low compared to all the characteristic times scales, i.e., $\tau_{\text{vibrational}} = \min(\tau_{\text{diffusive}}, \tau_{\text{buoyancy}}, \tau_{\text{viscous}})$. Secondly, vibrational amplitude is small enough so that the components corresponding to the rapid variation in velocity can be omitted, i.e., $l/[\theta(\overline{n}_1 - \overline{n}_2)(\Delta\rho/\rho_0)] \gg \hat{b}$. Here $(\overline{n}_1 - \overline{n}_2)$ is the density difference (of the cell concentration). Utilizing these assumptions, the variable quantities are decomposed as the sum of mean (slowly varying) and rapidly oscillating (with time period $\tau = 2\pi/\omega$) components to obtain the suitable equations for vibrational suspension. By using this time-averaging technique, we obtain the following systems of equations:

$$\nabla.\overline{\mathbf{v}} = 0 \tag{4}$$

$$\rho_0 c_a\left[\partial\overline{\mathbf{v}}/\partial t\right] = -\nabla\overline{p} + \tilde{\mu}\nabla^2\overline{\mathbf{v}} - \left(\mu/K\right)\overline{\mathbf{v}} + \overline{n}\theta\Delta\rho\mathbf{g} + \left(\rho_0/2\right)\left[\left(\Delta\rho/\rho_0\right)\hat{b}\omega\right]^2\left[\theta^2\left(\overline{\mathbf{w}}.\nabla\right)\left(\overline{n}\mathbf{k} - \overline{\mathbf{w}}\right)\right] \tag{5}$$

$$\phi\left[\partial\overline{n}/\partial t\right] = -\nabla.\left(\overline{n}\,\overline{\mathbf{v}} + \overline{n}\mathbf{k}q_c - D_c\nabla\overline{n}\right) \tag{6}$$

The last term in the right-hand side of Eq. (5) is the average body force of vibrational nature and the vector $\overline{\mathbf{w}}$ is the solenoidal part of $\overline{n}\mathbf{k}$. Here this vector $\overline{\mathbf{w}}$ satisfies the Helmholtz decomposition [7, 23]:

$$\mathbf{curl}\overline{\mathbf{w}} = \nabla\overline{n} \times \mathbf{k}, \quad \text{div}\overline{\mathbf{w}} = 0 \tag{7}$$

The rigid boundary conditions at lower and upper layer are taken as:

$$\left.\text{at } z = 0, 1: \quad \overline{\mathbf{v}} = 0, \quad \left(\overline{n}\,\overline{\mathbf{v}} + \overline{n}q_c\mathbf{k} - D_c\nabla\overline{n}\right).\mathbf{k} = 0, \quad \overline{\mathbf{w}}.\mathbf{k} = 0\right\} \tag{8}$$

2.1 Basic State Solution

In this state, we assume that the parameters' velocity, pressure, and number density vary in z-direction only. The time-independent quiescent solution of Equations [4–7] are:

$$\left.\begin{aligned} \overline{\mathbf{v}}^b = \mathbf{0}, \quad \overline{n}^b(z) = \nu\exp\left(q_c z/D_c\right), \quad Pe = q_c l/D_c \quad \overline{\mathbf{w}}^b = \mathbf{0}, \\ \overline{p}^b(z) = p_0 + \left[\exp(\text{Pe}) - \exp\left(q_c z/D_c\right)\right]g\nu\theta\Delta\rho\left(D_c/q_c\right) \end{aligned}\right\} \tag{9}$$

where the superscript 'b' denotes the basic state, $\overline{n}_{av}$ is the average concentration of cells, integration constant $\nu = \overline{n}_{av}Pe/\left(\exp(Pe) - 1\right)$ represents the basic number density at the bottom of the layer, and Pe is the bioconvection Péclet number which represents the ratio of the mean cell swimming speed to the speed of bulk fluid motions.

2.2 Linear Stability Analysis

For the linear stability analysis, applying the small perturbation $\overline{\mathbf{v}} = \overline{\mathbf{v}}^*$, $\overline{p} = \overline{p}^b + \overline{p}^*$, $\overline{n} = \overline{n}^b + \overline{n}^*$, $\overline{\mathbf{w}} = \overline{\mathbf{w}}^*$ to the basic state , we have:

$$\nabla.\overline{\mathbf{v}}^* = 0 \tag{10}$$

$$c_a\rho_0\left(\partial\overline{\mathbf{v}}^*/\partial t\right) = -\nabla\overline{p}^* + \tilde{\mu}\nabla^2\overline{\mathbf{v}}^* - \left(\mu/K\right)\overline{\mathbf{v}}^* + \mathbf{g}\overline{n}^*\theta\Delta\rho \tag{11}$$
$$+\left(\rho_0/2\right)\left[\left(\Delta\rho/\rho_0\right)\hat{b}\omega\right]^2\left[\theta^2\left(\overline{\mathbf{w}}^*.\nabla\right)\left(\overline{n}^*\mathbf{k} - \overline{\mathbf{w}}^*\right)\right]$$

$$\phi\left[\partial\overline{\mathbf{n}}^*/\partial t\right] = -v_z\left(\partial\overline{n}^b/\partial z\right) - q_c\left(\partial\overline{n}^*/\partial z\right) + D_c\nabla^2\overline{n}^* \tag{12}$$

$$\mathbf{curl}\overline{\mathbf{w}}^* = \nabla\overline{n}^* \times \mathbf{k} \tag{13}$$

Here $\overline{\mathbf{w}}^* = (w_x, w_y, w_z)$, $\overline{\mathbf{v}}^* = (v_x, v_y, v_z)$, and $\overline{n}^*$, which represents vibrational body force, perturbations to velocity and number density of micro-organisms, respectively.
Operating **k.curlcurl** on Eq. (11) and **curl** on Eq. (13) we get:

$$c_a\rho_0\left(\partial/\partial t\right)\left(\nabla^2 v_z\right) = -\theta\Delta\rho\nabla_1\overline{n}^* g + \left(\rho_0/2\right)\left[\left(\Delta\rho/\rho_0\right)\hat{b}\omega\right]^2 \tag{14}$$
$$\times\left[\theta^2\left(\partial n_b/\partial z\right)\nabla_1\left(w_z\right)\right] + \tilde{\mu}\left[\partial^4 v_z/\partial x^4 + \partial^4 v_z/\partial y^4 + \partial^4 v_z/\partial z^4\right.$$
$$\left.+2\left(\partial^4 v_z/\partial x^2\partial y^2 + \partial^4 v_z/\partial y^2\partial z^2 + \partial^4 v_z/\partial z^2\partial x^2\right)\right] - \left(\mu/K\right)\nabla^2 v_z$$

$$\nabla^2 w_z = \nabla_1\overline{n}^* \tag{15}$$

Here $\nabla_1 = \partial^2/\partial x^2 + \partial^2/\partial y^2$ is the two-dimensional Laplacian operator. To analyze the disturbances into normal modes, the perturbation quantities are taken as follows:

$$\left(v_z,\quad \overline{n}^*,\quad w_z\right) = \left(V_z,\quad N_z,\quad W_z\right)f(x, y)\exp\left(\sigma t\right) \tag{16}$$

Here, $\left(\nabla_1 f + \alpha^2\right)f(x, y) = 0$, and '$\alpha$' is the horizontal wave number. Introducing the following dimensionless quantities:

$$z' = z/l,\quad \alpha' = \alpha l,\quad V_z{}' = \nu\theta q_c l^2 V_z/D_c^2,\quad Pe = q_c l/D_c,\quad W_z' = W_z\theta,\quad N_z' = N\theta\} \tag{17}$$

We get the non-dimensionalized system of equations as:

$$\left(\tilde{\mu}/\mu\right)V_z^{\iota IV} - \left[2\alpha^{\iota^2}\left(\tilde{\mu}/\mu\right) + \left(1/Da\right) + \sigma\left(C_a\rho_0 l^2/\mu\right)\right]V_z^{\iota''} + \left[\alpha^{\iota^4}\left(\tilde{\mu}/\mu\right) + \left(\alpha^{\iota^2}/Da\right) + \sigma\left(\alpha^{\iota^2}C_a\rho_0 l^2/\mu\right)\right]V_z^{\iota} + \alpha^{\iota^2}(RbPe)N_z^{\iota} - \alpha^{\iota^2}(R\upsilon)\exp\left(z^{\iota}Pe\right)W_z^{\iota} = 0 \tag{18}$$

$$N_z^{\iota''} - PeN_z^{\iota'} - \left[\alpha^{\iota^2} + \sigma\left(l^2\phi/D_c\right)\right]N_z^{\iota} - V_z^{\iota}\exp\left(z^{\iota}Pe\right) = 0 \tag{19}$$

$$W_z^{\iota''} + \alpha^{\iota^2}\left(N_z^{\iota} - W_z^{\iota}\right) = 0 \tag{20}$$

Here, $Da = K/l^2$ is the Darcy number, $\left(RbPe\right) = q_c g\upsilon\theta\Delta\rho l^4/\mu D_c^2$ is the modified bioconvection Rayleigh number, Pe is the Péclet number, $R\upsilon = \rho_0\left[\theta\hat{b}\omega\upsilon q_c l^2(\Delta\rho/\rho_0)\right]^2/2\mu D_c{}^3$ is the vibrational Rayleigh number, σ is the growth rate. The principal of exchange of stabilities [24] is valid for this problem; therefore, in Eqs. (18)–(20) σ is set to zero for the onset of stationary convection.

$$Da_*\left[V_z^{\iota IV} - 2\alpha^{\iota 2}V_z^{\iota''} + \alpha^{\iota 4}V_z^{\iota}\right] + \left[\alpha^{\iota 2}V_z^{\iota} - V_z^{\iota''}\right] + \alpha^{\iota^2}R_b N_z^{\iota} - \alpha^{\iota^2}R_\upsilon\exp\left(z^{\iota}Pe\right)W_z^{\iota} = 0 \tag{21}$$

$$N_z^{\iota''} - PeN_z^{\iota'} - \alpha^{\iota^2}N_z^{\iota} - V_z^{\iota}\exp\left(z^{\iota}Pe\right) = 0 \tag{22}$$

$$W_z^{\iota''} + \alpha^{\iota^2}\left(N_z^{\iota} - W_z^{\iota}\right) = 0 \tag{23}$$

$Da_* = K\mu/l^2\tilde{\mu}$ is the modified Darcy number, $R_b = \left(DaRbPe\right)$ is the bioconvection Rayleigh–Darcy number, $R_\upsilon = \left(R\upsilon Da\right)$ is the vibrational Rayleigh–Darcy number. To obtain an approximate solution to the system of Eqs. (21)–(23), we apply a Galerkin-type weighted residuals method [25]. And select a trial solution (which satisfy the boundary conditions) and write

$$\left.V_z^{\iota} = \sum_{j=1}^{M}A_j\left(V_z^{\iota}\right)_j,\quad N_z^{\iota} = \sum_{j=1}^{M}B_j\left(N_z^{\iota}\right)_j,\quad W_z^{\iota} = \sum_{j=1}^{M}C_j\left(W_z^{\iota}\right)_j\right\} \tag{24}$$

Substituting Eq. (24) into system of Eqs. (21)–(23) and applying the standard Galerkin procedure, we get a system of 3M algebraic equations in 3M variables $A_j,\ B_j,\ C_j,\ (j = 1, 2, 3\ldots M)$. A non-trivial solution of this system (vanishing of the determinant of coefficients) leads to an interesting eigenvalue equation.

The dimensionless boundary conditions are taken as follows:

$$\text{at}\ z^{\iota} = 0,\ 1;\quad V_z^{\iota} = 0,\quad PeN_z^{\iota} = DN_z^{\iota},\quad W_z^{\iota} = 0 \tag{25}$$

and the trial solutions satisfying the boundaries Eq. (25) are chosen as:

$$\left.\begin{array}{r}\left(V_z^{\iota}\right)_1 = z^{\iota}\left(1 - z^{\iota}\right),\quad \left(N_z^{\iota}\right)_1 = 2 - Pe\left(1 - 2z^{\iota}\right) - \left(Pe\right)^2\left(z^{\iota} - z^{\iota^2}\right),\\ \text{and}\ \left(W_z^{\iota}\right)_1 = \left(z^{\iota} - z^{\iota^2}\right)\end{array}\right\} \tag{26}$$

Substituting Eq. (26) into Eqs. (21)–(23) and applying the Galerkin method [25], the eigenvalue problem takes the form:

$$\left(R_b\right)_{cr} = \underset{\alpha' \geq 0}{\text{Min}}\Bigg\{\Bigg(\Bigg\{\left[10(Pe)^4 + \alpha'^2\left(120 - 10(Pe)^2 + (Pe)^4\right)\right]\left(10 + \alpha'^2\right)$$
$$\times\left[Da_*\left(\alpha'^4 + 20\alpha'^2\right) + \left(10 + \alpha'^2\right)\right]\Bigg\} + \Bigg\{900\xi_1\alpha'^4\left(10 - (Pe)^2\right)R_v$$
$$\times\Bigg[\left(24/(Pe)^5 + 2/(Pe)^3\right)\left(\exp(Pe) - 1\right) - \left(12/(Pe)^4\right)$$
$$\times\left(\exp(Pe) + 1\right)\Bigg]\Bigg\}\Bigg)\Bigg/\left[30\xi_1\alpha'^2\left(10 + \alpha'^2\right)\left(10 - (Pe)^2\right)\right]\Bigg\} \tag{27}$$

where

$$\xi_1 = \left(8/(Pe)^2\right)\left(\exp(Pe) + 1\right) - \left(16/(Pe)^3 + 1/Pe\right)\left(\exp(Pe) - 1\right) \tag{28}$$

In the absence of vertical vibration, Eq. (27) collapses to:

$$\left(R_b\right)_{cr} = \underset{\alpha' \geq 0}{\text{Min}}\Bigg(\Bigg\{\left[Da_*\left(\alpha'^4 + 20\alpha'^2\right) + \left(10 + \alpha'^2\right)\right] \tag{29}$$
$$\times\left[10(Pe)^4 + \alpha'^2\left(120 - 10(Pe)^2 + (Pe)^4\right)\right]\Bigg\}\Bigg/\left[30\xi_1\alpha'^2\left(10 - (Pe)^2\right)\right]\Bigg)$$

When Pe tends to zero, above expression reduces to:

$$\left(R_b\right)_{cr} = \underset{\alpha' \geq 0}{\text{Min}}\left(\left\{2Da_*\left(\alpha'^4 + 20\alpha'^2\right) + \left(20 + 2\alpha'^2\right)\right\}\Big/\left(5\xi_1\right)_{Pe\to 0}\right) \tag{30}$$

Here, Eq. (30) represents the expression for the critical bioconvection Rayleigh–Darcy number for high-porosity porous layer case. And for this case $\left(R_b\right)_{cr}$ is obtained as 12 for the corresponding critical value of wave number α'_{cr} which is zero when the value of Da_* tending to 0.

As a special case, the permeability K approaches to zero, this Eq. (30) set off as:

$$\left(R_b\right)_{cr} = \underset{\alpha' \geq 0}{\text{Min}}\left\{2\left(10 + \alpha'^2\right)\Big/\left[5\xi_1\right]_{Pe\to 0}\right\} = \underset{\alpha' \geq 0}{\text{Min}}\left\{6\left(10 + \alpha'^2\right)/5\right\} \tag{31}$$

The above expression Eq. (31) holds for low porosity(Darcy model) and matches completely with the results obtained by Nield and coworkers for the case of fluid layer confined between two rigid boundaries [6]. Also this result coincides exactly with the known results for two impermeable boundaries obtained by Virendra kumar [15, 17].

For the case of stress-free upper and rigid lower surfaces, the analysis is same as that for rigid–rigid boundary case with the exception of boundary condition [in Eq. (25)] as: at $z' = 0$; $V'_z = 0$, $DV'_z = 0$ and $z' = 1$; $V'_z = 0$, $D^2V'_z = 0$ and trial solution (in Eq. (26)) for $\left(V'_z\right)_1$ as $z'^2(1 - z')(3 - 2z')$. An analytical relation of $\left(R_b\right)_{cr}$ is obtained [see in Appendix A.1]. And for limiting case, modified Darcy number $Da_* \to 0$, when Pe approaches to 0, the critical bioconvection Rayleigh–Darcy number is $\left(R_b\right)_{cr} = 7.619$ for the corresponding critical value of $\alpha'_{cr} = 0$. This $\left(R_b\right)_{cr}$ value is less in comparison with rigid–rigid case, means the suspension is less stable in the system with rigid-free boundaries.

2.3 Results and Discussion

For the fixed values of the dimensionless parameters $\alpha^{\iota} = 1.9$, $(R_b) = 100$, and Pe (0–3) [9, 17], desired control parameter values are computed and the results are depicted with graphs. Figures 2 and 3 illustrate the effect of the bioconvection Péclet number, Pe, on the modified critical bioconvection Rayleigh–Darcy number, $(R_b)_{cr}$, for different values of vibrational Ralyeigh–Darcy number and modified Darcy number respectively. For different values of R_v (0 to 5000) and modified Darcy number, Da_* (0.0001 to 1), the bioconvection Rayleigh–Darcy number $(R_b)_{cr}$ increases exponentially as Pe increases. Similar trends were attained in earlier reported study of Kuznetsov in the absence of porous media [9]. The vibration Rayleigh–Darcy number characterizes the effect of high-frequency and low-amplitude vertical vibration across the non-Darcy porous fluid layer, as R_v increases, magnitude of $(R_b)_{cr}$ increases, means vibrations have stabilizing effect on bioconvection. As modified Darcy number enlarges, the magnitude of $(R_b)_{cr}$ enlarges, which means this modified Darcy number makes the suspension stable.

In Figs. 4 and 5, the variation of α^{ι}_{cr} against bioconvection Péclet number, Pe, is examined graphically for distinct values of vibrational Rayleigh–Darcy number, modified Darcy number, respectively. From Fig. 4, it is observed that $\alpha^{\iota}_{c}r$ first increases, after certain range it takes on a maximum value, then decreases and also $\alpha^{\iota}_{c}r$ enlarges as R_v enlarges, shows the stronger vibrations correlate to larger critical wave number. For different values of modified Darcy number, the significant changes in the critical wave number have shown in Fig. 5. It is noticed that, with an increase in Péclet number (in a certain range, i.e., 0–2) critical wave number increases. Hence, the parameter Pe reduce the size of cells in this range. The magnitude of $\alpha^{\iota}_{c}r$ enlarges for decreased modified Darcy values.

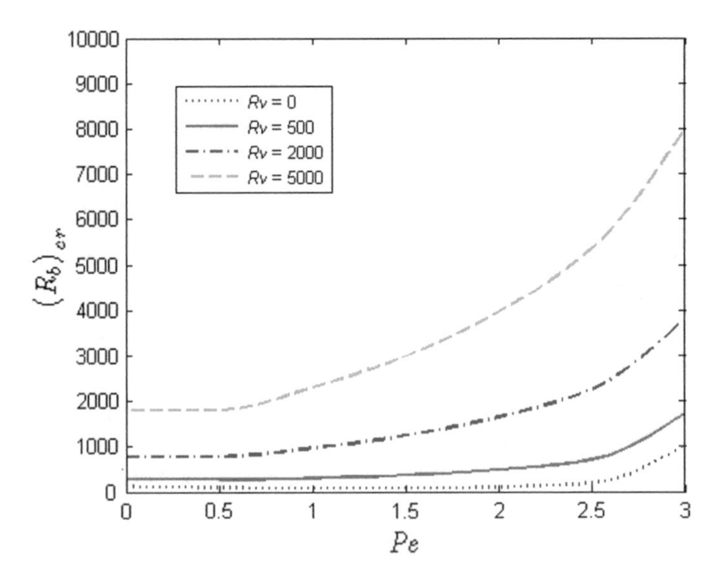

Fig. 2 Dependence of critical bioconvection Rayleigh–Darcy number $(R_b)_{cr}$ on Péclet number (Pe) for different values of vibrational Rayleigh–Darcy number (R_v)

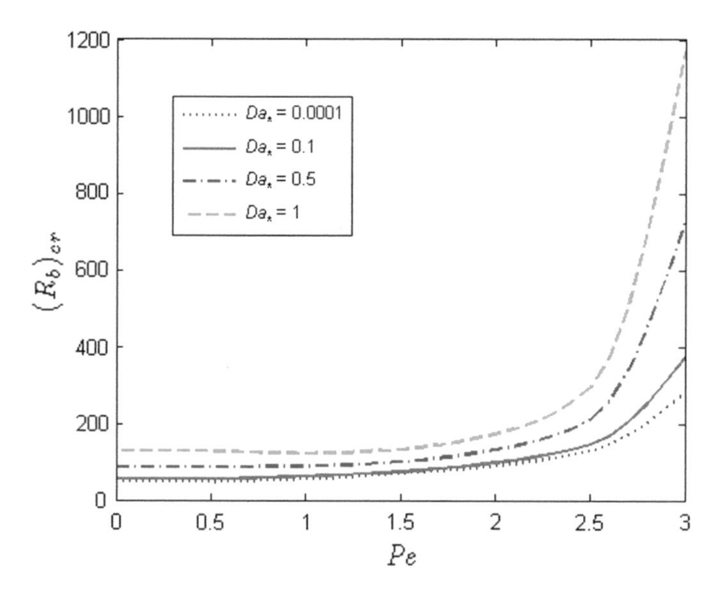

Fig. 3 Dependence of critical bioconvection Rayleigh–Darcy number $(R_b)_{cr}$ on Péclet number (Pe) for different values of modified Darcy number Da_*

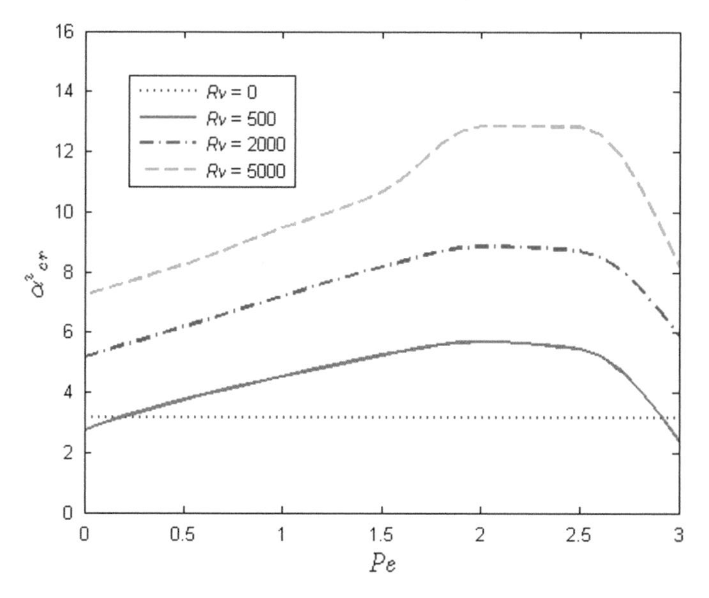

Fig. 4 Dependence of critical wave number(α^{ι}_{cr}) on Péclet number (Pe) for different values of vibrational Rayleigh–Darcy number (R_v)

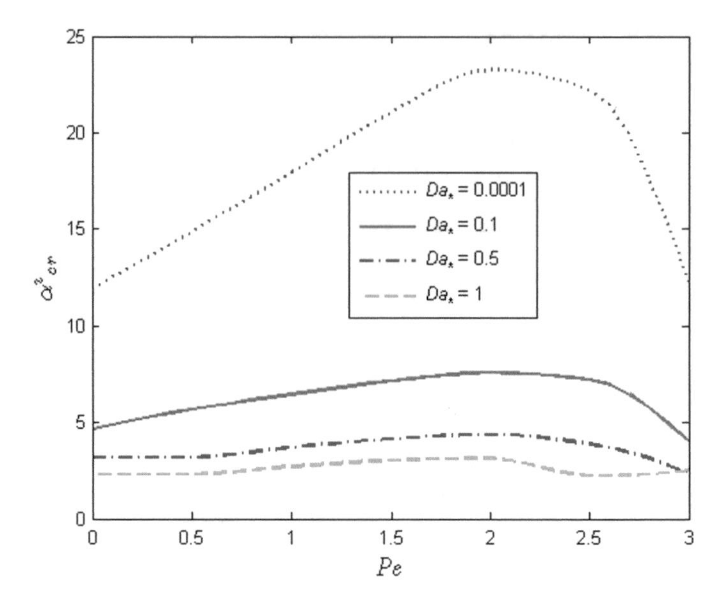

Fig. 5 Dependence of critical wave number(α^{l}_{cr}) on Péclet number (Pe) for different values of modified Darcy number Da_*

3 Conclusions

This paper analyzes the effect of high-frequency and low-amplitude vertical vibration of the onset of bioconvection in a suspension of gravitactic swimmers in fluid saturated with a horizontal non-Darcy porous layer of finite depth. The system is solved analytically using the Galerkin technique, and an expression for the critical bioconvection Rayleigh number for the non-oscillatory convection is obtained. The numerical results are in a good match-up with formerly published results. The main conclusions are drawn. The influence of the vibration effect stabilizes the suspension, and high-frequency vibrations can be applied to control (suppress) bioconvection. The impact of the modified Darcy number is to stabilize the system. Due to the presence of a non-Darcy porous medium, the magnitude of the parameters, bioconvection Rayleigh–Darcy number, wave number is less in comparison with its absence. The suspension with rigid-rigid boundaries is more stable than the system with rigid-free boundaries.

Appendix

An analytic expression for the dependence of critical bioconvection Rayleigh–Darcy number for rigid-free boundaries case is given below:

$$
\begin{aligned}
\left(R_b\right)_{cr} = \min_{\alpha' \geq 0}\Bigg\{ \Bigg(\Bigg\{ \left[Da_*\left(19\alpha'^4 + 432\alpha'^2 + 4536\right) + \left(19\alpha'^2 + 216\right)\right]\left(10 + \alpha'^2\right) \\
\times\left[10(Pe)^4 + \alpha'^2\left(120 - 10(Pe)^2 + (Pe)^4\right)\right]\Bigg\} + \Bigg\{ 37800\psi_1\alpha'^4\left(10 - (Pe)^2\right)R_v \\
\times\Bigg[2\exp(Pe)/(Pe)^3 - \left(6\exp(Pe) - 18\right)/(Pe)^4 - \left(72\exp(Pe) - 192\right)/(Pe)^5 \\
+\left(600\exp(Pe) + 840\right)/(Pe)^6 - 1440\left(\exp(Pe) - 1\right)/(Pe)^7 \Bigg]\Bigg\}\Bigg) \\
\Bigg/\left[90\psi_1\alpha'^2\left(10 + \alpha'^2\right)\left(126 + 7Pe - 13(Pe)^2\right)\right]\Bigg\}
\end{aligned}
\tag{32}
$$

where $\psi_1 = -\exp(Pe)/Pe + \left(4\exp(Pe) - 12\right)/(Pe)^2 + \left(54\exp(pe) - 138\right)/(Pe)^3 - \left(444\exp(Pe) + 612\right)/(Pe)^4 + 1056\left(\exp(Pe) - 1\right)/(Pe)^5$ and $\left(\psi_1\right)_{Pe \to 0} = 3/10$

References

1. Childress S, Levandowsky M, Spiegel EA (1975) Pattern formation in a suspension of swimming microorganisms: equations and stability theory. J Fluid Mech 63:591–613
2. Hill NA, Pedley TJ, Kessler JO (1989) Growth of bioconvection patterns in a suspension of gyrotactic micro-organisms in a layer of finite depth. J Fluid Mech 208:509–543
3. Pedley TJ (1992) Hydrodynamic phenomena in a suspension of swimming microorganisms. Annu Rev Fluid Mech 24:313–358
4. Stewart TL, Fogler HS (2001) Biomass plug development and propagation in porous media. Biotech Bioeng 72:353–63
5. Kuznetsov AV, Jiang N (2003) Bioconvection of negatively geotactic microorganisms in a porous medium: the effect of cell deposition and declogging. Int J Numer Methods Heat Fluid Flow 13:341–364
6. Nield DA, Kuznetsov AV, Avramenko AA (2004) The onset of bioconvection in a horizontal porous-medium layer. Trans Porous Med 54:335–344
7. Cisse I, Bardan G, Mojtabi A (2004) Rayleigh benard instability of a fluid under high frequency vibration. Int J Heat Mass Trans 47:4101–4112
8. Bahloul A, Nguyen-Quang T, Nguyen TH (2005) Bioconvection of gravitactic microorganisms in a fluid layer. Int Commun Heat Mass Transfer 32:64–71
9. Kuznetsov AV (2005) The onset of bioconvection in a suspension of negatively geotactic microorganisms with high-frequency vertical vibration. Int Commun Heat Mass Trans 32:1119–1127
10. Kuznetsov AV (2006) Investigation of the onset of bioconvection in a suspension of oxytactic microorganisms subjected to high-frequency vertical vibration. Theor Comput Fluid Dyn 20:73–87
11. Alloui Z, Nguyen TH, Bilgen E (2006) Stability analysis of thermo-bioconvection in suspensions of gravitactic microorganisms in a fluid layer. Int Commun Heat Mass Transfer 33:1198–1206
12. Tri Nguyen-Quang, Lepalec Georges (2008) Gravitactic bioconvection in a fluid-saturated porous medium with double diffusion. J Porous Med 11:751–764
13. Nguyen-Quang T, Nguyen H, Guichard F, Nicolau A, Szatmari G, LePalec G, Dusser M, Lafossee J, Bonnet JL, Bohatier J (2009) Two-dimensional gravitactic bioconvection in a protozoan (Tetrahymena pyriformis) culture. Zoolog Sci 26:54–65

14. Kuznetsov AV, Nield DA (2010) Thermal instability in a porous medium layer saturated by a nanofluid: Brinkman model. Transp Porous Med 81:409–422
15. Sharma YD, Kumar V (2012) Effect of high-frequency vertical vibration in a suspension of negatively geotactic microorganisms saturating a porous medium. Int J Appl Math Mech 8:47–58
16. Sharma YD, Kumar V (2012) The effect of high-frequency vertical vibration in a suspension of gyrotactic microorganisms. Mech Res Commun 44:40–46
17. Kumar V, Sharma YD (2014) Instability analysis of gyrotactic microorganisms: a combined effect of high frequency vertical vibration and porous media. Transp Porous Med 102:153–165
18. Saini S, Sharma YD (2018) Numerical study of nanofluid thermo-bioconvection containing gravitactic microorganisms in porous media: effect of vertical throughflow. Adv Powder Technol 29:2725–2732
19. Saini S, Sharma YD (2018) Analysis of onset of bio-thermal convection in a fluid containing gravitactic microorganisms by the energy method. Chinese J Phys 56:2031–2038
20. Zhao M, Wang S, Wang H, Mahabaleshwar US (2019) Darcy-brinkman bio-thermal convection in a suspension of gyrotactic microorganisms in a porous medium. Neural Comput Applic 31:1061–1067
21. Kumar V, Srikanth K (2021) An overstability analysis of vertically vibrated suspension of active swimmers subjected to thermal stratification. SN Appl Sci 3:612
22. Nield DA, Bejan A (2006) Convection in porous media, 3rd edn. Springer, New York
23. Gershuni GZ, Lyubimov DU (1998) Thermal vibrational convection. Wiley, New York
24. Chandrasekhar S (1981) Hydrodynamic and hydromagnetic stability. Dover, New York
25. Finlayson BA (1972) The method of weighted residuals and variational principles. Academic Press, NewYork

Impact of Two Temperatures on a Generalized Thermoelastic Plate with Thermal Loading

Ankit Bajpai and P. K. Sharma

Abstract This study investigates the effects of two temperatures on a generalized thermoelastic plate in light of generalized thermoelasticity. The plate is infinite in x- and y-direction and has a finite thickness in the z-direction. The origin of the coordinate system is taken on the middle plane of the plate. Various field quantities are taken as functions of x, z, and t only. The boundary of the plate is rigidly fixed and subjected to thermal loading. The governing equations are non-dimensionalized. To solve the governing equations, potential functions are introduced, and harmonic solutions are obtained. With the help of obtained solutions, the stress and displacement components, and conductive and thermodynamic temperatures are determined analytically in the closed-form. Using boundary conditions, the constants in the solutions are obtained. To show the results graphically, numerical results are computed for copper material. The variation of stress components and conductive and thermodynamic temperatures, is presented graphically for different values of two-temperature parameters and compared with one temperature thermoelasticity.

Keywords Generalized thermoelasticity · Two temperature · Harmonic solution

1 Introduction

The theory of thermoelasticity is the coupling of thermal and mechanical fields. In the classical theory of thermoelasticity, there are two shortcomings. The first one is that the heat conduction equation has no elastic term. So, elastic changes do not affect the temperature. The second one is that it has a parabolic type heat conduction equation providing infinite speed for heatwaves' deliverance. Biot [1] introduced the coupled theory of thermoelasticity, eliminating the first drawback. To overcome the second shortcoming, generalized theories of thermoelasticity have been developed by many researchers predicting the finite speed of heatwaves. Lord and Shulman [2] proposed

A. Bajpai (✉) · P. K. Sharma
Department of Mathematics & Scientific Computing, National Institute of Technology Hamirpur, Hamirpur, HP 177005, India
e-mail: ankitbajpai@nith.ac.in

S. S. Ray et al. (eds.), *Applied Analysis, Computation and Mathematical Modelling in Engineering*, Lecture Notes in Electrical Engineering 897,
https://doi.org/10.1007/978-981-19-1824-7_5

the first generalized theory of thermoelasticity by introducing one relaxation time. Green and Lindsay [3] have submitted another approach of thermoelasticity involving two relaxation times.

In problems of ultra-short laser heating, where very high heat flux generates in the body for a brief time interval (about 10^{-7}), the two-temperature model's role becomes more significant. It explains more realistic results compared to one temperature theory. For heat conduction in deformable bodies, Chen and Gurtin [4] and Chen et al. [5] suggested two different temperatures, viz. conductive and thermodynamic temperatures that arise due to thermal and mechanical processes, respectively, separated by heat supply for time-independent situations, and hence when heat supply vanishes, both temperatures will be equal. But for time-dependent problems, two temperatures are in general distinct even though heat supply is zero. Warren and Chen [6] explained that, in two-temperature theory, the propagation speed increases and discontinuities in strain and conductive and thermodynamic temperatures become smooth. Thereafter, for many years, this theory was underestimated and ignored. But, in recent time, two-temperature theory (2TT) has been noticed by many researchers. They further obtained advancement in two-temperature theory and explained their applications, primarily describing the continuity of stress function as it is discontinuous for one temperature thermoelasticity (1TT) [7]. Various authors have presented a good number of the problems of two-temperature thermoelasticity by considering various models and boundary conditions [8–17].

The present study is motivated by the broad applications of two-temperature theory in pulsed laser technologies in material processing and nondestructive detecting, and explaining the continuity of the stress function [7]. Also, the plate structures are broadly utilized in many engineering fields, for example, aerospace, mechanical, and automotive engineering disciplines. The plate theory is a significant piece of transport engineering, whereby the utilization of plate and shell structures is common, particularly in aerospace engineering.

In this study, authors formulated a thermoelastic plate problem to analyze the impact of two temperatures on symmetric and skew-symmetric modes of various field quantities due to thermal loading. The Youssef model [18] of the generalized thermoelasticity with two temperatures has been used to investigate the deformation in thermoelastic plate. For numerical computations, copper material is taken, and numerical results are obtained using MATLAB programming. The numerical results obtained are presented graphically for symmetric as well as skew-symmetric cases in light of Lord and Shulman's theory with and without two temperatures. Stress and displacement components, and conductive and thermodynamic temperatures are displayed graphically to show the two-temperature effect. The results are compared graphically with the one temperature thermoelasticity (1TT).

2 Mathematical Modeling and Solution of the Problem

In this section, basic equations, formulation of the problem, and its solution will be presented.

2.1 Basic Equations

In the framework of generalized two-temperature thermoelasticity following Youssef [18], governing equations and constitutive relations in the absence of body forces and heat sources are

$$(\lambda + \mu)u_{j,\mathrm{ij}} + \mu u_{i,\mathrm{jj}} - \gamma\left(\theta + \nu\frac{\partial\theta}{\partial t}\right)_{,i} = \rho\frac{\partial^2 u_i}{\partial t^2} \tag{1}$$

$$K\phi_{,ii} = \rho C_E\left(\frac{\partial\theta}{\partial t} + \tau_0\frac{\partial^2\theta}{\partial t^2}\right) + \gamma T_0\left(\frac{\partial u_{i,i}}{\partial t} + n_0\tau_0\frac{\partial^2 u_{i,i}}{\partial t^2}\right) \tag{2}$$

$$\sigma_{\mathrm{ij}} = 2\mu e_{\mathrm{ij}} + \lambda e_{\mathrm{kk}}\delta_{\mathrm{ij}} - \gamma\left(\theta + \nu\frac{\partial\theta}{\partial t}\right)\delta_{\mathrm{ij}} \tag{3}$$

$$e_{\mathrm{ij}} = \frac{1}{2}\left(u_{i,j} + u_{j,i}\right) \tag{4}$$

The relation between two temperatures:

$$\phi - \theta = a^*\phi_{,\mathrm{ii}}, \theta = |T - T_0| \text{ with } \frac{\theta}{T_0} \ll 1 \tag{5}$$

where $u_i (i = 1, 2, 3)$—displacement components, σ_{ij}—stress components, e_{ij}—strain components, θ—thermodynamic temperature, ϕ—conductive temperature, T_0—reference temperature, λ, μ—Lame's parameters, K—thermal conductivity, ρ—mass density, C_E—specific heat at constant strain, $\gamma = (3\lambda + 2\mu)\alpha_t$, α_t-linear thermal expansion, "a^*"—two-temperature parameter (2TP), τ_0, ν—relaxation times, and n_0—parameter.

By putting $\tau_0 = \nu = 0$, the governing equations correspond to coupled thermoelasticity; $n_0 = 1, \nu = 0$ correspond to LS theory and $n_0 = 0$ correspond to GL theory of thermoelasticity. On taking $a^* = 0$, the respective thermoelastic models in 1TT can be obtained.

2.2 Formulation of the Problem

Consider an infinite elastic plate of finite width "2d," which is homogeneous, isotropic, and thermally conducting with initial uniform temperature T_0. Initially, the plate is assumed to be unstrained and unstressed. The middle plane of the plate coincides with the x–y plane such that $-d \leq z \leq d$ and $-\infty < x, y < \infty$ as shown in figure below. The origin of the coordinate system is taken at any point of the middle plane. The boundary surfaces $z = \pm d$ are considered to be rigidly fixed with thermal loading.

We consider the x–z plane as the plane of incidence and restrict our analysis to this plane so that various quantities are functions of only x, z, and t. Hence, the displacement components and temperatures are given by

$$u = u(x, z, t), v = 0, w = w(x, z, t), \phi = \phi(x, z, t) \ \& \ \theta = \theta(x, z, t) \tag{6}$$

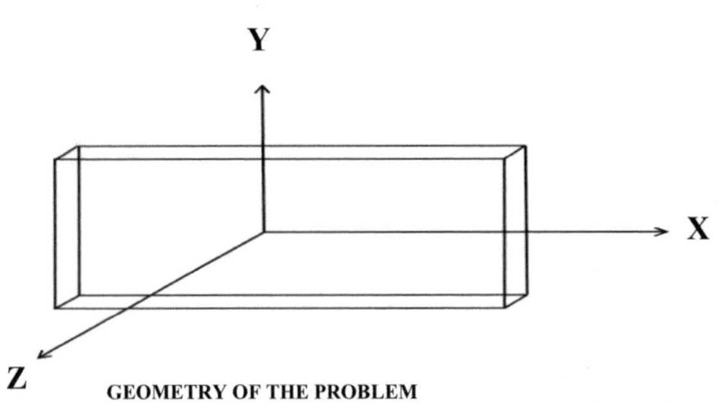

GEOMETRY OF THE PROBLEM

Equations (1)–(5) along with (6) can be written in nondimensional form (after hiding primes) as

$$\frac{\partial^2 u}{\partial x^2} + (1 - \delta^2)\frac{\partial^2 w}{\partial x \partial z} + \delta^2 \frac{\partial^2 u}{\partial z^2} - \frac{\partial}{\partial x}\left(\theta + \nu \frac{\partial \theta}{\partial t}\right) = \frac{\partial^2 u}{\partial t^2} \tag{7}$$

$$\frac{\partial^2 w}{\partial z^2} + (1 - \delta^2)\frac{\partial^2 u}{\partial x \partial z} + \delta^2 \frac{\partial^2 w}{\partial x^2} - \frac{\partial}{\partial z}\left(\theta + \nu \frac{\partial \theta}{\partial t}\right) = \frac{\partial^2 w}{\partial t^2} \tag{8}$$

$$\frac{\partial^2 \phi}{\partial x^2} + \frac{\partial^2 \phi}{\partial z^2} = \left(\frac{\partial \theta}{\partial t} + \tau_0 \frac{\partial^2 \theta}{\partial t^2}\right) + \in \left(\frac{\partial}{\partial t} + n_0 \tau_0 \frac{\partial^2}{\partial t^2}\right)\left(\frac{\partial u}{\partial x} + \frac{\partial w}{\partial z}\right) \tag{9}$$

$$\sigma_{zz} = \frac{\partial w}{\partial z} + (1 - 2\delta^2)\frac{\partial u}{\partial x} - \left(\theta + \nu \frac{\partial \theta}{\partial t}\right) \tag{10}$$

$$\sigma_{xz} = \delta^2 \left(\frac{\partial u}{\partial z} + \frac{\partial w}{\partial x}\right) \tag{11}$$

where nondimensional quantities are introduced as

$$
\begin{aligned}
&(x', z') = \frac{\omega^*}{c_1}(x, z), \quad (u', w') = \frac{\omega^* c_1 \rho}{\gamma T_0}(u, w), \quad t' = \omega^* t, \\
&\nu' = \omega^* \nu, \tau_0' = \omega^* \tau_0, a^{*\prime} = \frac{a^* \omega^{*2}}{c_1^2}, \\
&\theta' = \frac{\theta}{T_0}, \ \phi' = \frac{\phi}{T_0}, \ \sigma_{ij}' = \frac{\sigma_{ij}}{\gamma T_0}, \ c_1^2 = \frac{(\lambda + 2\mu)}{\rho}, \\
&\omega^* = \frac{C_E(\lambda + 2\mu)}{K}, \ \delta^2 = \frac{\mu}{(\lambda + 2\mu)}, \ \in = \frac{T_0 \gamma^2}{\rho C_E(\lambda + 2\mu)}.
\end{aligned}
\tag{12}
$$

2.3 Solution of the Problem

To simplify Eqs. (7)–(9), potential functions Φ and Ψ are introduced by Helmholtz decomposition theorem as [19]

$$
u = \frac{\partial \Phi}{\partial x} + \frac{\partial \Psi}{\partial z}, \quad w = \frac{\partial \Phi}{\partial z} - \frac{\partial \Psi}{\partial x} \tag{13}
$$

Equations (7)–(9) with (13) can be obtained as

$$
\nabla^2 \phi = (\dot{\phi} - a^* \nabla^2 \dot{\phi}) + \tau_0 (\ddot{\phi} - a^* \nabla^2 \ddot{\phi}) + \in [\nabla^2 \dot{\Phi} + n_0 \tau_0 (\nabla^2 \ddot{\Phi})] \tag{14}
$$

$$
\nabla^2 \Phi - \frac{\partial^2 \Phi}{\partial t^2} = \theta + \nu \frac{\partial \theta}{\partial t} \tag{15}
$$

$$
\nabla^2 \Psi - \frac{1}{\delta^2} \frac{\partial^2 \Psi}{\partial t^2} = 0 \tag{16}
$$

We take the solutions following [19] as

$$
(\Phi, \Psi, \phi) = [f(z), g(z), h(z)] \exp[i\xi(x - ct)] \tag{17}
$$

where $c = \frac{\omega}{\xi}$—phase velocity, ω—frequency, and ξ—wave number.

Equations (14)–(16) along with (13) and (17), after some simplifications, yield

$$
\begin{aligned}
u(x, z, t) =& [i\xi(C_3 \cos m_1 z + C_4 \sin m_1 z + C_5 \cos m_2 z + C_6 \sin m_2 z) \\
&- \beta C_7 \sin \beta z + \beta C_8 \cos \beta z] \exp[i\xi(x - ct)]
\end{aligned}
\tag{18}
$$

$$
\begin{aligned}
w(x, z, t) =& [-m_1 C_3 \sin m_1 z + m_1 C_4 \cos m_1 z - m_2 C_5 \sin m_2 z + m_2 C_6 \cos m_2 z \\
&- i\xi(C_7 \cos \beta z + C_8 \sin \beta z)] \exp[i\xi(x - ct)]
\end{aligned}
\tag{19}
$$

$$\phi(x,z,t) = \left[\frac{1}{\alpha_1}(C_3\cos m_1 z + C_4 \sin m_1 z) + \frac{1}{\alpha_2}(C_5 \cos m_2 + C_6 \sin m_2 z)\right] \exp[i\xi(x-ct)] \tag{20}$$

$$\theta(x,z,t) = \begin{bmatrix} \dfrac{\{1+a^*g_1\}}{\alpha_1}(C_3\cos m_1 z + C_4 \sin m_1 z) + \\ \dfrac{\{1+a^*g_2\}}{\alpha_2}(C_5 \cos m_2 + C_6 \sin m_2 z) \end{bmatrix} \exp[i\xi(x-ct)] \tag{21}$$

where

$$\alpha_j = \frac{-i\omega t_2\{1+a^*g_j\}}{\alpha^2 - m_j^2}, g_j = \left(m_j^2 + \xi^2\right), (j = 1, 2),$$

$$\alpha^2 = \xi^2\left(c^2 - 1\right), \beta^2 = \xi^2\left(\frac{c^2}{\delta^2} - 1\right),$$

$$m_1^2 = \frac{1}{2}\left(A + \sqrt{A^2 - 4B}\right), m_2^2 = \frac{1}{2}\left(A - \sqrt{A^2 - 4B}\right),$$

$$A = \frac{2\xi^2 - \omega^2\left(1 + t_0 - i \in \omega t_1 t_2 + a^*\left(\omega^2 t_0 + 2i\xi^2 \omega t_1 t_2 - 2\xi^2 t_0\right)\right)}{-1 + a^*\omega^2(t_0 - i \in \omega t_1 t_2)},$$

$$B = \frac{-\xi^4 - \omega^4 t_0 + \omega^2\xi^2\left(1 + t_0 - i \in \omega t_1 t_2 + a^*\left(-\omega^2 t_0 - i \in \xi^2 \omega t_1 t_2 + \xi^2 t_0\right)\right)}{-1 + a^*\omega^2(t_0 - i \in \omega t_1 t_2)},$$

$$t_0 = \tau_0 + i\omega^{-1}, t_1 = \nu + i\omega^{-1}, t_2 = n_0\tau_0 + i\omega^{-1}.$$

Terms corresponding to two-temperature parameter "a^*" stand for 2TT. If the two-temperature parameter "$a^* = 0$," one will get the results for 1TT.

2.4 Boundary Conditions

The boundary conditions are taken as under.

Boundary conditions in nondimensional form at $z = \pm$ d are given as.

I. Mechanical boundary conditions: The surfaces of plate are considered rigidly fixed, hence

$$u = w = 0, \tag{22}$$

II. Thermal boundary conditions: Thermal load is applied on surfaces as

$$\phi = G_1 e^{i\xi(x-ct)}, \tag{23}$$

G_1 is the constant temperature applied on the boundary.

3 Amplitudes of Stress and Displacement Components, and Conductive and Thermodynamic Temperatures

Invoking boundary conditions (22)–(23) with the help of solutions given in (18)–(21) and relations (10)–(11), the analytical expressions of stress and displacement components, and conductive and thermodynamic temperatures for symmetric as well as skew-symmetric cases are obtained as

$$\left.\begin{array}{l}(\sigma_{zz})_{sym}(x,z,t)=[q_1C_3\cos m_1z+q_2C_5\cos m_2z+p_1C_8\beta\cos\beta z]\exp[i\xi(x-\mathrm{ct})]\\(\sigma_{zz})_{sksym}(x,z,t)=[q_1C_4\sin m_1z+q_2C_6\sin m_2z+p_1C_7\beta\sin\beta z]\exp[i\xi(x-\mathrm{ct})]\end{array}\right\}\tag{24}$$

$$\left.\begin{array}{l}(\sigma_{xz})_{sym}(x,z,t)=[q_3C_3\cos m_1z+q_4C_5\cos m_2z+p_2C_8\beta\cos\beta z]\exp[i\xi(x-\mathrm{ct})]\\(\sigma_{xz})_{sksym}(x,z,t)=[q_3C_4\sin m_1z+q_4C_6\sin m_2z+p_2C_7\beta\sin\beta z]\exp[i\xi(x-\mathrm{ct})]\end{array}\right\}\tag{25}$$

$$\left.\begin{array}{l}(u)_{sym}(x,z,t)=[i\xi(C_3\cos m_1z+C_5\cos m_2z+\beta C_8\cos\beta z]\exp[i\xi(x-\mathrm{ct})]\\(u)_{sksym}(x,z,t)=[i\xi(C_4\sin m_1z+C_6\sin m_2z)-\beta C_7\sin\beta z]\exp[i\xi(x-\mathrm{ct})]\end{array}\right\}\tag{26}$$

$$\left.\begin{array}{l}(w)_{sym}(x,z,t)=[-m_1C_3\sin m_1z-m_2C_5\sin m_2z-i\xi C_8\sin\beta z]\exp[i\xi(x-\mathrm{ct})]\\(w)_{sksym}(x,z,t)=[m_1C_4\cos m_1z+m_2C_6\cos m_2z-i\xi C_7\cos\beta z]\exp[i\xi(x-ct)]\end{array}\right\}\tag{27}$$

$$\left.\begin{array}{l}(\phi)_{sym}(x,z,t)=\left[\frac{1}{\alpha_1}(C_3\cos m_1z)+\frac{1}{\alpha_2}(C_5\cos m_2z)\right]\exp[i\xi(x-\mathrm{ct})]\\(\phi)_{sksym}(x,z,t)=\left[\frac{1}{\alpha_1}(C_4\sin m_1z)+\frac{1}{\alpha_2}(C_6\sin m_2z)\right]\exp[i\xi(x-\mathrm{ct})]\end{array}\right\}\tag{28}$$

$$\left.\begin{array}{l}(\theta)_{sym}(x,z,t)=\left[\frac{\{1+a^*g_1\}}{\alpha_1}(C_3\cos m_1z)+\frac{\{1+a^*g_2\}}{\alpha_2}(C_5\cos m_2z)\right]\exp[i\xi(x-\mathrm{ct})]\\(\theta)_{sksym}(x,z,t)=\left[\frac{\{1+a^*g_1\}}{\alpha_1}(C_4\sin m_1z)+\frac{\{1+a^*g_2\}}{\alpha_2}(C_6\sin m_2z)\right]\exp[i\xi(x-\mathrm{ct})]\end{array}\right\}\tag{29}$$

where constants $C_i(i=3,4,5,6,7,8)$ are obtained by using boundary conditions (22) and (23) such that

$$C_i=\frac{\Delta_i}{\Delta},\ (i=3,4,5,6,7,8)$$

where,

$$\Delta = \det(A),\ A = \begin{pmatrix} i\xi cs_1 & i\xi sn_1 & i\xi cs_2 & i\xi sn_2 & -\beta sn_3 & \beta cs_3 \\ i\xi cs_1 & -i\xi sn_1 & i\xi cs_2 & -i\xi sn_2 & \beta sn_3 & \beta cs_3 \\ -m_1 sn_1 & m_1 cs_1 & -m_2 sn_2 & m_2 cs_2 & -i\xi cs_3 & -i\xi sn_3 \\ m_1 sn_1 & m_1 cs_1 & m_2 sn_2 & m_2 cs_2 & -i\xi cs_3 & i\xi sn_3 \\ \frac{1}{\alpha_1} cs_1 & \frac{1}{\alpha_1} sn_1 & \frac{1}{\alpha_2} cs_2 & \frac{1}{\alpha_2} sn_2 & 0 & 0 \\ \frac{1}{\alpha_1} cs_1 & -\frac{1}{\alpha_1} sn_1 & \frac{1}{\alpha_2} cs_2 & -\frac{1}{\alpha_2} sn_2 & 0 & 0 \end{pmatrix},$$

Δ_i = determinant of matrix A when ith column of A is replaced by column vector, $B = \begin{pmatrix} 0\ 0\ 0\ 0\ G_1\ G_1 \end{pmatrix}^T$

$$\begin{aligned} \mathrm{cs}_i &= \cos m_i d,\ \mathrm{sn}_i = \sin m_i d, \quad i = 1, 2 \\ \mathrm{cs}_3 &= \cos \beta d,\ \mathrm{sn}_3 = \sin \beta d, \\ q_j &= -m_j^2 \alpha_j - \xi^2 (1 - 2\delta^2) \alpha_j - \left(1 + a(m_j^2 + \xi^2)\right)(1 - i\xi \mathrm{c}\nu), \quad (j = 1, 2) \\ q_k &= 2i\xi m_{k-2} \alpha_{k-2}, \quad (k = 3, 4) \\ p_1 &= 2i\xi \beta \delta^2, \\ p_2 &= (\xi^2 - \beta^2). \end{aligned}$$

sksym = Skew symmetric, sym = Symmetric.

4 Numerical Results and Discussion

In order to portray theoretical results presented in preceding sections, we have chosen copper material (following [12]) for evaluation of numerical results and physical data for which is as given below

$$\begin{aligned} &\lambda = 7.76 \times 10^{10} \mathrm{Kg\,m^{-1}} s^{-2},\ \mu = 3.86 \times 10^{10} \mathrm{Kg\,m^{-1} s^{-2}}, \\ &\varepsilon = 0.0168,\ \rho = 8954 \mathrm{Kg\,m^{-3}}, \\ &T_0 = 293\,\mathrm{K},\ C_E = 383.1 \mathrm{J\,Kg^{-1} K^{-1}}, \\ &K = 386 \mathrm{W\,m^{-1} K^{-1}},\ \alpha_t = 1.78 \times 10^{-5} \mathrm{K^{-1}},\ \omega = -0.3 \mathrm{s^{-1}}, \\ &\tau_0 = 0.003\,\mathrm{s},\ \nu = 0\,\mathrm{s},\ t = 0.1\,\mathrm{s},\ x = 1\,\mathrm{m},\ \xi = 1 \mathrm{m^{-1}},\ G_1 = 1. \end{aligned}$$

The variation of amplitudes of stress component σ_{zz}, displacement component w, thermodynamic temperature (T), and conductive temperature (ϕ) for skew-symmetric and symmetric modes of vibration with the thickness of the plate z in Figs. 1, 2, 3, 4, 5, and 6 for $x = 1$ and with the length of the plate x in Figs. 7 and 8 for $z = 1$ are demonstrated for three different values of two-temperature parameter (2TP), viz.$a^* = 0, 0.5, 0.9$ in the context of two-temperature LS theory. The solid

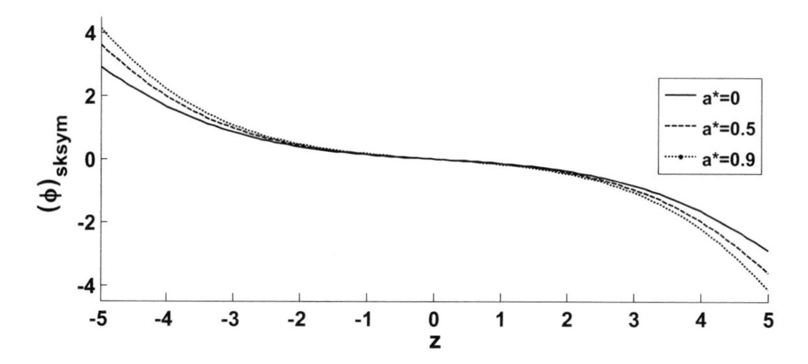

Fig. 1 Variation of skew symmetric conductive temperature with plate thickness

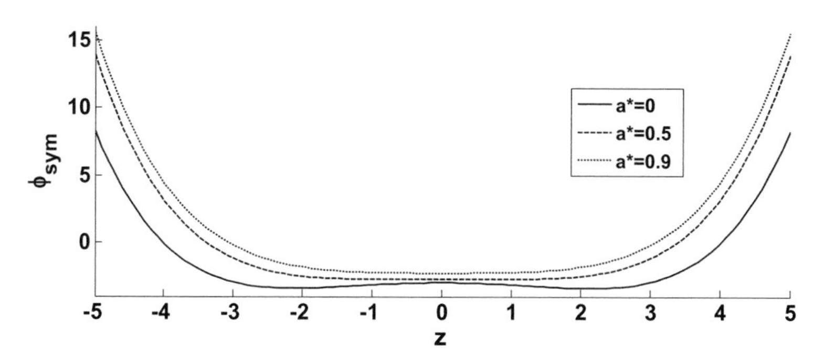

Fig. 2 Variation of symmetric conductive temperature with plate thickness

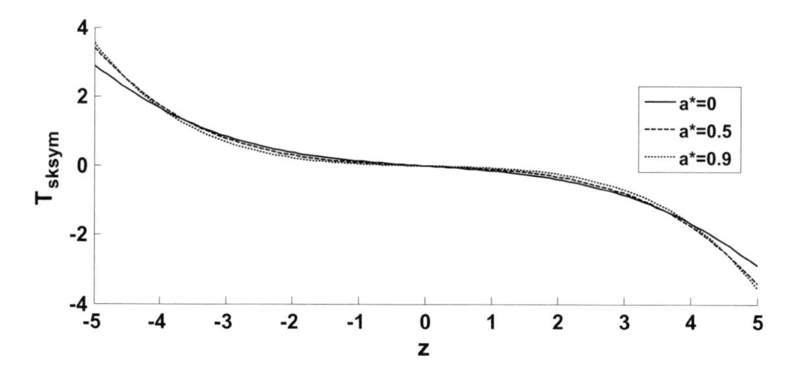

Fig. 3 Variation of skew symmetric thermodynamic temperature with plate thickness

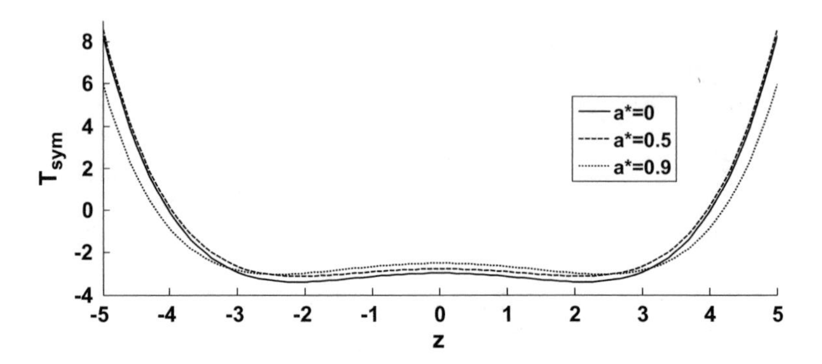

Fig. 4 Variation of symmetric thermodynamic temperature with plate thickness

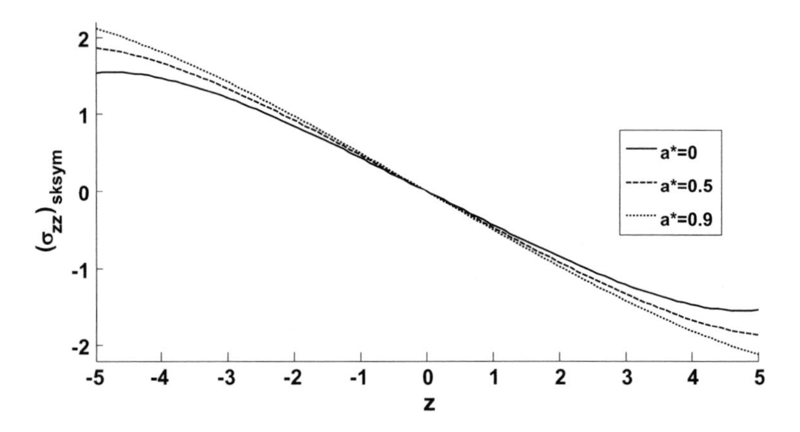

Fig. 5 Variation of skew symmetric normal stress component with plate thickness

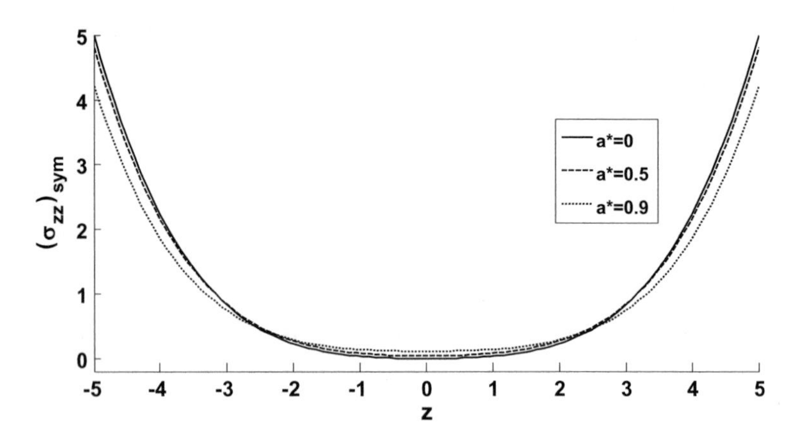

Fig. 6 Variation of symmetric normal stress component with plate thickness

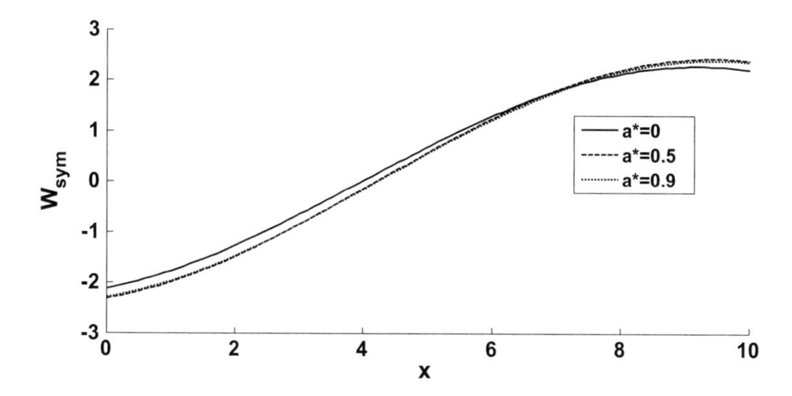

Fig. 7 Variation of symmetric displacement W with plate length x

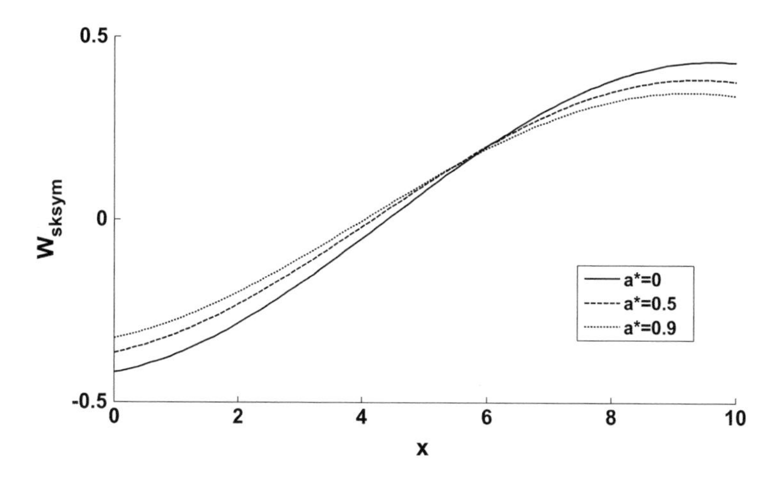

Fig. 8 Variation of skew symmetric displacement W with plate length x

lines (—), dashed lines (- - -), and dotted lines (• • •) correspond to 1TT ($a^* = 0$) and 2TT ($a^* = 0.5$, $a^* = 0.9$), respectively.

Figures 1, 2, 3, and 4 represent the skew-symmetric and symmetric conductive and thermodynamic temperature variation along the plate's thickness due to thermal loading, respectively. The symmetric and skew-symmetric components of conductive and thermodynamic temperature have similar behavior with different magnitudes. 2TT has a maximum impact near the boundary surfaces. Its impact reduces on moving away from the boundary to the middle of the plate. Figures 5 and 6 show the symmetric variation and skew-symmetric normal stress components with plate thickness. It is clear from the figures that normal stress components for symmetric and skew-symmetric cases have maximum impact of two-temperature parameter near the load. Figures 7 and 8 depict the skew-symmetric and symmetric displacement component

W variation along the length of the plate x. Displacement W has a similar pattern for both symmetric and skew-symmetric case, but magnitude for both cases is different. Similar to other field quantities, 2TT has its maximum impact near the loading boundaries and reduces on moving away from the boundaries.

5 Conclusion

The effect of two-temperature parameter on various quantities in both modes is maximum on the boundary and decreases on moving toward the plate's middle plane except skew-symmetric normal stress and conductive temperature. A similar pattern of variations in these quantities is observed in 1TT and 2TT. However, various quantities show higher magnitudes in thermal load environment. Skew-symmetric thermodynamic temperature is less affected by two-temperature parameters as compared to symmetric one. Conductive and thermodynamic temperatures and normal stress have their minima in the middle of the plate in symmetric modes. The two-temperature parameter is observed to have more effects in the symmetric case than the skew-symmetric one near the plate's boundary. The two-temperature parameter leads to a change in the magnitude of various computed quantities; however, trends of variation remain unaffected.

References

1. Biot M (1956) Thermoelasticity and irreversible thermodynamics. J Appl Phys 27:240–253
2. Lord HW, Shulman Y (1967) A generalized dynamical theory of thermoelasticity. J Mech Phys Solid 15:299–309
3. Green AE, Lindsay KA (1972) Thermoelasticity. J Elast 2:1–7
4. Chen PJ, Gurtin ME (1968) On a theory of heat conduction involving two temperatures. Z Angew Math Phys 19:614–627
5. Chen PJ, Gurtin ME, Willams WO (1969) On the thermodynamics of non-simple elastic material with two temperatures. Z Angew Math Phys 20:107–112
6. Warren WE, Chen PJ (1973) Wave propagation in the two temperature theory of thermoelasticity. Acta Mech 16:21–31
7. Youssef HM, Al-Harby AH (2007) State-space approach of two-temperature generalized thermoelasticity of infinite body with a spherical cavity subjected to different types of thermal loading. Arch Appl Mech 77(9):675–687
8. Abbas IA, Abdalla AN, Alzahrani FS, Spagnuolo M (2016) Wave propagation in a generalized thermoelastic plate using eigenvalue approach. J Therm Stress 39:1367–1377
9. Kumar R, Manthena VR, Lamba NK, Kedar GD (2017) Generalized thermoelastic axi-symmetric deformation problem in a thick circular plate with dual phase lags and two temperatures. Mater Phys Mech 32:123–132
10. Khamis AK, El-Bary AA, Youssef HM, Nasr AM (2019) Two-temperature high-order lagging effect of living tissue subjected to moving heat source. Microsys Tech. 25:4731–4740
11. Kumar R, Miglani A, Rani R (2018) Generalized two temperatures thermoelasticity of micro polar porous circular plate with three phase lag model. J Mech 34:779–789

12. Xue ZN, Yu YJ, Tian XG (2017) Effects of two-temperature parameter and thermal nonlocal parameter on transient responses of a half-space subjected to ramp-type heating. Waves Random Complex Media 27(3):440–457
13. Abo-Dahab SM, Lotfy K (2017) Two-temperature plane strain problem in a semiconducting medium under photothermal theory. Waves Random Complex Media 27(1):67–91
14. Kumar R, Devi S, Abo-Dahab SM (2019) Propagation of Rayleigh waves in modified couple stress generalized thermoelastic with a three-phase-lag model. Waves Random Complex Media. https://doi.org/10.1080/17455030.2019.1588482
15. Said SM, Othman MIA (2019) Generalized electro–magneto-thermoelasticity with two-temperature and internal heat source in a finite conducting medium under three theories. Waves Random Complex Media. https://doi.org/10.1080/17455030.2019.1637552
16. Deswal S, Kumar S, Jain K (2020) Plane wave propagation in a fiber-reinforced diffusive magneto-thermoelastic half spce with two-temperature. Waves Random Complex Media. https://doi.org/10.1080/17455030.2020.1758832
17. Abbas IA (2017) Free vibration of a thermoelastic hollow cylinder under two-temperature generalized thermoelastic theory. Mech Based Design Struc Mach 45:395–405
18. Youssef HM (2006) Theory of two-temperature generalized thermoelasticity. IMA J Appl Math 71:383–390
19. Sharma JN, Singh D, Kumar R (2000) Generalized thermoelastic waves in homogeneous isotropic plates. J Acoust Soc Am 108:848–851

Numerical Investigation of Baffle Spacing in a Shell and Tube Heat Exchanger with Segmental Baffle

Ravi Gugulothu, Narsimhulu Sanke, Farid Ahmed, Naga Sarada Somanchi, and M. T. Naik

Abstract Baffle spacing has a decisive effect on heat transfer and pumping power. The development of baffle spacing significantly dominates the turbulence created inside the shell and tube heat exchanger and heat transfer. The study will focus on the impact of baffle spacing in both global and local thermo-hydraulic characteristics. The shell side flow rate varied from 0.18 to 0.31 kg/s, whereas the tube side flow rate varied from 0.11 to 0.18 kg/s. The fluid was assumed to be incompressible Newtonian fluid. Finite volume method was implemented to predict the thermal and flow behavior inside the heat exchanger. The simulations were carried out under steady-state assumption. The results show that with the decrease in baffle spacing, the amount of heat transfer rate increases. On the contrary, the increase in pressure drop was observed with the increase in baffle spacing.

Nomenclature

STHE	Shell and Tube Heat Exchanger
P	Tube pitch
d	Tube diameter
D	Shell diameter

R. Gugulothu (✉) · N. Sanke
Department of Mechanical Engineering, University College of Engineering, Osmania University, Hyderabad, India
e-mail: ravi.gugulothu@gmail.com

N. Sanke
e-mail: nsanke@osmania.ac.in

F. Ahmed
Department of Nuclear Science and Engineering, Military Institute of Science and Technology, Dhaka, Bangladesh

N. S. Somanchi · M. T. Naik
Department of Mechanical Engineering, JNTUH College of Engineering Hyderabad, Hyderabad, India

S. S. Ray et al. (eds.), *Applied Analysis, Computation and Mathematical Modelling in Engineering*, Lecture Notes in Electrical Engineering 897,
https://doi.org/10.1007/978-981-19-1824-7_6

S	Baffle spacing
H	Baffle Cut/height
L	Length
PDEs	Partial Differential Equations
Nu	Nusselt number
De	Effective diameter
Re	Reynolds number
TEMA	Tubular Exchanger Manufacturers Association

Greek symbols

ΔP	Pressure drop (Pa)
U	Overall Heat Transfer Coefficient (W/m^2K)

Subscripts

s	Shell
t	Tube
0	Outer
i	Inner
avg	Average

1 Introduction

Heat exchanger devices are used to transfer heat between two or more fluids which are at different temperatures and pressures, separated by a solid wall [1]. In heat exchanger apparatus, shell and tube heat exchanger (STHE) is playing key role (energy generation, oil refining industries, waste to heat recovery systems, etc.) due to robust construction of geometrical structure, ease of upgrade, and maintenance [2, 3]. To calculate the performance of STHE, baffle geometry and arrangement are essential components which are indicated in Fig. 1. Segmental baffle STHE is a commonly used one, and it will lead the fluid flow in tortuous and zigzag manner in the shell body [4–6]. These baffles are improving level of mixing to enhance the heat transfer of STHE which is given in Figs. 1 and 2.

Ramananda Rao et al. [7] studied the shell and tube heat exchanger by choosing the parameters toward minimizing the cost of heat exchanger for any heat duty, author considered tube pitch (P_t) should be 1.25–1.5d_0, minimum tube sheet thickness is 3.2 mm, length to diameter ratio should be 3 to 15 and baffle spacing should be 20% of

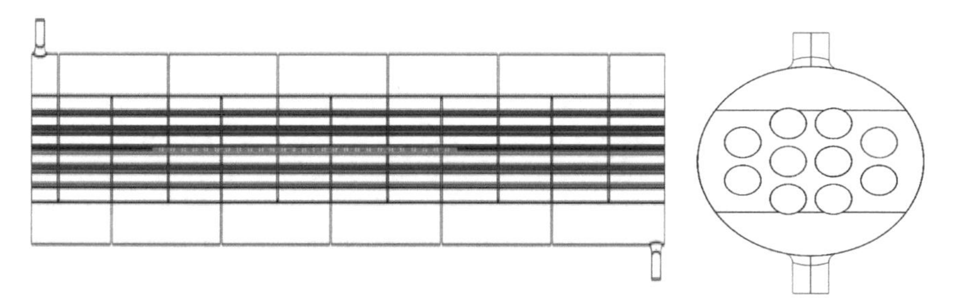

Fig. 1 Segmental baffle shell and tube heat exchanger

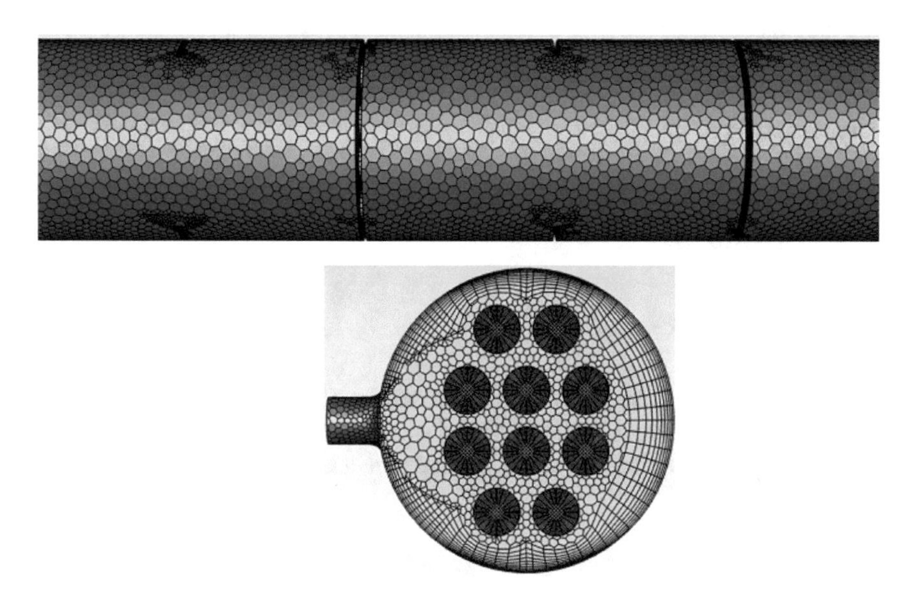

Fig. 2 Mesh geometry

D_i to 100% of D_i, it should not be less than 50 mm and finally concluded that these design parameters shown limited pressure drop and space and length constraints. Khalifeh Soltan et al. [8] listed out 20 and 100%D_i. Saffar Avval et al. [9] presented an optimization program to calculate an optimum baffle spacing and the number of sealing strips for STHE, all most negligible effect found on the optimum pumping power and heat transfer, while the baffle spacing has a noticeable effect by author.

Gaddis et al. [10] presented a procedure to evaluate shell side pressure drop in segmental baffle shell and tube heat exchanger. Author varied the ratio of baffle spacing to shell inside diameter (S/D_i) and the ratio of baffle cut to shell inside diameter (H/D_i) within the range and concluded that $0.2 \leq \left(\frac{S}{D_i}\right) \leq 1$ and $0.15 \leq \left(\frac{H}{D_i}\right) \leq 0.4$. Iyer et al. [11] successfully established and designed optimal STHE. Kallannavar et al. [12] experimentally studied that STHE with different tube layout as 30^0, 45^0, and 60^0. Among these three tube layouts, 30^0 degree tube layout proved better heat transfer rate. Emad et al. [13] experimentally studied STHE with four different segmental baffle (conventional single segmental baffle, staggered single segmental baffle, flower segmental baffle, and hybrid segmental baffle) configurations to enhance the thermal, hydraulic, and thermodynamic performances. Hybrid segmental baffle configuration STHE is the best among the studied STHE. Gugulothu et al. [14] numerically studied different hydrodynamic characteristics and local parameters of a 3D geometry of shell and tube heat exchangers with segmental baffle.

The objective of the study is to investigate the heat transfer rate of different configured baffled spacing and provide the potential configuration through optimization study. For the investigation, the baffle spacing varied from 50 to 100 mm. The shell side flow rate varied from 0.18 to 0.31 kg/s, whereas the tube side flow rate varied from 0.11 to 0.18 kg/s. The study will put a remark to choose the best configuration of baffle spacing and flow rate considering both the variation of heat transfer rate and pressure drop.

2 Simulation and Modeling

2.1 Governing Equations

In the present investigation, the heat transfer and fluid flow phenomena on the shell side and tube side are studied in shell and tube heat exchanger with segmental baffle. The conservation equation, continuity, momentum, and energy are presented here.

Continuity equation:

$$\frac{\partial(\rho f_v)}{\partial t} + \frac{\partial\left(\rho f_s u_j\right)}{\partial x_j} = 0 \tag{1}$$

Momentum equation:

$$\frac{\partial(\rho f_v u_i)}{\partial t} + \frac{\partial\left(\rho f_s u_i u_j\right)}{\partial x_j} = -f_s \frac{\partial p_i}{\partial x_i} + \frac{\partial\left(f_s \mu_{\text{eff}}\left(\frac{\partial u_i}{\partial x_j} + \frac{\partial u_j}{\partial x_i}\right)\right)}{\partial x_j} - 0.5 f_i \rho |u_i| u_i \tag{2}$$

Energy equation:

$$\frac{\partial\left(\rho c_p f_v T\right)}{\partial t} + \frac{\partial\left(\rho c_p T f_s u_j\right)}{\partial x_j} = \frac{\partial\left(f_s \lambda_{\text{eff}}\left(\frac{\partial T}{\partial x_j}\right)\right)}{\partial x_j} + \Omega K (T_t - T) \tag{3}$$

where u is velocity

2.2 Computational Domain

The STHE as shown in Fig. 1 is a single shell and tube pass with four tubes. The geometry of STHE is shell inner diameter (D_i)=100 mm, length (Ls)=1150 mm, and tube bundle consists with ten tubes of 19 mm outer diameter in triangular arrangement with 25-mm tube pitch.

2.3 Boundary Conditions

The shell and tube heat exchanger consists of two loops carrying the coolant. The cold fluid flows in the shell with a temperature of 303.15 K. The hot fluid flows in four tubes at a temperature of 338.15 K. Several assumptions are made to simplify the geometry and study the numerical computational domain. Those are insulated shell walls, neglected fluid flow leakage, and fully developed fluid flow.

2.4 Mesh Selection and Sensitivity Analysis

STHX is a complex structure, so unstructured tetra-hydrate mesh was selected which is a proper grid system to mesh numerical models, i.e., shown in Fig. 2. Very fine mesh is done nearby tube wall region to obtain more accurate results. The finite volume method with SIMPLE pressure velocity coupling algorithm is used to solve the PDEs. PRESTO method with double precision solver was implemented to solve the convective terms in the governing equations [15–17].

2.5 Data Reduction

Tube outer diameter (d_0) is 16 mm taken,

$$\text{Shell side Nusselt number } \mathrm{Nu}_s = 1.86\mathrm{Re}_s^{0.33} \, \mathrm{Pr}_s^{0.33} \left(\frac{D_i}{L}\right)^{0.33} \tag{4}$$

$$\text{Shell side Reynolds number } \mathrm{Re}_s = \frac{D_e V_s}{\vartheta_s} \tag{5}$$

$$\text{Effective diameter (m) } D_e = \frac{P_t^2 - \frac{\pi d_0^2}{4}}{\frac{\pi d_0}{4}} \quad (6)$$

$$\text{According to TEMA standards } P_t = 1.25 d_0 \text{ in mm} \quad (7)$$

$$\text{Shell side Prandtl number } Pr_s = \frac{\mu_s C_{p_s}}{k_s} \quad (8)$$

$$\text{Free flow area } A_s = \frac{\pi}{4}\left(D_i^2 - N_t d_0^2\right) \quad (9)$$

$$\text{Number of tubes } N_t = K_1 \left(\frac{D_s}{d_0}\right)^{n_1} \quad (10)$$

No. of passes	Triangular pitch		Square and rotated square pitch	
	K_1	N_1	K_1	n_1
1	0.319	2.142	0.215	2.207
2	0.249	2.207	0.156	2.291
4	0.175	2.285	0.158	2.263
6	0.0743	2.499	0.0402	2.617
8	0.0365	2.675	0.0331	2.643

$$\text{Shell outer diameter } D_s = \frac{D_{\text{otl}}}{0.95} + \delta_{\text{clearence}} \quad (11)$$

$$\text{Tube bundle outer diameter } D_{\text{otl}} = \left(\frac{N_t}{K_1}\right)^{\frac{1}{n_1}} d_0 \quad (12)$$

According to Ravi et al., baffle space (BS) is 20–100%D_i and it should not be less than 51 mm. In this research paper, geometry is created with different baffle spacing from 0.2 to $1D_i$ and simulation is conducted.

$$\text{Heat exchanger surface area}(\text{m}^2) A_0 = \pi N_t d_0 L \quad (13)$$

$$\text{Shell side Nusselt number } \text{Nu}_s = 1.86 \text{Re}_s^{0.33} Pr_s^{0.33} \left(\frac{D_i}{L}\right)^{0.33} \quad (14)$$

$$\text{Tube side Reynolds number } \text{Re}_t = \frac{d_i V_t}{\vartheta_t} \quad (15)$$

$$\text{Tube side Prandtl number } Pr_t = \frac{\mu_t C_{p_t}}{k_t} \tag{16}$$

3 Results and Discussions

3.1 Validation of Results

In order to validate the numerical model, the results of the present study were compared with the experimental study of Kamel et al. [18] following the same boundary conditions. The results of the comparison study are shown in Fig. 3. It is well recognized from the figure that the present study aligns precisely with a good agreement along with the study of Kamel et al. [18]. The maximum errors encountered during the simulation are found to be 14.6% at 0.18 kg/s flow rate, whereas the minimum error of 6.5% was encountered at 0.11 kg/s flow rate. Hence, following the results of the assessment, it could be said that the present numerical model could capture the thermo-hydraulic characteristics of the computational geometry accurately.

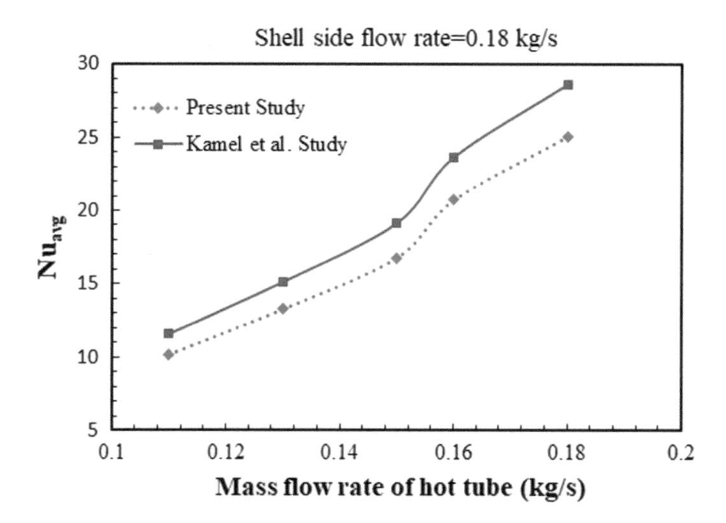

Fig. 3 Comparison of the present study with experimental investigations

3.2 Pressure Drop

The pressure distribution of heat exchangers can reflect the resistance characteristic and heat transfer performance. Figure 4 presents the pressure contour with superimposed fluid velocity. It is well recognized from the figure that along with the flow direction, the pressure drops linearly.

Figure 5 shows the temperature distribution of all the studied six baffle space shell and tube heat exchangers with the segmental baffle. The temperature distribution is shallow behind the baffles, and local heat transfer is very poor. In Fig. 5, tube side temperature will gradually decrease from 338.15 to 322.618 K by increasing the

Fig. 4 Shell side pressure of $m_s = 0.18$ kg/s and $m_t = 0.11$ to 0.18 kg/s

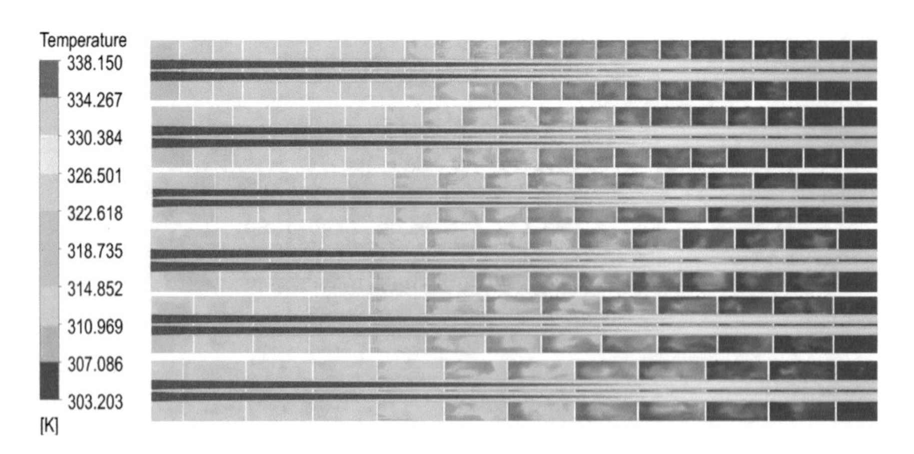

Fig. 5 Temperature in x–y plane of $m_s = 0.18$ kg/s and $m_t = 0.11$ to 0.18 kg/s

shell side temperature until 314 K. It is noticed that the temperature is higher and gradually reduces near the shell wall center.

The velocity contours of computational flow domain are shown in Fig. 6 for

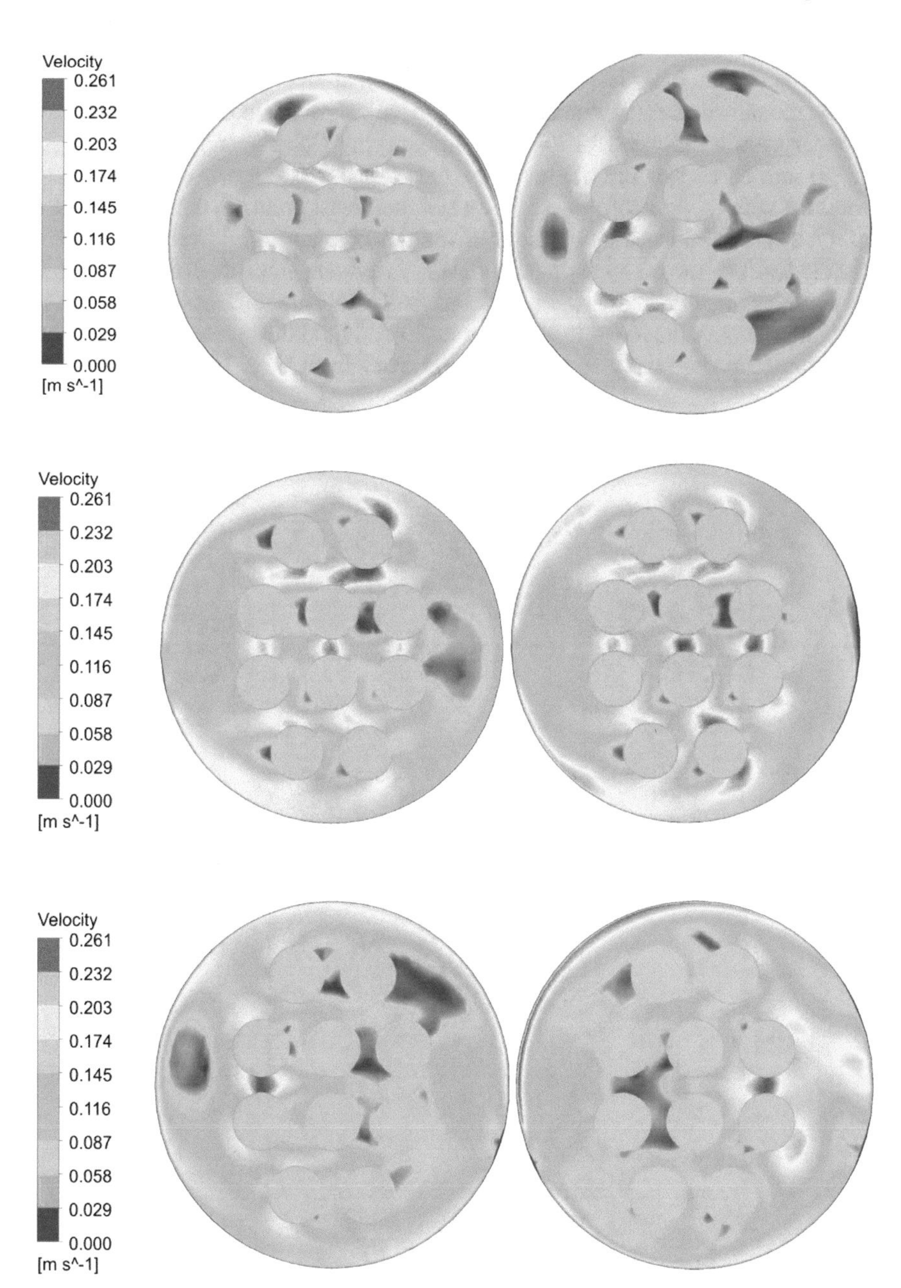

Fig. 6 Velocity at $x = 0.24$ m of radial plane of $m_s = 0.18$ kg/s and $m_t = 0.11$ to 0.18 kg/s

the radial velocity at x=0.24 m distance. These contours are demonstrated in radial directions, and the authors noticed that due to the presence of the baffles, the vortex formations is observed which initiates the secondary flow in the flow regime. Such phenomenon of inducing secondary flow with the help of initiating and increasing turbulent intensity improves the heat transfer rate of the flow domain. Due to having lower space within the baffles, 50-mm baffle spacing induces higher secondary flow, which gradually decreases with the increase in baffle spacing and the least amount of secondary flow is evaluated for 100-mm baffle spacing. Consequently, it is recognized that 50-mm baffle spacing improves the heat transfer rate most, which gradually decreases with the increase in baffle spacing and least amount of improvement is observed for 100-mm baffle spacing. In addition, the upsurge in velocity gradient near the walls of the computational geometry is evident since the leakage zones in the heat exchanger decreases the velocity at shell center and increases at shell walls.

Figure 7 represents the average heat transfer and fluid flow characteristics across tube bundles in shell body at different baffle space from 50 to 100 mm. According

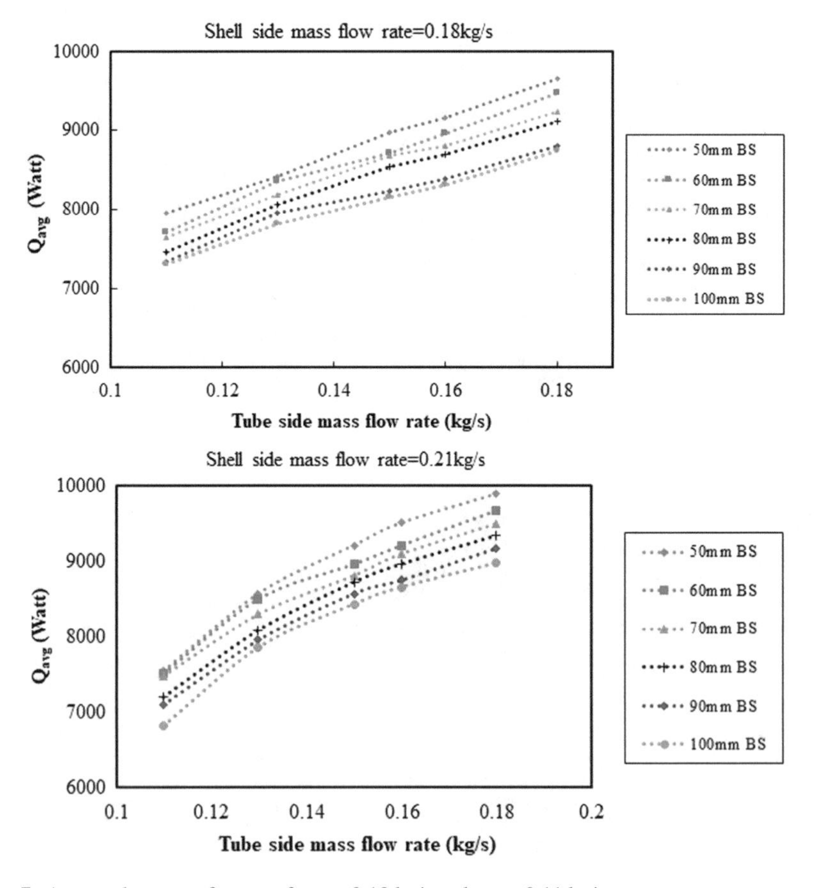

Fig. 7 Average heat transfer rate of $m_s = 0.18$ kg/s and $m_t = 0.11$ kg/s

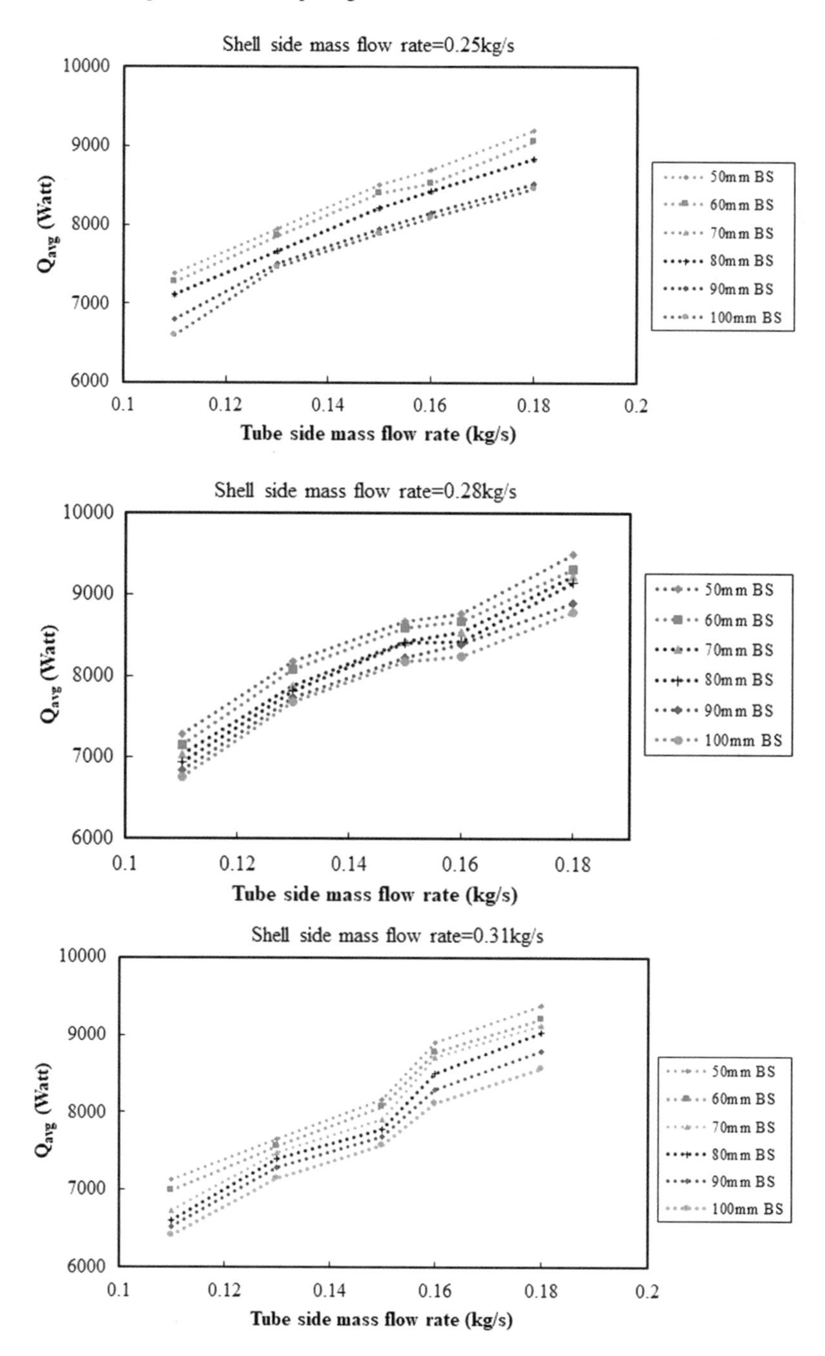

Fig. 7 (continued)

to Fig. 7, heat transfer rate is increasing along with the increase in mass flow rate and baffle spacing. The fluctuation of the flow pattern increases with the decrease in baffle spacing, which eventually increases the turbulent intensity. Since the turbulent intensity plays as a predominant parameter in enhancement of heat transfer, lowering the baffle spacing induces higher heat transfer rate. As a result, the highest heat transfer rate was observed at 50-mm baffle spacing in Fig. 7, which progressively decreases with the increase in baffle spacing and achieved lowest value of heat transfer rate at 100-mm baffle spacing.

Figure 8 indicates the pressure drop of all the studied shell and tube heat exchanger with different baffle spaces from 50 to 100 mm, i.e., 20–1%D_i. From Fig. 8, it is identified that with the decrease in baffle spacing, the pressure drop increases. The maximum pressure drop was observed for 50-mm baffle spacing, which gradually decreases with the increase in baffle spacing, and lowest value of pressure drop was recognized for 100-mm baffle spacing.

4 Conclusion

In this research paper, work is done with different baffle space from 20 to 100%D_i with segmental baffle shell and tube heat exchanger. In this work, 20%D_i baffle space showed more pressure, high heat transfer, and 100%D_i baffle space shown as less pressure and less heat transfer. Moreover, 70- and 80-mm baffle spacing in segmental baffle has shown the optimal solutions. The study will privilege the heat exchanger industries to implement the optimized configuration of baffle spacing, taking both heat transfer and pressure drop into consideration.

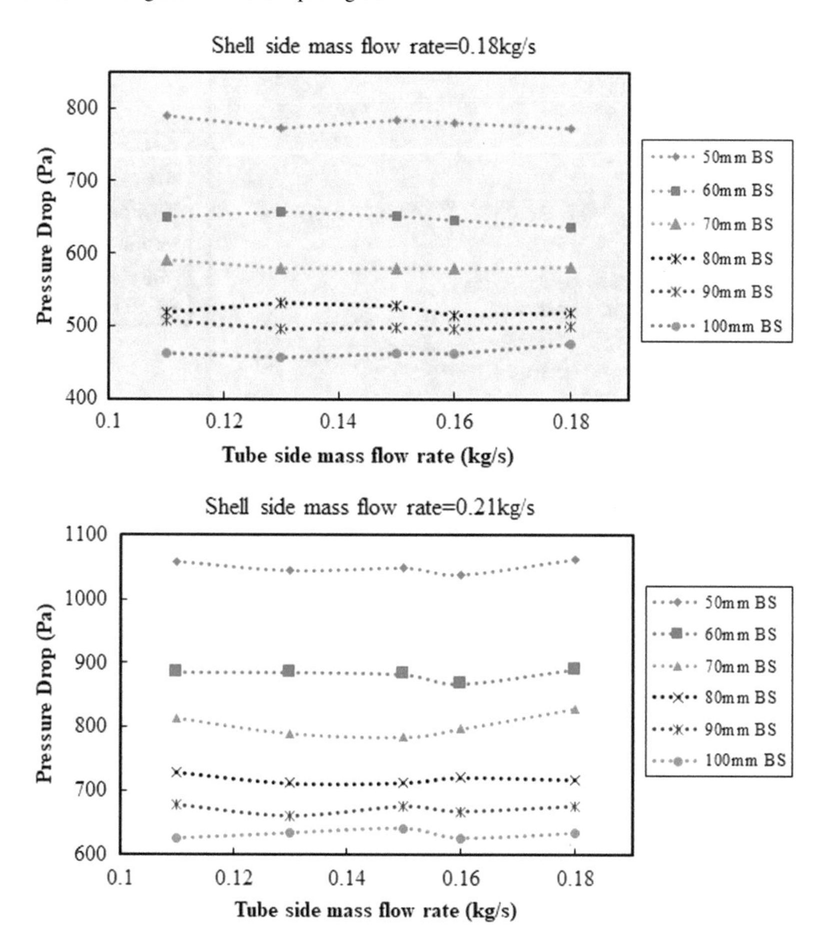

Fig. 8 Pressure drop of $m_s = 0.18$ kg/s and $m_t = 0.11$ kg/s

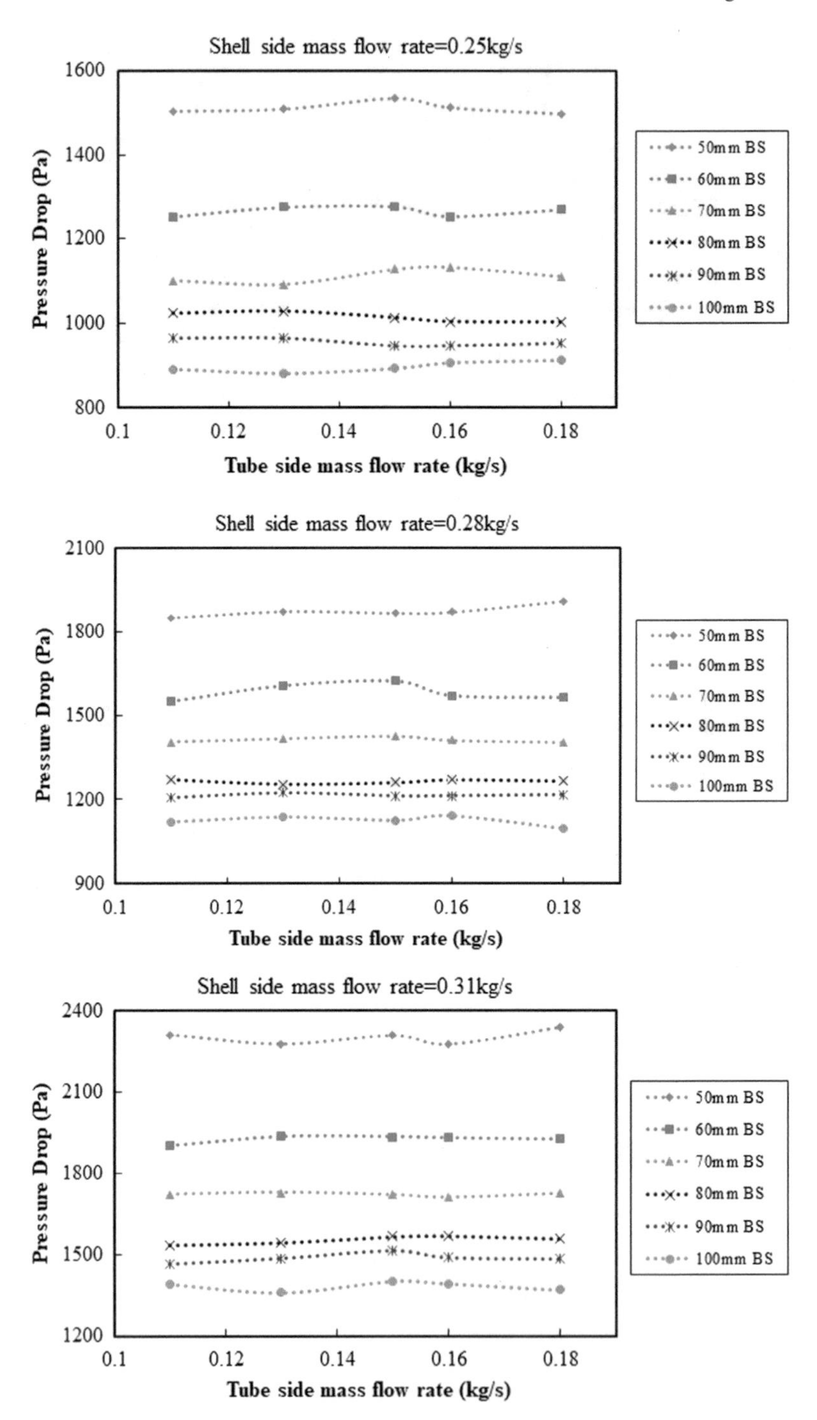

Fig. 8 (continued)

References

1. Hadidi A, Hadidi M, Nazari A (2013) A new design approach for shell and tube heat exchangers using imperialist competitive algorithm (IAC) from economic point of view. Energy Convers Manage 67(2013):66–74
2. Gugulothu R, Somanchi NS, Reddy VK, Tirkey JV (2016) A review on design of baffles for shell and tube heat exchangers. In: Proceeding of ACGT-2016, Indian Institute of Technology Bombay, during 14–16th Nov 2016
3. Gugulothu R, Sanke N, Gupta AVSSKS (2018) Numerical study of heat transfer characteristics in shell and tube heat exchanger. In: Lecture notes in mechanical engineering. Springer Nature Singapore Pte Ltd. 2019, pp 375–383
4. Gugulothu R, Vijaya Kumar Reddy K, Somanchi NS, Adithya EL (2017) A review on enhancement of heat transfer techniques. In: 5th international conference of materials processing and characterization (ICMPC-2016), Materials Today: Proceedings, vol 4, pp 1051–1056
5. Vijaya Kumar Reddy K, Sudheer Prem Kumar B, Gugulothu R, Anuja K, Vijaya Rao P (2017) CFD analysis of a helically coiled tube in tube heat exchanger. In: 5th international conference of materials processing and characterization (ICMPC-2016), materials today: proceedings, vol 4, pp 2341–2349
6. Gugulothu R, Sanke N, Ahmed F, Jilugu RK (2020) Numerical study on shell and tube heat exchanger with segmental baffle. In: International joint conference on advances in computational intelligence (IJCACI-2020), at Daffodil International University Dhaka, Bangladesh, during 20–21st Nov 2020
7. Ramananda Rao R, Shrinivasa U, Srinivasan J (1991) Synthesis of cost optimal shell and tube heat exchangers. Heat Transfer Eng 12(3):47–55
8. Khalifeh Soltan B, Saffar Avval M, Damangir E (2004) Short communication: minimizing capital and operating costs of shell and tube condensers using optimum baffle spacing. Appl Therm Eng 24(2004):2801–2810
9. Saffar Avval M, Damangir E (1995) A general correlation for determining optimum baffle spacing for all types of shell and tube exchangers. Int J Heat Mass Transfer 38(13):2501–2506
10. Gaddis ES, Gnielinski V (1997) Pressure drop on the shell side of shell and tube heat exchangers with segmental baffles. Chem Eng Process 36:149–159
11. Iyer VH, Mahesh S, Malpani R, Sapre M, Kulkarni AJ (2019) Adaptive range genetic algorithm: a hybrid optimization approach and its application in the design and economic optimization of shell and tube heat exchanger. Eng Appl Artif Intell 85:444–461
12. Kallannavar S, Mashyal S, Rajangale M (2019) Effect of tube layout on the performance of shell and tube heat exchangers. In: Mater Today: Proc
13. El Said EMS, Abou Al Sood MM (2019) Shell and tube heat exchanger with new segmental baffles configurations: a comparative experimental investigation. Appl Therm Eng 150:803–810
14. Gugulothu R, Sanke N, Ahmed F, Jilugu RK (2021) Numerical study on shell and tube heat exchanger with segmental baffle. In: Proceedings of international joint conference on advances in computational intelligence, algorithms for intelligent systems. Springer Nature Singapore Pte. Ltd. pp 309–318
15. Ahmed F, Abir MA, Bhowmik PK, Deshpande V, Mollah AS, Kumar D, Alam S (2021) Computational assessment of thermo-hydraulic performance of Al_2O_3 water nanofluid in hexagonal rod-bundles subchannel. Prog Nuclear Energy 135:103700
16. Ahmed F, Abir Md A, Fuad M, Akter F, Bhowmik PK, Alam S, Kumar D (2021) Numerical investigation of the thermo hydraulic performance of water based nanofluids in a dimpled channel flow using Al_2O_3, CuO and hybrid Al_2O_3-CuO as nanoparticles. In: Heat transfer, pp 1–26
17. Fluent ANSYS (2015) Ansys fluent. Academic Research. Release, 14
18. Shirvan KM, Mamourian M, Esfahabi JA (2018) Experimental investigation on thermal performance and economic analysis of cosine wave tube structure in a shell and tube heat exchanger. Energy Conver Manag 175:86–98

19. Yang JF, Zeng M, Wang QW (2015) Numerical investigation on shell side performances of combined parallel and serial two shell pass shell and tube heat exchangers with continuous helical baffles. Appl Energy 139:163–174
20. Gugulothu R, Sanke N (2022) Use of segmental baffle in shell and tube heat exchanger for nano emulsions. Heat Trans 51(3):2645–2666. https://doi.org/10.1002/htj.22418
21. Gugulothu R, Sanke N (2022) Effect of helical baffles and water-based Al_2O_3, CuO, and SiO_2 nanoparticles in the enhancement of thermal performance for shell and tube heat exchanger. Heat Trans. https://doi.org/10.1002/htj.22474

New Analytical Exact Solutions of Time Fractional (2+1)-Dimensional Calogero–Bogoyavlenskii–Schiff (CBS) Equations

A. K. Sahoo and A. K. Gupta

Abstract In this article, Kudryashov and modified Kudryashov techniques were implemented to acquire new exact solutions of the time fractional (2+1)-dimensional CBS equation. The solutions thus attained have been stated explicitly and graphical models have been illustrated by choosing appropriate values to the parameters to visualize the mechanism of the considered nonlinear fractional differential equation (FDE). The considered methods are very powerful and effective enough to utilize for establishing solutions of various nonlinear FDEs applied in mathematical physics.

Keywords Conformable fractional derivative · Fractional Calogero–Bogoyavlenskii–Schiff (CBS) equation · Generalized Kudryashov method · Modified Kudryashov method

1 Introduction

In present century, fractional calculus has been established as a vibrant field of study in engineering and science, as many researchers are paying interest to it due to its diversified implementation in various fields like heat transfer, signal processing, biology, robotics, electronics, genetic algorithms, control systems, etc. Many mathematical models are governed by fractional order differential equations. These ideas were attributed to Leibniz, Liouville, Riemann, Caputo, etc. The basic literature related to fractional derivatives and integrals is discussed in refs [1, 2].

In examining the physical phenomena, the study of nonlinear FDEs executes a significant role. Various methods have already been implemented to attain exact

A. K. Sahoo · A. K. Gupta (✉)
Department of Mathematics, School of Applied Sciences, KIIT Deemed to be University, Bhubaneswar, Odisha 751024, India
e-mail: arun27.nit@gmail.com

A. K. Sahoo
e-mail: ajaysahoo20@gmail.com

S. S. Ray et al. (eds.), *Applied Analysis, Computation and Mathematical Modelling in Engineering*, Lecture Notes in Electrical Engineering 897,
https://doi.org/10.1007/978-981-19-1824-7_7

solutions of FPDEs [3–5]. Few distinguished efforts have been noticed by the scientists for analyzing FDEs arising in mathematical physics. In this manuscript, the conformal time fractional CBS equation has been contemplated for the first time.

Consider the conformable time fractional (2+1)-dimensional CBS equation

$$T_t^{\alpha} u_x - 4u_x u_{xy} - 2u_{xx}u_y + u_{xxxy} = 0 \tag{1.1}$$

where α denotes the order of conformable derivative. Bogoyavlenskii constructed the CBS equation by utilizing the modified Lax formalism [6, 7]. Later, it was derived by Schiff [8] using self-dual Yang Mills equation. The fundamental interaction of a Riemann wave along the y-axis with a long wave along the x-axis is governed by the CBS equation.

Methods such as Cole Hopf transformation [9], singular manifold method [10, 11], similarity transformations method [12], the Tanh–Coth method [13], the Tanh function method [14], the improved Tanh–Coth method [15], sine–cosine approach [16], the extended homoclinic test approach [17], Hirota's bilinear method [18], and the Lie transformation method [10, 19] had been employed to acquire solutions of CBS equation. But the extensive study of conformable time fractional CBS equation is just the opening.

The acquired exact solutions for fractional (2+1)-dimensional CBS equation have been documented first time ever in this manuscript.

This paper is systematized as follows. In Sect. 2, introduction to conformable fractional derivative has been provided. The algorithm of generalized Kudryashov and modified Kudryashov technique is provided in Sect. 3. The Kudryashov techniques have been implemented to fractional CBS equation which is discussed in Sect. 4. In Sect. 5, results are furnished, and Sect. 6 concludes the article.

2 Conformable Fractional Derivative

The conformable fractional derivative of a function $u(t)$ of order α is defined by [20–22]

$$T_\alpha(u(t)) = \lim_{\tau \to 0} \frac{u\left(t + \tau t^{1-\alpha}\right) - u(t)}{\tau} \quad \text{for all } t > 0, \alpha \in (0, 1) \tag{2.1}$$

Some properties of conformal fractional derivatives are as follows [22]

(i) $T_\alpha(au + bv) = aT_\alpha(u) + bT_\alpha(v)$.
(ii) $T_\alpha(t^\mu) = \mu t^{\mu-\alpha}$, for all $\mu \in \mathbb{R}$.
(iii) $T_\alpha(u * v)(t) = t^{1-\alpha} v'(t) u'(v(t))$.

3 Algorithm of the Proposed Method

The generalized Kudryashov and modified Kudryashov techniques have been employed to calculate the exact solutions of fractional (2+1)-dimensional CBS equations. A number of researchers have demonstrated the generalization, reliability, and efficacy of this approach [23, 24]. The basic ideas of generalized Kudryashov and modified Kudryashov methods are discussed in Refs. [22, 25, 26].

4 Exact Solution of Time Fractional (2+1)-Dimensional CBS Equation

This segment consists of the solutions of time fractional (2+1)-dimensional CBS equations by utilizing the generalized Kudryashov and modified Kudryashov techniques to calculate the exact solution for the same.

4.1 *Application of Generalized Kudryashov Technique*

By utilizing the wave transformation [27] $u(x, y, t) = U(\xi)$, $\xi = x + y - \lambda(\frac{t^{\alpha}}{\alpha})$, Eq. (1.1) can be reduced as

$$-\lambda U_{\xi\xi} - 6U_{\xi}U_{\xi\xi} + U_{\xi\xi\xi\xi} = 0 \tag{4.1}$$

where λ is the constant that is to be evaluated later. Now, the value of N is found to be 1 by equating the nonlinear term and highest order derivative term of Eq. (4.1).

Therefore,

$$U(\xi) = \varphi_0 + \varphi_1\psi(\xi) \tag{4.2}$$

where $\psi(\xi) = \frac{1}{1 \pm e^{\xi}}$.

φ_0, φ_1 are constants to be determined.

Next, system of algebraic equations was acquired from Eq. (4.2), substituting the derivatives of $U(\xi)$ and considering the ansatz. The same powers of ψ^i, $(i = 1, 2, 3, \ldots)$ are collected and equate to zero, to acquire the following system of nonlinear equations:

Coefficient of ψ^1:

$$\varphi_1 - \lambda\varphi_1 = 0, \tag{4.3}$$

Coefficient of ψ^2:

$$-15\varphi_1 + 6\varphi_1^2 + 3\varphi_1\lambda = 0, \tag{4.4}$$

Coefficient of ψ^3:

$$50\varphi_1 - 24\varphi_1^2 - 2\varphi_1\lambda = 0, \tag{4.5}$$

Coefficient of ψ^4 :

$$-60\varphi_1 + 30\varphi_1^2 = 0, \tag{4.6}$$

Coefficient of ψ^5:

$$24\varphi_1 - 12\varphi_1^2 = 0, \tag{4.7}$$

On solving Eqs. (4.3)–(4.7), the following nontrivial solutions can be acquired.

Case I: For $\varphi_1 = 2, \lambda = 1$ and $\psi(\xi) = \frac{1}{1+e^{\xi}}$, the new exact solution of the conformable time fractional CBS equation given in Eq. (1.1) can be identified as

$$u_1(x, y, t) = \frac{2 + \varphi_0 + \varphi_0 \cos h\left[x + y - \frac{t^\alpha}{\alpha}\right] + \varphi_0 \sin h\left[x + y - \frac{t^\alpha}{\alpha}\right]}{1 + \cos h\left[x + y - \frac{t^\alpha}{\alpha}\right] + \sin h\left[x + y - \frac{t^\alpha}{\alpha}\right]} \tag{4.8}$$

Case II: For $\varphi_1 = 2, \lambda = 1$ and $\psi(\xi) = \frac{1}{1-e^{\xi}}$, the exact solution of Eq. (1.1) is given by

$$u_2(x, y, t) = \frac{-2 - \varphi_0 + \varphi_0 \cos h\left[x + y - \frac{t^\alpha}{\alpha}\right] + \varphi_0 \sin h\left[x + y - \frac{t^\alpha}{\alpha}\right]}{-1 + \cos h\left[x + y - \frac{t^\alpha}{\alpha}\right] + \sin h\left[x + y - \frac{t^\alpha}{\alpha}\right]} \tag{4.9}$$

4.2 *Application of Modified Kudryashov Technique*

Equation (4.2) can be attained in the aforementioned manner. Next, considering the ansatz $\psi(\xi) = \frac{1}{1 \pm a^{\xi}}$ in Eq. (4.2) and substituting the derivatives of $U(\xi)$ in Eq. (4.1), the following equations can be acquired.

Coefficient of ψ^1:

$$-\lambda\varphi_1(\ln(a))^2 + \varphi_1(\ln(a))^4 = 0, \tag{4.10}$$

Coefficient of ψ^2:

$$3\varphi_1\lambda(\ln(a))^2 + 6\varphi_1^2(\ln(a))^3 - 15\varphi_1(\ln(a))^4 = 0, \tag{4.11}$$

Coefficient of ψ^3:

$$-2\varphi_1\lambda(\ln(a))^2 - 24\varphi_1^2(\ln(a))^3 + 50\varphi_1(\ln(a))^4 = 0, \tag{4.12}$$

Coefficient of ψ^4 :

$$30\varphi_1^2(\ln(a))^3 - 60\varphi_1(\ln(a))^4 = 0, \tag{4.13}$$

Coefficient of ψ^5:

$$-12\varphi_1^2(\ln(a))^3 + 24\varphi_1(\ln(a))^4 = 0, \tag{4.14}$$

On solving Eqs. (4.10)–(4.14), the following nontrivial solutions can be attained.

Case I: For $\varphi_1 = 2\ln(a)$, $\lambda = (\ln(a))^2$ and $\psi(\xi) = \frac{1}{1+a^{\xi}}$, the new exact solution of the conformable time fractional CBS equation given in Eq. (1.1) can be identified as

$$u_3(x, y, t) = \frac{\varphi_0 a^{x+y} + \varphi_0 a^{\frac{t^{\alpha}(\ln(a))^2}{\alpha}} + 2a^{\frac{t^{\alpha}(\ln(a))^2}{\alpha}}\ln(a)}{a^{x+y} + a^{\frac{t^{\alpha}(\ln(a))^2}{\alpha}}}. \tag{4.15}$$

Case II: For $\varphi_1 = 2\ln(a)$, $\lambda = (\ln(a))^2$ and $\psi(\xi) = \frac{1}{1-a^{\xi}}$, the exact solution of Eq. (1.1) is given by

$$u_4(x, y, t) = \frac{\varphi_0 a^{x+y} - \varphi_0 a^{\frac{t^{\alpha}(\ln(a))^2}{\alpha}} - 2a^{\frac{t^{\alpha}(\ln(a))^2}{\alpha}}\ln(a)}{a^{x+y} - a^{\frac{t^{\alpha}(\ln(a))^2}{\alpha}}} \tag{4.16}$$

5 Results

The exact solutions of conformable time fractional (2+1)-dimensional CBS equation have been attained for the first time by employing the generalized Kudryashov and modified Kudryashov techniques with the aid of wave transform. The following 3D graphical models of the obtained solutions $u(x, y, t)$ have been illustrated in Figs. 1, 2, 3, 4, 5, 6, 7, and 8 for various values α and by choosing appropriate values to the parameters.

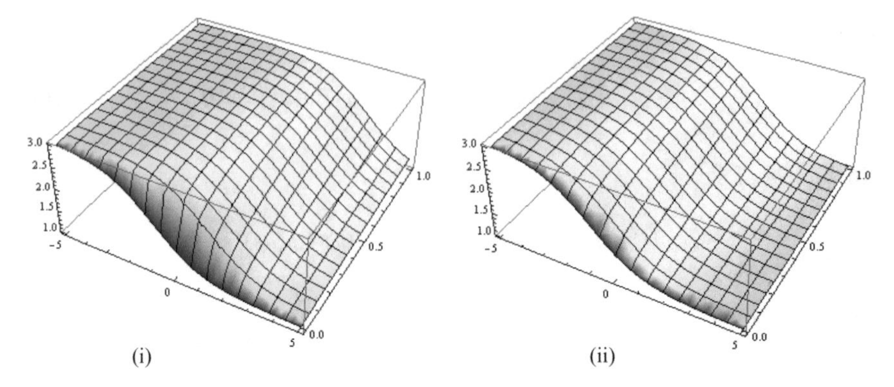

Fig. 1 3D plot of $u_1(x, y, t)$ for time fractional CBS equation given in Eq. (4.8) when $\varphi_0 = 1$, $y = 1$, at (i) $\alpha = 0.25$, (ii) $\alpha = 0.5$

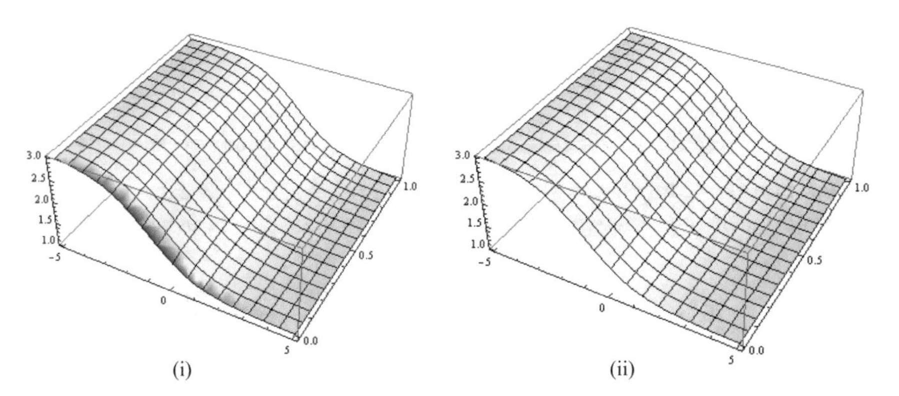

Fig. 2 3D plot of $u_1(x, y, t)$ for time fractional CBS equation given in Eq. (4.8) when $\varphi_0 = 1$, $y = 1$, at (i) $\alpha = 0.75$, (ii) $\alpha = 1$

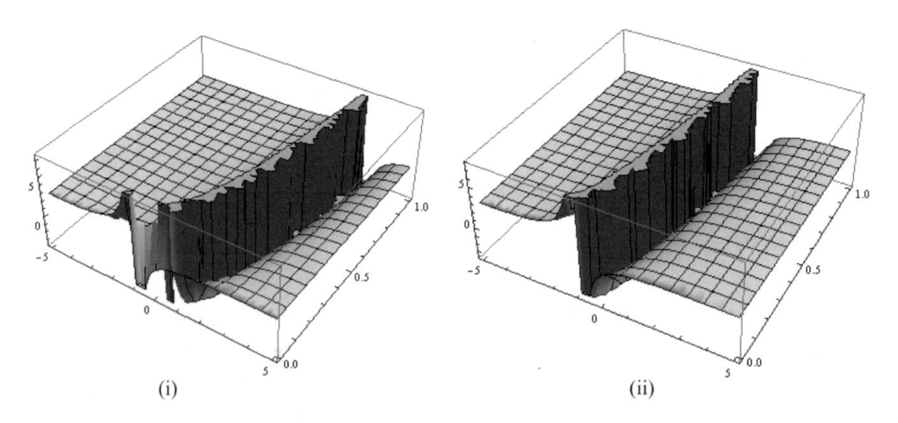

Fig. 3 3D plot of $u_2(x, y, t)$ for time fractional CBS equation given in Eq. (4.9) when $\varphi_0 = 1$, $y = 1$, at (i) $\alpha = 0.25$, (ii) $\alpha = 0.5$

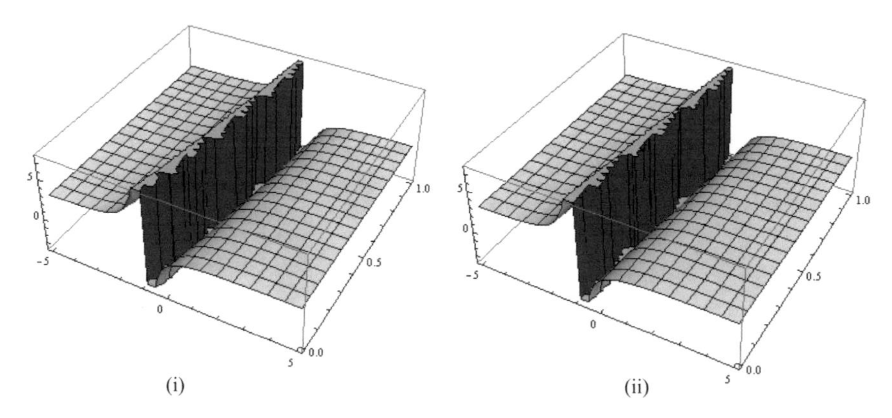

Fig. 4 3D plot of $u_2(x, y, t)$ for time fractional CBS equation given in Eq. (4.9) when $\varphi_0 = 1$, $y = 1$, at (i) $\alpha = 0.75$, (ii) $\alpha = 1$

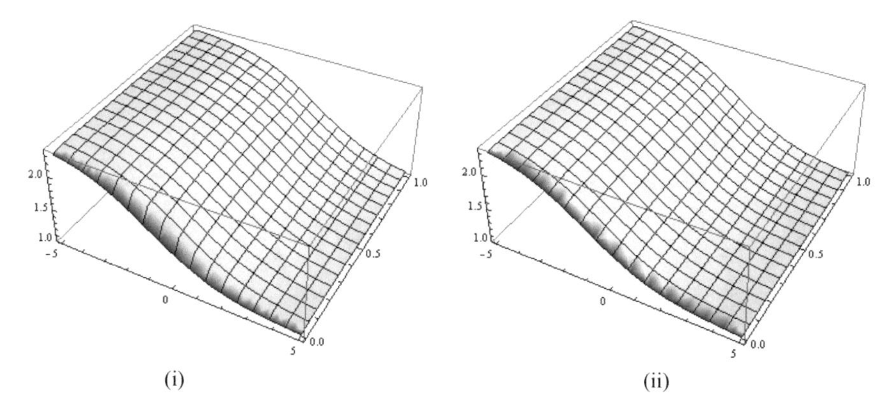

Fig. 5 3D plot of $u_3(x, y, t)$ for time fractional CBS equation given in Eq. (4.15) when $\varphi_0 = 1$, $y = 1$, $a = 2$ at (i) $\alpha = 0.25$, (ii) $\alpha = 0.5$

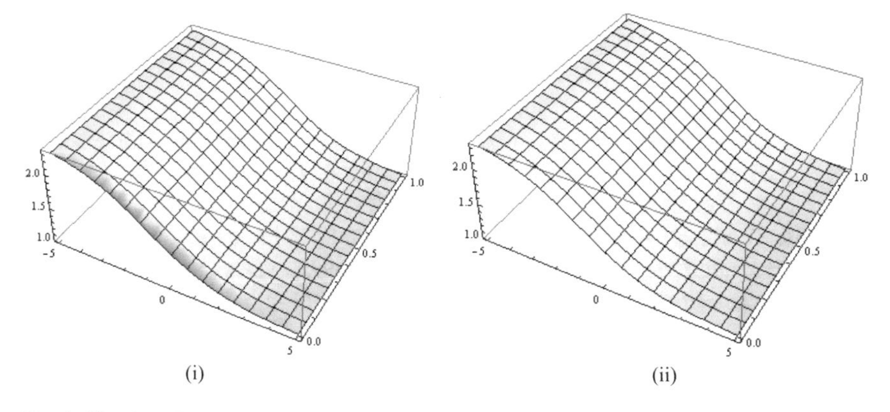

Fig. 6 **3D** plot of $u_3(x, y, t)$ for time fractional CBS equation given in Eq. (4.15) when $\varphi_0 = 1$, $y = 1$, $a = 2$ at (i)$\alpha = 0.75$, (ii) $\alpha = 1$

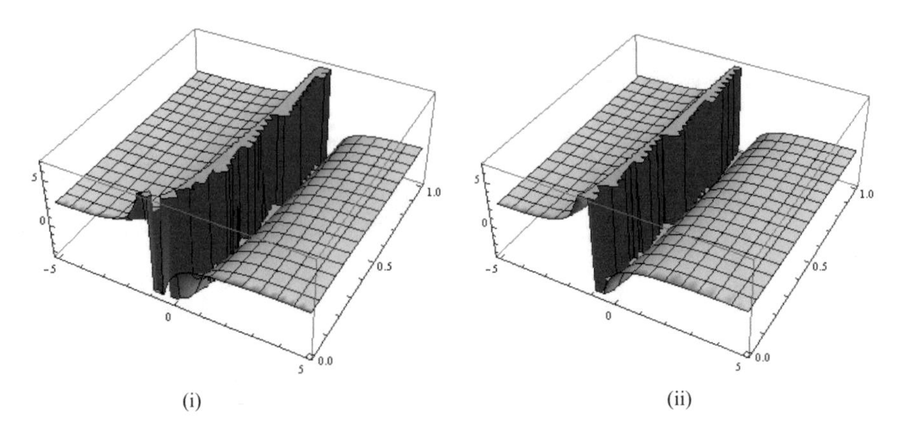

Fig. 7 3D plot of $u_4(x, y, t)$ for time fractional CBS equation given in Eq. (4.16) when $\varphi_0 = 1$, $y = 1$, $a = 2$ at (i)$\alpha = 0.25$, (ii) $\alpha = 0.5$

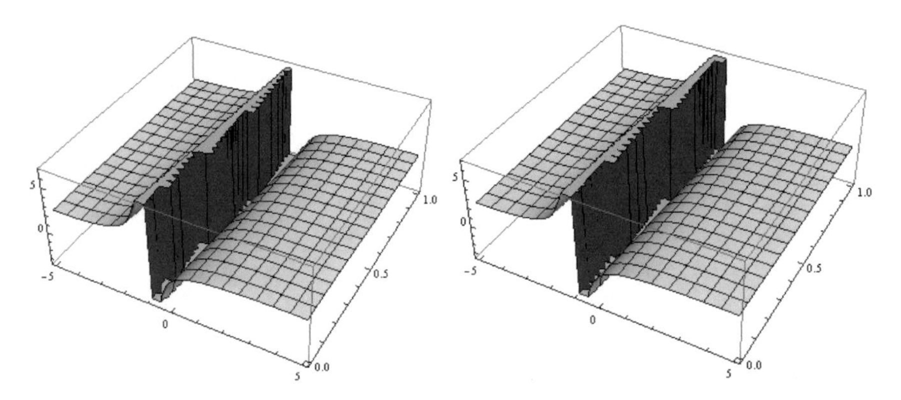

Fig. 8 3D plot of $u_4(x, y, t)$ for time fractional CBS equation given in Eq. (4.16) when $\varphi_0 = 1$, $y = 1$, $a = 2$ at (i)$\alpha = 0.75$, (ii) $\alpha = 1$

6 Conclusion

Here, the solitary wave solutions of conformable time fractional (2+1)-dimensional CBS equations are investigated. The generalized Kudryashov and modified Kudryashov techniques have successfully implemented to formally derive these solutions. To visualize the mechanism of fractional CBS equation, the graphs of the acquired solutions are furnished by selecting suitable values to the parameters. This discourse evident that the focused methods are efficacious for analytical solutions of higher order FDEs. It can also be manifested that the methods are thoroughly dependable and significant to discover new exact solutions.

References

1. Podlubny I (1999) Fractional differential equations. Academic Press, New York, USA
2. Das S (2011) Functional fractional calculus. Springer, New York
3. Saha Ray S, Gupta AK (2018) Wavelet methods for solving partial differential equations and fractional differential equations. CRC Press, Boca Raton
4. Gupta AK, Saha Ray S (2017) On the solitary wave solution of fractional Kudryashov-Sinelshchikov equation describing nonlinear wave processes in a liquid containing gas bubbles. Appl Math Comput 298:1–12
5. Saha Ray S (2017) On conservation laws by Lie symmetry analysis for (2+1)-dimensional Bogoyavlensky-Konopelchenko equation in wave propagation. Comput Math Appl 74(6):1158–1165
6. Bogoyavlenskii OI (1990) Overturning solitons in new two dimensional integrable equations. Math USSR-Izvestiya 34(2):245–259
7. Bogoyavlenskii OI (1990) Breaking solitons in (2+1)-dimensional integrable equations. Russ Math Surv 45:1–86
8. Calogero F (1995) Integrable nonlinear evolution equations and dynamical systems in multidimensions. Acta Appl Math 39:229–244. https://doi.org/10.1007/BF00994635
9. Wazwaz AM (2010) The (2+1) and (3+1) dimensional CBS equations: multiple soliton solutions and multiple singular soliton solutions. Zeitschrift für Naturforschung 65(a):173–181
10. Peng Y, Shen M (2006) On exact solutions of the Bogoyavlenskii equation. J Phys 67(3):449–456
11. Saleh R, Kassem M, Mabrouk S (2017) Exact solutions of Calgero-Bogoyavlenskii-Schiff equation using the singular manifold method after Lie reductions. Math Meth Appl Sci 40(16):5851–5862
12. Kumar R (2016) Application of Lie-group theory for solving Calogero-Bogoyavlenskii-Schiff equation. IOSR J Math 12:144–147
13. Wazwaz AM (2008) New solutions of distinct physical structures to high-dimensional nonlinear evolution equations. Appl Math Comput 196:363–370
14. Moatimid G, El-Shiekh R, Abdul-Ghani AN (2013) Exact solutions for Calogero-Bogoyavlenskii-Schiff equation using symmetry method. Appl Math Comput 220:455–462
15. Gomez C (2015) Exact solution of the Bogoyavlenskii equation using the improved Tanh-Coth method. Appl Math Sci 9:4443–4447
16. Najafi M, Arbabi S, Najafi M (2012) New exact solutions of (2+1) dimensional Bogoyavlenskii equation by the sine-cosine method. Int J Basic Appl Sci 1(4):490–497
17. Darvishi M, Najafi M, Najafi M (2010) New application of EHTA for the generalized (2+1) dimensional nonlinear evolution equations. World Acad Sci Eng Technol 37:494–500
18. Wazwaz AM (2008) Multiple-soliton solutions for the Calogero-Bogoyavlenskii-Schiff, Jimbo-Miwa and YTSF equations. Appl Math Comput 203:592–597
19. Gandarias M, Bruzon M (2008) Travelling wave solutions for a CBS equation in (2+1) dimensions. In: Proceedings of American conference on applied mathematics. World Scientific and Engineering Academy and Society, Harvard, pp 153–158
20. Khalil R, Al-Horani M, Yousefi A, Sababheh M (2014) A new definition of fractional derivative. J Comput Appl Math 264:65–70
21. Abdeljawad T (2015) On conformable fractional calculus. J Comput Appl Math 279:57–66
22. Hosseini K, Bekir A, Ansari R (2017) New exact solutions of the conformable time-fractional Cahn-Allen and Cahn-Hilliard equations using the modified Kudryashov method. Optik 132:203–209
23. Kudryashov NA (2009) On new travelling wave solutions of the KdV and the KdV-Burgers equation. Commun Nonlinear Sci Numer Simul 14:1891–1900
24. Ryabov PN, Sinelshchikov DI, Kochanov MB (2011) Application of the Kudryashov method for finding exact solutions of the high order nonlinear evolution equations. Appl Math Comput 218:3965–3972

25. Gupta AK (2019) On the exact solution of time-fractional (2+1) dimensional Konoplechenko-Dubrovsky equation. Int J Appl Comput Math 5. https://doi.org/10.1007/s40819-019-0678-z
26. Hosseini K, Mayeli P, Ansari R (2017) Modified Kudryashov method for solving the conformable time-fractional Klein-Gordon equations with quadratic and cubic nonlinearities. Optik-Int J Light Electron Opt 130:737–742
27. Hosseini K, Ansari R (2017) New exact solutions of nonlinear conformable time-fractional Boussinesq equations using the modified Kudryashov method. Waves Random Complex Media 27(4):628–636

Nonlinear Convective Flow of Power-law Fluid over an Inclined Plate with Double Dispersion Effects and Convective Thermal Boundary Condition

P. Naveen, Ch. RamReddy, and D. Srinivasacharya

Abstract This study explores the impact of double dispersion effects on the nonlinear convective flow of power-law fluid along an inclined plate. Besides, the density differences with concentration and temperature are assumed to be larger and also convective thermal condition is considered at the boundary. Governing nonlinear partial differential equations are solved numerically using the successive linearization method (SLM) together with the local non-similarity technique. Accuracy and convergence of obtained results of successive linearization method are confirmed through error analysis. Also, present results are validated with previously published works in a special case. The present study enables us to discuss the influence of pertinent governing parameters on the heat and mass transfer rates of the fluid flow at the wall. This kind of investigation is useful in the mechanism of combustion, aerosol technology, high-temperature polymeric mixtures and solar collectors, which operate at moderate to very high temperatures and concentrations.

Keywords Power-law fluid · Nonlinear Boussinesq approximation · Convective boundary condition · Successive linearization method · Double dispersion effects

1 Introduction

Analysis of heat and mass transfer of non-Newtonian power-law fluid (Ostwald-de Waele type) flow through porous media acquired huge attention by many researchers [1–7] due its comprehensive applications in energy and geophysical industries. In the fields of oil reservoir, ceramic processing and heat storage beds, the double dispersion effects are more predominant with the consideration of inertial effects

P. Naveen (✉)
Department of Mathematics, School of Advanced Sciences, Vellore Institute of Technology, Vellore 632014, India
e-mail: naveen.p@vit.ac.in

Ch. RamReddy · D. Srinivasacharya
Department of Mathematics, National Institute of Technology Warangal, Warangal 506004, India

S. S. Ray et al. (eds.), *Applied Analysis, Computation and Mathematical Modelling in Engineering*, Lecture Notes in Electrical Engineering 897,
https://doi.org/10.1007/978-981-19-1824-7_8

in the flow region of porous medium (refer Nield and Bejan [8]). Moreover, the fluid flow through intricate paths activates dispersion effects in porous media at pore level. With this consideration, the importance of the thermal and solutal dispersion on the flow of fluid through a porous medium are exhibited by many authors. A lot of research has been accounted-for on this point with different fluids as depicted in the articles of researchers [9–13].

In addition to the above said point, double dispersion effects are more prevalent in fluid flow regime when the temperature–concentration-dependent density relation is nonlinear (also known as, nonlinear Boussinesq approximation [14, 15]) in the buoyancy term. Since, most of the thermal equipment works at moderate and very high temperatures and concentrations, this leads to have nonlinearity in buoyancy with temperature–concentration-dependent density relation [16, 17].

Heat transfer analysis with the convective thermal boundary condition is beneficial consideration and has very important applications in the of fields nuclear plants, gas turbines, heat exchangers related industries. In view of these applications, Munir [18] (on viscous fluid flow), Hayat [19] (on power-law fluid flow) and RamReddy and Naveen [20] (on micropolar fluid flow) considered this thermal condition at the boundary for the study of fluid flow behaviour over different geometries.

In the present study, the heat and mass transport phenomena of power-law fluid past an inclined plate with a convective thermal boundary condition is examined. The double dispersion effects and nonlinear Boussinesq approximations are included in order to investigate their effect over fluid flow.

2 Governing Equations

Consider, the steady, 2D, laminar mixed convective flow of incompressible power-law fluid along an inclined plate in a non-Darcy porous medium. The inclination angle is measured in terms of Ω about vertical direction. By left convection, the infinite plate is either heated or cooled from a fluid with temperature T_f. The solutal concentration over the wall is C_w, and ambient porous medium concentration and temperature are taken to be C_∞ and T_∞, respectively.

By employing nonlinear Boussinesq approximation and with usual boundary layer conditions, the governing equations for the power-law fluid flow in a non-Darcy porous medium (Forchheimer model) [2, 21, 22] are given by

$$\frac{\partial u}{\partial x} + \frac{\partial v}{\partial y} = 0 \tag{1}$$

$$\frac{\partial u^n}{\partial y} + \frac{b\sqrt{K_p}}{\nu}\frac{\partial u^2}{\partial y} = \frac{K_p g^*}{\nu}\left\{[\beta_0 + 2\beta_1(T - T_\infty)]\frac{\partial T}{\partial y} + [\beta_2 + 2\beta_3(C - C_\infty)]\frac{\partial C}{\partial y}\right\}\cos\Omega \tag{2}$$

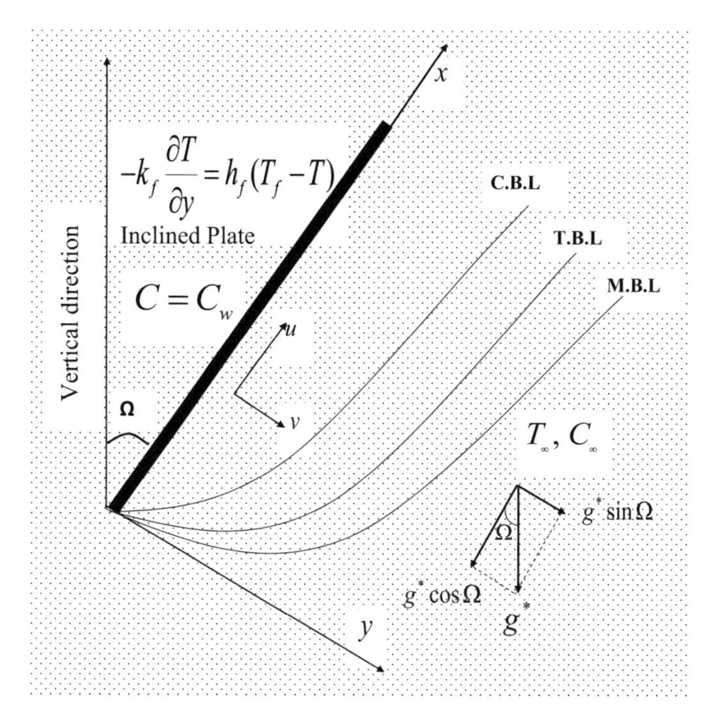

Fig. 1 Schematic diagram of the problem

$$u\frac{\partial T}{\partial x} + v\frac{\partial T}{\partial y} = \frac{\partial}{\partial y}\left[(\alpha + \chi\, d\, u)\frac{\partial T}{\partial y}\right] \tag{3}$$

$$u\frac{\partial C}{\partial x} + v\frac{\partial C}{\partial y} = \frac{\partial}{\partial y}\left[(D + \zeta\, d\, u)\frac{\partial C}{\partial y}\right] \tag{4}$$

The associated boundary conditions are

$$\begin{aligned} v = 0, \quad -k_f\frac{\partial T}{\partial y} = h_f(T_f - T), \quad C = C_w \quad &\text{at} \quad y = 0 \\ u = u_\infty, \quad T = T_\infty, \quad C = C_\infty \quad &\text{as} \quad y \to \infty \end{aligned} \tag{5}$$

where u and v are the velocity components in x and y directions, T is the temperature, C is the concentration, b is the empirical constant associated with the Forchheimer porous inertia term, n is the power-law index, Ω is the inclination angle, g^* is the acceleration due to gravity, K_p is the permeability, ν is the kinematic viscosity, k_f is the thermal conductivity of the fluid, h_f is the convective heat transfer coefficient, α is the molecular thermal diffusivity, D is the molecular solutal diffusivity, d is the pore diameter, χ is the thermal dispersion coefficient and ζ is the solutal dispersion coefficient, respectively.

Further, the first- and second-order expansions of thermal coefficients are denoted by β_0 and β_1. Like that, the first- and second-order expansions of solutal coefficients are defined by β_2 and β_3, respectively.

Now introduce stream function $\psi(x, y)$ as $u = \frac{\partial \psi}{\partial y}$ and $v = -\frac{\partial \psi}{\partial x}$, which is automatically satisfied the continuity equation (1).

Here, we define non-similarity transformations [23–25] in the following form

$$\begin{gathered} \xi = \frac{x}{L},\quad Pe = \frac{u_\infty L}{\alpha}, \eta = \frac{y}{L} Pe^{\frac{1}{2}} \xi^{\frac{-1}{2}},\quad \psi(\xi, \eta) = \alpha\, \xi^{\frac{1}{2}} Pe^{\frac{1}{2}} f(\xi, \eta) \\ T(\xi, \eta) = T_\infty + \left(T_f - T_\infty\right) \theta(\xi, \eta),\quad C(\xi, \eta) = C_\infty + (C_w - C_\infty)\, \phi(\xi, \eta) \end{gathered} \tag{6}$$

where ξ is the stream-wise coordinate, L is the characteristics length and $Pe = \frac{u_\infty L}{\alpha}$ is the global Peclet's number. f, θ and ϕ are the dimensionless stream function, temperature and concentrations, respectively.

Substituting the transformations (6) into (1)–(4), we obtain the non-dimensional governing equations,

$$n\left(f'\right)^{n-1} f'' + 2F_0 Pe f' f'' = \lambda^n \left[(1 + 2\alpha_1\theta)\theta' + \mathcal{B}(1 + 2\alpha_2\phi)\phi'\right] \cos\Omega \tag{7}$$

$$\theta'' + Pe_\chi \left(f'\theta'\right)' + \frac{1}{2} f\theta' = \xi\left(f'\frac{\partial\theta}{\partial\xi} - \frac{\partial f}{\partial\xi}\theta'\right) \tag{8}$$

$$\frac{1}{Le}\phi'' + Pe_\zeta \left(f'\phi'\right)' + \frac{1}{2} f\phi' = \xi\left(f'\frac{\partial\phi}{\partial\xi} - \frac{\partial f}{\partial\xi}\phi'\right) \tag{9}$$

The resultant boundary conditions from Eq. (5)

$$\begin{gathered} f(\xi, 0) = -2\xi\left(\frac{\partial f}{\partial \xi}\right)_{\eta=0},\quad \theta'(\xi, 0) = -Bi\,\xi^{\frac{1}{2}}\left[1 - \theta(\xi, 0)\right], \phi(\xi, 0) = 1, \\ f'(\xi, \infty) = 1,\ \theta(\xi, \infty) = 0,\ \phi(\xi, \infty) = 0. \end{gathered} \tag{10}$$

In the above equations, Ra denotes the global Rayleigh number, λ denotes the mixed convection parameter, $F_0 Pe$ denotes the Forchheimer number, $\mathcal{B}$ denotes the and Buoyancy ratio, Ω denotes the angle of inclination, Pe_d denotes the pore diameter-dependent Péclet number, Pe_χ denotes the thermal dispersion parameter, Le denotes the diffusivity ratio, Pe_ζ denotes the solutal dispersion parameter, α_1 denotes the nonlinear density–temperature parameter (NDT), Bi denotes the Biot number, and α_2 denotes the nonlinear density–concentration parameter (NDC).

Mathematical expressions for the parameters are given below, $Ra = (\frac{L}{\alpha}) \left(\frac{[K_p g^* \beta_0 (T_f - T_\infty)]}{\nu}\right)^{1/n}$, $\lambda = \frac{Ra}{Pe}$, $F_0 Pe = f_0 (Pe_d)^{2-n} \left(f_0 = \left[\frac{(b\sqrt{K_p})}{\nu}\right] (\frac{\alpha}{d})^{2-n}\right)$, $\mathcal{B} =$

$\frac{\beta_2(C_w-C_\infty)}{(\beta_0(T_f-T_\infty))}$, $Pe_d = \frac{(u_\infty d)}{\alpha}$, $Pe_\chi = \frac{[\chi\, d\, u_\infty]}{\alpha}$, $Le = \frac{\alpha}{D}$, $Pe_\zeta = \frac{[\zeta\, d\, u_\infty]}{\alpha}$, $\alpha_1 = \frac{\beta_1(T_f-T_\infty)}{\beta_0}$, $\alpha_2 = \frac{\beta_3(C_w-C_\infty)}{\beta_2}$ and $Bi = \frac{h_f L}{(k_f\, Pe^{1/2})}$.

Here, the physical quantities of interest, the non-dimensional Nusselt and the Sherwood numbers are given by

$$NuPe^{\frac{-1}{2}} = -\xi^{\frac{1}{2}}\left[1 + Pe_\chi\, f'(\xi, 0)\right]\theta'(\xi, 0), \;\; ShPe^{\frac{-1}{2}} = -\xi^{\frac{1}{2}}\left[1 + Pe_\zeta\, f'(\xi, 0)\right]\phi'(\xi, 0).$$

3 Numerical Solution

For the solutions of partial differential equations (7)–(9) together with (10), the following steps were followed

- Firstly, the above-said PDEq equations are transformed into ODEqs [26] by introducing auxiliary variables.
- Next, a novel successive linearization method is used to linearize the resultant equations.
- Lastly, the linearized equations are solved with the Chebyshev collocation method [27–29].

The detailed procedure of solution methodology to solve equations (7)–(9) together with (10) is presented in the following Sects. 3.1 to 3.3.

3.1 Local Non-similarity Procedure

The preliminary approximate solution can be found from local similarity equations for a particular case of $\xi << 1$; the terms containing $\xi \frac{\partial}{\partial \xi}$ are supposed to be negligible. Then, the first-level truncation or local similarity equations from (7)–(10) are

$$\left[n\,(f')^{n-1} + 2F_0 Pe f'\right] f'' - \lambda^n\left[(1+2\alpha_1\theta)\theta' + \mathcal{B}(1+2\alpha_2\phi)\phi'\right]\cos\Omega = 0 \tag{11}$$

$$\theta'' + Pe_\chi\left(f'\theta'\right)' + \frac{1}{2} f\theta' = 0 \tag{12}$$

$$\frac{1}{Le}\phi'' + Pe_\zeta\left(f'\phi'\right)' + \frac{1}{2} f\phi' = 0 \tag{13}$$

The corresponding boundary conditions are

$$f(\xi,0)=0,\ \theta'(\xi,0)=-Bi\,\xi^{\frac{1}{2}}\left[1-\theta(\xi,0)\right],\phi(\xi,0)=1,$$
$$f'(\xi,\infty)=1,\ \theta(\xi,\infty)=0,\ \phi(\xi,\infty)=0. \tag{14}$$

The local non-similarity ordinary nonlinear differential equations in the second-level truncation is discovered by introducing new variables to recall the omitted expressions from the first-level truncation, i.e. take $U=\dfrac{\partial f}{\partial \xi}$, $V=\dfrac{\partial \theta}{\partial \xi}$, $W=\dfrac{\partial \phi}{\partial \xi}$. Thus, the second-level truncation is

$$\left[n\left(f'\right)^{n-1}+2F_0Pef'\right]f''-\lambda^n\left[(1+2\alpha_1\theta)\theta'+\mathcal{B}(1+2\alpha_2\phi)\phi'\right]\cos\Omega=0 \tag{15}$$

$$\theta''+Pe_\chi\left(f'\theta'\right)'+\frac{1}{2}f\theta'=\xi\left(Vf'-U\theta'\right) \tag{16}$$

$$\frac{1}{Le}\phi''+Pe_\zeta\left(f'\phi'\right)'+\frac{1}{2}f\phi'=\xi\left(Wf'-U\phi'\right) \tag{17}$$

The corresponding boundary conditions are

$$f(\xi,0)=-2\xi\,U(\xi,\eta),\ \theta'(\xi,0)=-Bi\,\xi^{\frac{1}{2}}\left[1-\theta(\xi,0)\right],\ \phi(\xi,0)=1,$$
$$f'(\xi,\infty)=1,\ \theta(\xi,\infty)=0,\ \phi(\xi,\infty)=0. \tag{18}$$

The two-level local non-similarity technique is accomplished with a third level of truncation; for this, we differentiate equations (15)–(18) with respect to ξ and omit the partial derivatives of U, V, W. Then, the resultant equations are

$$n(n-1)\left(f'\right)^{n-2}f''U'+n\left(f'\right)^{n-1}U''+2F_0Pe(f''U'+U''f')-$$
$$\lambda^n\left[V'+2\alpha_1(V\theta'+\theta V')+\mathcal{B}(W'+2\alpha_2(W\phi'+\phi W'))\right]\cos\Omega=0 \tag{19}$$

$$V''+\frac{3}{2}U\theta'+\frac{1}{2}V'f+Pe_\chi\left[U''\theta'+f''V'+U'\theta''+f'V''\right]-Vf'=\xi\left(U'V-UV'\right) \tag{20}$$

$$\frac{1}{Le}W''+\frac{3}{2}U\phi'+\frac{1}{2}W'f+Pe_\zeta\left[U''\phi'+f''W'+U'\phi''+f'W''\right]-Wf'=\xi\left(U'W-UW'\right) \tag{21}$$

The corresponding boundary conditions are

$$U(\xi,0)=0,\ V'(\xi,0)=Bi\xi^{\frac{1}{2}}V(\xi,0)+\frac{1}{2}Bi\xi^{\frac{-1}{2}}[\theta(\xi,0)-1],\ W(\xi,0)=0,$$
$$U'(\xi,\infty)=0,\ V(\xi,\infty)=0,\ W(\xi,\infty)=0 \tag{22}$$

The coupled nonlinear ordinary differential equations (15)–(17) and (19)–(21) along with the boundary conditions (18) and (22) are evaluated using successive linearization method. First, it linearize the non-similarity equation, and then, it utilizes Chebyshev collocation method for the approximate solution.

3.2 Successive Linearization

Let us consider an independent vector $\mathbb{Q}(\eta) = [f(\eta), \theta(\eta), \phi(\eta), U(\eta), V(\eta), W(\eta)]$ and assume that it can be represented as

$$\mathbb{Q}(\eta) = \mathbb{Q}_k(\eta) + \sum_{m=0}^{k-1} \mathbb{Q}_m(\eta) \tag{23}$$

where $\mathbb{Q}_k(\eta)$, $k = 1, 2, 3\ldots$ are unknown vectors, and those are determined by recursively evaluating the linearized version of the non-similarity equations and presuming that $\mathbb{Q}_m(\eta)$, $(0 \le m \le k-1)$ are expected from antecedent iterations. The initial guesses $\mathbb{Q}_0(\eta)$ is selected so that it satisfy the boundary conditions (18) and (22). By imposing Eq. (23) in Eqs.(15)–(22) and considering only linear terms, we obtain the linearized equations to be evaluated which are

$$\tilde{p}_{1,k-1} f_k'' + \tilde{p}_{2,k-1} f_k' + \tilde{p}_{3,k-1}\theta_k' + \tilde{p}_{4,k-1}\theta_k + \tilde{p}_{5,k-1}\phi_k' + \tilde{p}_{6,k-1}\phi_k = \tilde{z}_{1,k-1} \tag{24}$$

$$\tilde{q}_{1,k-1} f_k + \tilde{q}_{2,k-1}\theta_k'' + \tilde{q}_{3,k-1}\theta_k' + \tilde{q}_{4,k-1}U_k + \tilde{q}_{5,k-1}V_k = \tilde{z}_{2,k-1} \tag{25}$$

$$\tilde{a}_{1,k-1} f_k + \tilde{a}_{2,k-1}\phi_k'' + \tilde{a}_{3,k-1}\phi_k' + \tilde{a}_{4,k-1}U_k + \tilde{a}_{5,k-1}W_k = \tilde{z}_{3,k-1} \tag{26}$$

$$\begin{aligned}\tilde{b}_{1,k-1} f_k'' + \tilde{b}_{2,k-1} f_k' + \tilde{b}_{3,k-1}\theta_k' + \tilde{b}_{4,k-1}\theta_k + \tilde{b}_{5,k-1}\phi_k' + \tilde{b}_{6,k-1}\phi_k + \tilde{b}_{7,k-1}U_k'' + \tilde{b}_{8,k-1}U_k' \\ + \tilde{b}_{9,k-1}V_k' + \tilde{b}_{10,k-1}V_k + \tilde{b}_{11,k-1}W_k' + \tilde{b}_{12,k-1}W_k = \tilde{z}_{4,k-1}\end{aligned} \tag{27}$$

$$\tilde{c}_{1,k-1} f_k + \tilde{c}_{2,k-1}\theta_k' + \tilde{c}_{3,k-1}U_k' + \tilde{c}_{4,k-1}U_k + \tilde{c}_{5,k-1}V_k'' + \tilde{c}_{6,k-1}H_k' + \tilde{c}_{7,k-1}V_k = \tilde{z}_{5,k-1} \tag{28}$$

$$\tilde{d}_{1,k-1} f_k + \tilde{d}_{2,k-1}\phi_k' + \tilde{d}_{3,k-1}U_k' + \tilde{d}_{4,k-1}U_k + \tilde{d}_{5,k-1}W_k'' + \tilde{d}_{6,k-1}W_k' + \tilde{d}_{7,k-1}W_k = \tilde{z}_{6,k-1} \tag{29}$$

The linearised boundary conditions are

$$\begin{aligned} f_k(0) = f_k'(0) = f_k'(\infty) = 0,\ Bi\,\xi^{\frac{1}{2}}\theta_k(0) + \theta_k'(0) = 0,\ \theta_k(\infty) = 0, \\ \phi_k(0) = \phi_k(\infty) = 0,\ U_k(0) = U_k'(0) = U_k'(\infty) = 0, \\ -\frac{1}{2}Bi\,\xi^{\frac{-1}{2}}\theta_k(0) + V_k'(0) - Bi\,\xi^{\frac{1}{2}}V_k(0) = 0,\ V_k(\infty) = 0,\ W_k(0) = W_k(\infty) = 0 \end{aligned} \tag{30}$$

Here, the coefficient parameters $\tilde{p}_{s,k-1}$, $\tilde{q}_{s,k-1}$, $\tilde{a}_{s,k-1}$, $\tilde{b}_{s,k-1}$, $\tilde{c}_{s,k-1}$, $\tilde{d}_{s,k-1}$, and $\tilde{z}_{s,k-1}$ which depend on the $\mathbb{Q}_0(\eta)$ and on the $\mathbb{Q}_k(\eta)$ derivatives.

3.3 Chebyshev Collocation Scheme

We solve linearized equations (24)–(29) by an established procedure, namely Chebyshev collocation scheme [30]. In the context of numerical implication, the original region $[0, \infty)$ is truncated to $[0, L]$ for large value of L, and further, the truncated region $[0, L]$ is transformed into $[-1, 1]$ using the following mapping

$$\frac{\eta}{L} = \frac{\tau + 1}{2}, \qquad -1 \le \tau \le 1 \tag{31}$$

In this procedure, The Chebyshev polynomials $T_m(\tau) = \cos[m \ \cos^{-1}\tau]$ are used to approximate the unknown functions $\mathbb{Q}_k(\eta)$ and these polynomials are collocated at $K + 1$ Gauss–Lobatto points in the interval $[-1, 1]$ and those are defined as

$$\tau_m = \cos\frac{\pi m}{K}, \qquad m = 0, 1, ..., K \tag{32}$$

The unknown function $\mathbb{Q}_k(\eta)$ is imprecise at the collocation points by

$$\mathbb{Q}_k(\tau) = \sum_{j=0}^{K} \mathbb{Q}_k(\tau_j) T_j(\tau_m), \qquad m = 0, 1, ...K \tag{33}$$

and

$$\frac{d^{\mathbb{S}}}{d\eta^{\mathbb{S}}} \mathbb{Q}_k(\tau) = \sum_{r=0}^{K} \mathbf{D}_{rm}^{\mathbb{S}} \mathbb{Q}_k(\tau_r), \qquad m = 0, 1, 2, ...K \tag{34}$$

where $\mathcal{D}$ is the Chebyshev spectral derivative matrix such that $\mathbf{D} = (2/L)\mathcal{D}$ and $\mathbb{S}$ is the order of differentiation. After employing Eqs.(31)–(34) into linearized form of Eqs. (24)–(29), the resultant solution is

$$\tilde{\mathbf{Y}}_k = \tilde{\mathbf{B}}_{k-1}^{-1} \tilde{\mathbf{Z}}_{k-1} \tag{35}$$

In Eq. (35), $\tilde{\mathbf{B}}_{k-1}$ is a $(6N + 6) \times (6N + 6)$ matrix, $\tilde{\mathbf{Y}}_k$ and $\tilde{\mathbf{Z}}_{k-1}$ are $(6N + 1) \times 1$ column vectors defined by

$$\begin{gathered} \tilde{\mathbf{B}}_{k-1} = \left[\tilde{\mathbf{B}}_{ij}\right], \ i, j = 1, 2, \ldots 6, \ \ \tilde{\mathbf{Y}}_k = \left[\tilde{\mathbb{F}}_k \ \tilde{\Theta}_k \ \tilde{\Phi}_k \ \tilde{\mathbb{U}}_k \ \tilde{\mathbb{V}}_k \ \tilde{\mathbb{W}}_k\right]^T, \\ \tilde{\mathbf{Z}}_{k-1} = \left[\tilde{\mathbf{z}}_{1,k-1} \ \tilde{\mathbf{z}}_{2,k-1} \ \tilde{\mathbf{z}}_{3,k-1} \ \tilde{\mathbf{z}}_{4,k-1} \ \tilde{\mathbf{z}}_{5,k-1} \ \tilde{\mathbf{z}}_{6,k-1}\right]^T \end{gathered} \tag{36}$$

3.4 Residual Error Analysis

It can be ensured the convergence of the proposed method by evaluating the norm of the difference between two consecutive iterations. This algorithm is accepted to have converged when the error norms are less than a given tolerance level. The error norms are given by

$$E_f = \max_{0 \le i \le Nx} \| f_{r+1,i} - f_{r,i} \|_\infty, \; E_\theta = \max_{0 \le i \le Nx} \| \theta_{r+1,i} - \theta_{r,i} \|_\infty,$$
$$E_\phi = \max_{0 \le i \le Nx} \| \phi_{r+1,i} - \phi_{r,i} \|_\infty \tag{37}$$

Norm of the residual errors of the governing Eqs. (7)–(9) across ξ at different iterations levels of the present numerical scheme are depicted by Fig. 2. This figure revels that the residual errors decrease with an increase of number of iterations, and this is an indication of convergence of the solutions. Also, observed that the residual errors are nearly uniform across ξ. This result proves that the accuracy of solution method does not depend on the length of ξ interval. Furthermore, the small residual errors, which are obtained after a few iterations, are a clear sign of the accuracy of the solution method. Hence, the residual error results validate the accuracy of generated results in this study.

4 Results and Discussion

In addition to the error analysis, present numerical results are also validated with the previously published works [9, 31] without nonlinear convection and double dispersion effects as appeared in Tables 1 and 2. Results are in good agreement, and the influence of the parameters $\alpha_1, \alpha_2, Pe_\chi, Pe_\zeta, \Omega$ and Bi are depicted by the Figs. 3, 4 and 5 for $f', \theta, \phi, Nu\, Pe^{\frac{-1}{2}}$ and $Sh\, Pe^{\frac{-1}{2}}$.

4.1 Influence of α_1 and α_2 With Viscosity Index n

Variations of fluid flow profiles for $\alpha_1 (0, 6)$, α_2 $(0, 5)$ and n (0.7, 1.0, 1.4) with $Pe_\chi = 0.5$, $Pe_\zeta = 0.2$, $Bi = 0.5$, $\xi = 0.5$ and $\Omega = 30°$ shown in Fig. 3a–c. It uncovers that the influence of n is extensive and increases both thermal and solutal boundary layer thickness, whereas it diminishes the thickness of momentum boundary layer. Regarding α_1, the velocity is predominant at the inclined plate surface and for η_{max} value, it achieves unity. As found in Fig. 3a, $\alpha_1 > 0$ infers that $T_f > T_\infty$; henceforth, some amount of the heat is induced to fluid region by the wall surface. Moreover, Fig. 3a displays the impact of the α_2 on the behaviour of velocity. The results of this figure repeat the same kind of behaviour just like α_1 in all three fluids. The thicknesses of thermal and solutal boundary layer diminish with the rise of both α_1

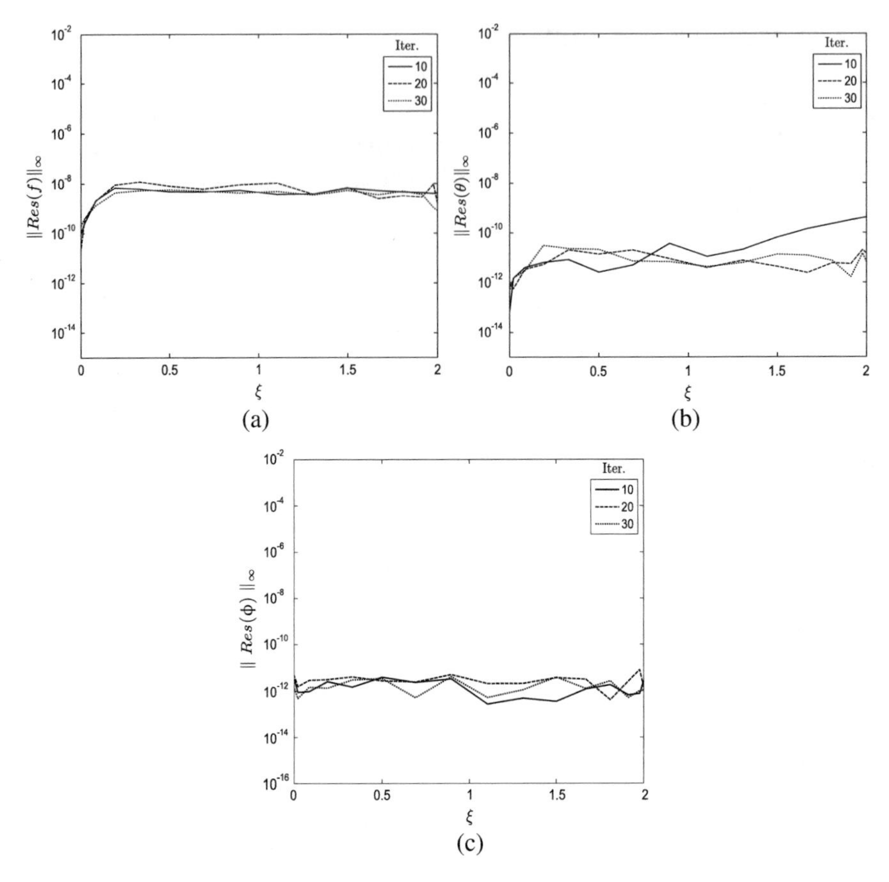

Fig. 2 Residual error over iterations when $\mathcal{B} = 0.5$, $F_0 Pe = 1, \alpha_1 = 1, \alpha_2 = 1, Bi = 0.5, n = 1,$ $Pe_\chi = 0.5$, $Pe_\zeta = 0.2$, $\Omega = 30°$

and α_2, as shown in Fig. 3b, c. Obviously, the nonlinear temperature and concentration differences between the wall and ambient medium are increased for larger values of α_1 and α_2, due to which higher velocity and smaller temperature concentration are obtained.

Figure 3d–e show the effect of $\alpha_1(0, 6)$ and $\alpha_2(0, 5)$ on the $NuPe^{-1/2}$ and $ShPe^{-1/2}$ against ξ. The rise of either α_1 or α_2 increases all the fluid profiles of the pseudo-plastic fluid flow. So, also, these two dimensionless quantities have a same change in the other two fluid flows of Newtonian and dilatant fluid. Evidently, these two heat and mass transfer rates are identically comparable with the works of Partha [16] in Newtonian fluid (for $n = 1$) case.

Table 1 Comparison of $-\theta'(0)$ against λ with the fixed values of $\mathcal{B} = 0$, $F_0 Pe = 0$, $\alpha_1 = 0$, $Bi \to \infty$, $\alpha_2 = 0$, $Pe_\chi = 0$, $Pe_\zeta = 0$ and $\Omega = 0$

	$n = 0.5$		$n = 1.0$		$n = 1.5$	
λ	[31]	Present	[31]	Present	[31]	Present
0	0.5641	0.5642	0.5641	0.5642	0.5641	0.5642
0.5	0.8209	0.8217	0.6473	0.6474	0.6034	0.6034
1.0	0.9303	0.9296	0.7205	0.7206	0.6634	0.6634
4.0	1.3010	1.3007	1.0250	1.0558	1.0180	1.0176
8.0	1.6100	1.6097	1.3540	1.3801	1.3800	1.4357
15.0	2.0010	2.0005	1.8120	1.8123	1.8620	1.8606

Table 2 Comparison of $f'(0)$, $-\theta'(0)$ and $-\phi'(0)$ against $Le, \mathcal{B}, \lambda$ with the fixed values of $F_0 Pe = 1, n = 1$, $Pe_\chi = 0, \alpha_2 = 0$, $Pe_\zeta = 0$, $Bi \to \infty$, $\alpha_1 = 0$ and $\Omega = 0$

				$Le = 1$		$Le = 10$			
		$f'(0)$		$-\theta'(0)$ & $-\phi'(0)$		$-\theta'(0)$		$-\phi'(0)$	
	λ	[9]	Present	[9]	Present	[9]	Present	[9]	Present
$\mathcal{B} = -0.5$	0	1.0	1.0	0.5645	0.5642	0.5642	0.5642	1.7841	1.7841
	1	1.1583	1.1583	0.5922	0.5922	0.6054	0.6054	1.9329	1.9329
	5	1.6794	1.6794	0.6793	0.6793	0.7244	0.7244	2.3534	2.3534
	10	2.1926	2.1926	0.7580	0.7580	0.8247	0.8247	2.7009	2.7009
	20	3.0	3.0	0.8706	0.8706	0.9617	0.9617	3.1686	3.1686
$\mathcal{B} = 1.0$	0	1.0	1.0	0.5642	0.5642	0.5642	0.5642	1.7841	1.7841
	1	1.5616	1.5616	0.6603	0.6603	0.6377	0.6377	2.1381	2.1381
	5	3.0	3.0	0.8706	0.8706	0.8083	0.8083	2.8864	2.8865
	10	4.217	4.2167	1.0203	1.0203	0.9358	0.9358	3.4061	3.4061
	20	6.0	6.0	1.2097	1.2097	1.1012	1.1012	4.0548	4.0548

4.2 *Influence of Ω and Bi With Viscosity Index n*

Effect of $\Omega\,(0°, 60°)$ and $Bi(0.1, 10)$ on the f', θ and ϕ are displayed in Fig. 4a–c for three instances of viscosity index ($n = 0.7, 1.0, 1.4$). As depicted in Fig. 4a, reduction in the buoyancy effect caused by Ω diminishes the velocity f'. Also, it is observed from Fig. 4a that the velocity of the power-law fluid enhances by magnifying Bi. With rising values of Ω values of θ and ϕ enhance, as show in Fig. 4b, c, and these results are qualitatively matched with the work of Chen [32].

Impact of Bi on the θ is collected through Fig. 4b for the case of wall and non-isothermal conditions. Since, convective thermal condition can be changed to wall condition for a larger value of Bi (i.e. $Bi \to \infty$) [33], the same was observed from Fig. 4b. At the wall surface, the temperature is accelerating when Bi changing from $Bi < 0.1$ (known as, thermally thin case) to $Bi > 0.1$ (known as, thermally thick case). Further, that the concentration decreasing function of Bi, as shown in Fig. 4c.

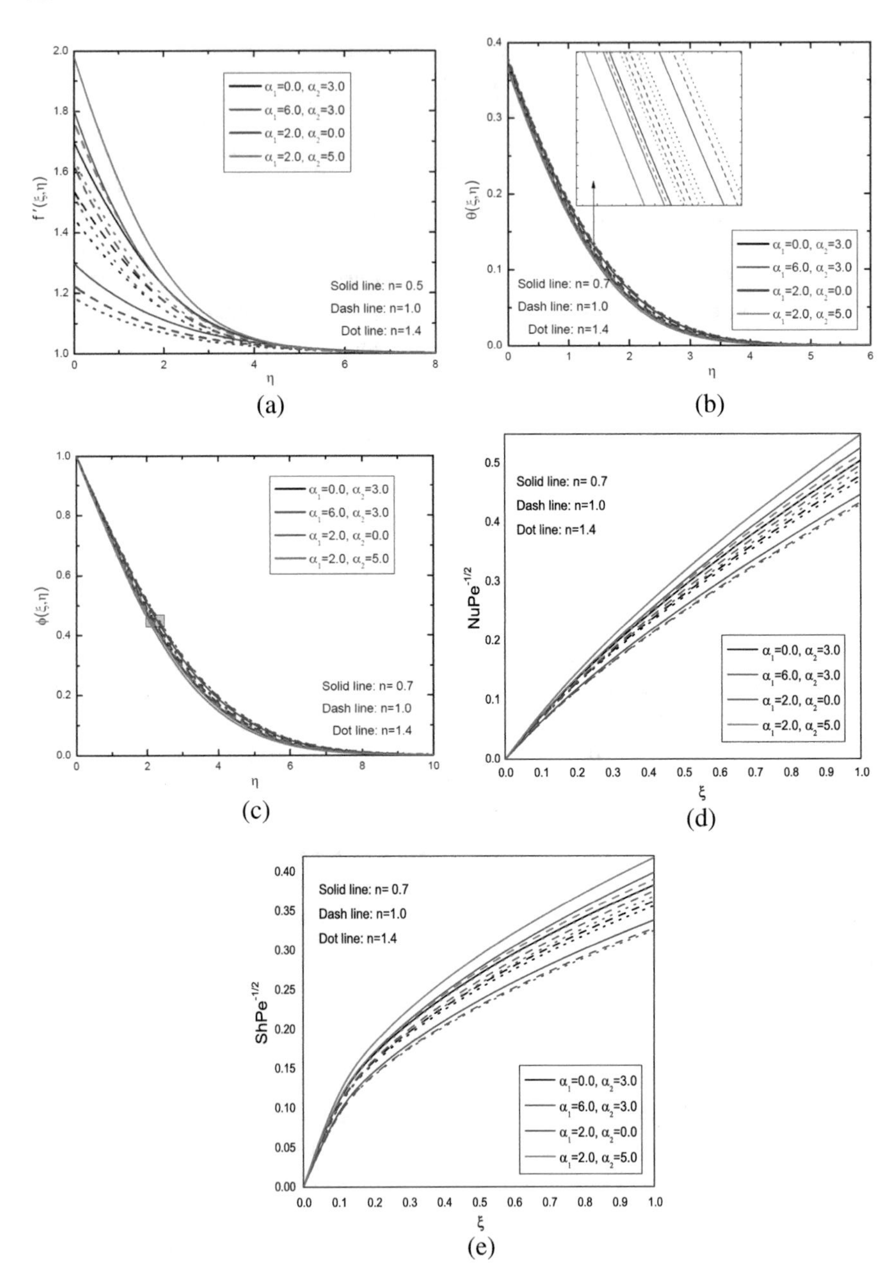

Fig. 3 Impact of α_1 and α_2 on the **(a)** f', **(b)** θ, **(c)** ϕ, **(d)** $Nu\,Pe^{\frac{-1}{2}}$, and **(e)** $Sh\,Pe^{\frac{-1}{2}}$ for three values of n with $Bi = 0.5$, $Pe_\chi = 0.5$, $Pe_\zeta = 0.2$, $\Omega = 30^\circ$

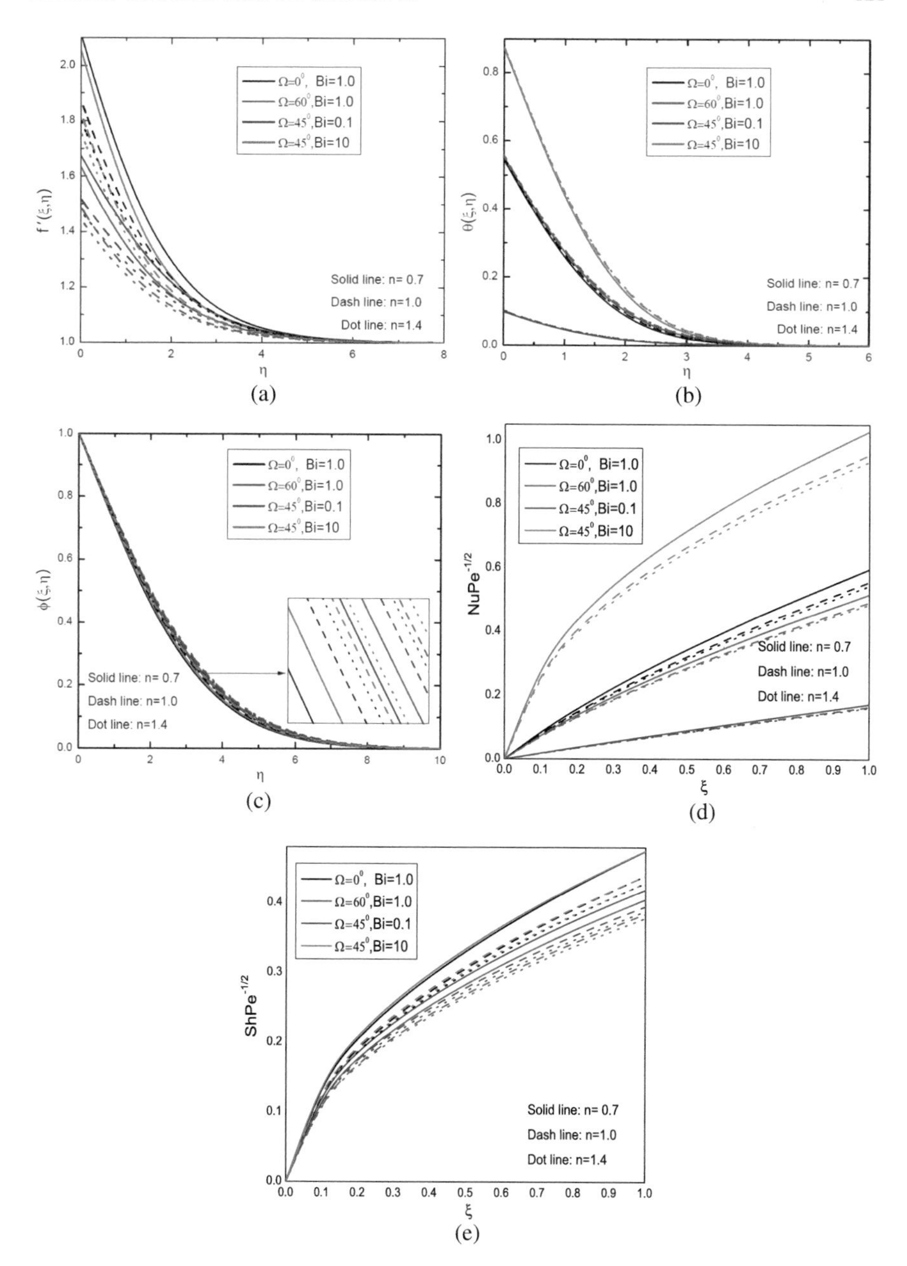

Fig. 4 Impact of Ω and Bi on the **(a)** f', **(b)** θ, **(c)** ϕ, **(d)** $Nu\ Pe^{\frac{-1}{2}}$, and **(e)** $Sh\ Pe^{\frac{-1}{2}}$ for three values of n with $\alpha_1 = 1$, $Pe_\chi = 0.5$, $\alpha_2 = 1$, $Pe_\zeta = 0.2$

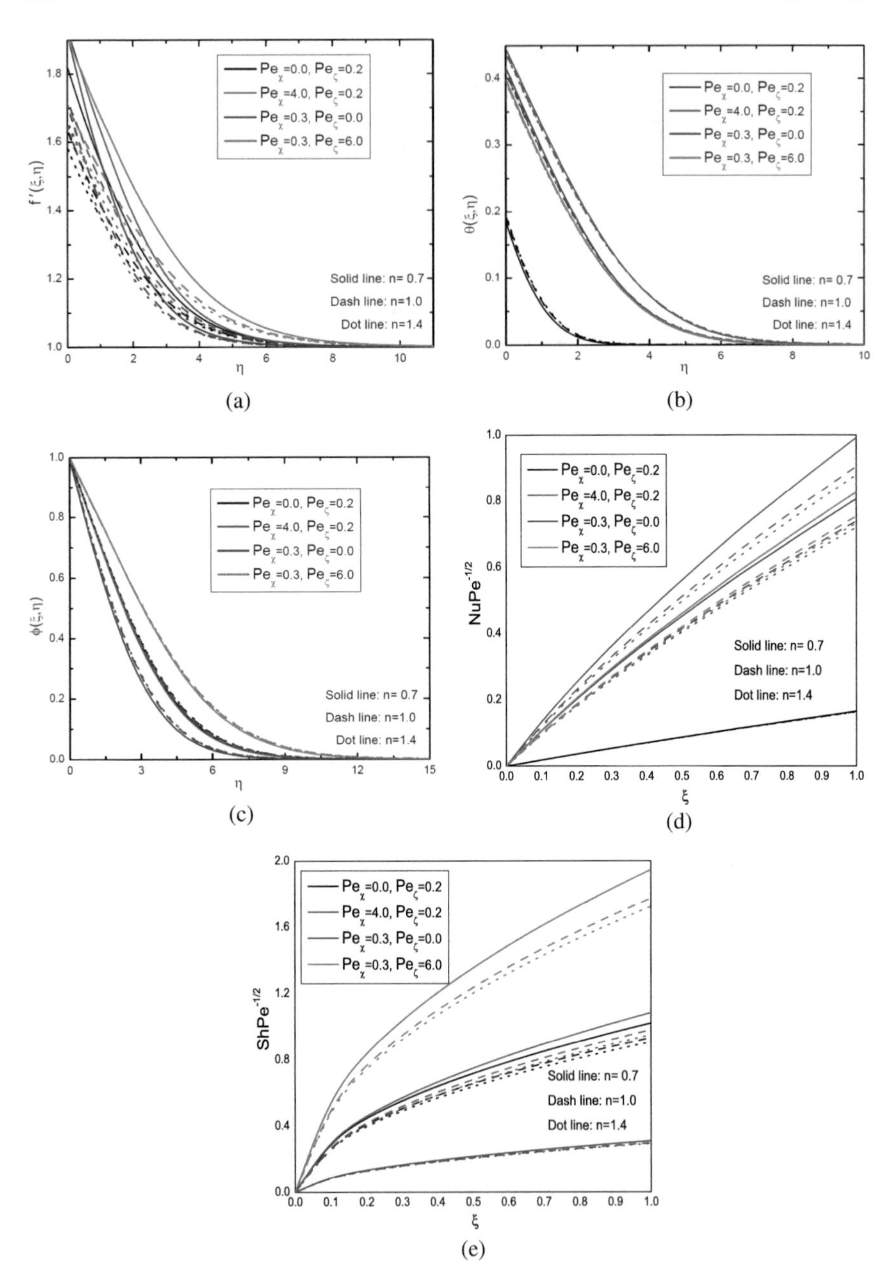

Fig. 5 Impact of Pe_χ and Pe_ζ on the **(a)** f', **(b)** θ, **(c)** ϕ, **(d)** $Nu\ Pe^{\frac{-1}{2}}$, and **(e)** $Sh\ Pe^{\frac{-1}{2}}$ for three values of n with $Bi = 0.3, \alpha_1 = 1, \alpha_2 = 1, \Omega = 30°$

Table 3 Parameters in the model and their values

Parameters	Symbol	Value range	Source
Biot number	Bi	(0, 10)	[19, 34]
Angle of inclination	Ω	$(0^0, 90^0)$	[35, 36]
NDT parameter	α_1	(0, 6)	[16]
Thermal dispersion parameter	Pe_χ	(0, 4)	[11, 13]
NDC parameter	α_2	(0, 5)	[16]
Solutal dispersion parameter	Pe_ζ	(0, 6)	[11, 13]
Power-law index	n	(0.5, 1.5)	[36]

Figure 4d–e exhibit the effect of $\Omega\,(0°, 60°)$ and $Bi\,(0.1, 10)$ on the $NuPe^{-1/2}$ and $ShPe^{-1/2}$ for three fluid cases with $Pe_\chi = 0.6$, $Pe_\zeta = 0.3, \alpha_1 = 1, \alpha_2 = 1$. If position of plate is changing from vertical to horizontal, there is a decrement in $g^* cos\Omega$ term, and this degrade the buoyancy. Hence, this reduction diminishes $NuPe^{-1/2}$ and $ShPe^{-1/2}$. However, $NuPe^{-1/2}$ and $ShPe^{-1/2}$ enhance by the increase of Bi and decrease with the viscosity index n (Table 3).

4.3 Influence of Pe_χ and Pe_ζ Viscosity Index n

Figure 5a–c revels the effect of Pe_χ (0, 4), Pe_ζ (0, 6) on f', θ and ϕ, for a fixed value of $\alpha_1 = 1, \alpha_2 = 1$, $Bi = 0.3, \xi = 0.5$ and $\Omega = 30°$. From Fig. 5a, it is significant that the thickness of the momentum boundary layer increases with the double dispersion parameters. Supplementing thermal dispersion effect into the energy equation gives more dominance in thermal conduction, and it improves thermal boundary layer thickness near to the surface of the inclined plate, as shown in Fig. 5b. On the other hand, increasing the solutal dispersion parameter leads to increase the thickness of concentration boundary layers, as depicted in Fig. 5c. However, in the absence or in the presence of double dispersion parameters, the temperature and concentration profiles are increased for viscosity index n.

The effect of $Pe_\chi\,(0, 4)$ and $Pe_\zeta = (0, 6)$ on Nusselt and Sherwood numbers is displayed in Fig. 5d–e. It is referred that thermal dispersion increases the heat transfer rate and solutal dispersion favours the mass transfer rate, as shown in Fig. 5d–e. However, these two transfer rates are more in pseudo-plastic fluids when compared with Newtonian and dilatant fluids.

5 Conclusions

In the present study, the characteristics power-law fluid flow along an inclined plate is studied with the consideration double dispersion effects and convective boundary condition at the wall. Influence of nonlinear convection parameters angle inclination is discussed. Major findings are listed below:

- Influence of α_2 is notable on the Nusselt and Sherwood number, when compared with the influence of α_1.
- Angle of inclination increases the thermal and solutal boundary layer thicknesses, whereas decreases the velocity, heat, and mass transfer rates of power-law fluid.
- Influence of Biot number is prominent in velocity, temperature, heat, and mass transfer rates.
- Heat transfer rate and temperatures of power-law fluid are magnified with by thermal dispersion parameter.
- Solutal dispersion parameter increases the concentration and mass transfer rates of power-law fluid.

Nomenclature

b	inertia coefficient which is taken to be a constant	(–)
Bi	Biot number	$\frac{h_f L}{k_f\, Pe^{\frac{1}{2}}}$
$\mathcal{B}$	buoyancy ratio	$\frac{\beta_2(C_w - C_\infty)}{\beta_0(T_f - T_\infty)}$
C	concentration	$\mathrm{kgmol/m^3}$
ϕ	dimensionless concentration	(–)
C_p	specific heat capacity	$\mathrm{J/(kg \cdot K)}$
C_w	wall concentration	$\mathrm{kgmol/m^3}$
C_∞	ambient concentration	$\mathrm{kgmol/m^3}$
D	solutal diffusivity	$\mathrm{m^2/s}$
f	dimensionless stream function	(–)
g^*	acceleration due to gravity	$\mathrm{m/s^2}$
h_f	convective heat transfer coefficient	$\mathrm{W/(m^2K)}$
k_f	thermal conductivity	$\mathrm{kg\,m/(K\,s^3)}$
K_p	permeability	$\mathrm{m^2}$
L	characteristic length	m
Le	diffusivity ratio	$\frac{\alpha}{D}$
n	power-law index	(–)
Nu_x	local Nusselt number	$\frac{-x}{(T_f - T_\infty)}\left[\frac{\partial T}{\partial y}\right]_{y=0}$
$Nu\, Pe^{\frac{-1}{2}}$	dimensionless Nusselt number	$-\xi^{\frac{1}{2}}\left[1 + Pe_\chi\, f'(\xi, 0)\right]\theta'(\xi, 0)$
Pe	global Peclet's number	$\frac{u_\infty L}{\alpha}$
Pe_d	pore diameter-dependent Peclet number	$\frac{u_\infty d}{\alpha}$
Pe_χ	Thermal dispersion parameter	$\frac{[\chi\, d\, u_\infty]}{\alpha}$
Pe_ζ	Solutal dispersion parameter	$\frac{[\zeta\, d\, u_\infty]}{\alpha}$
Pr	Prandtl number	$\frac{\nu}{\alpha}$
Ra	global Rayleigh number	$\left[\frac{L}{\alpha}\right]\left[\frac{K_p g^* \beta_0 (T_f - T_\infty)}{\nu}\right]^{1/n}$
$F_0 Pe$	Forchheimer number	$\left[\left(\frac{b\sqrt{K_p}}{\nu}\right)(\frac{\alpha}{d})^{2-n}\right](Pe_d)^{2-n}$
Sh_x	local Sherwood number	$\frac{-x}{(C_w - C_\infty)}\left[\frac{\partial C}{\partial y}\right]_{y=0}$
$Sh\, Pe^{\frac{-1}{2}}$	dimensionless Sherwood number	$-\xi^{\frac{1}{2}}\left[1 + Pe_\zeta\, f'(\xi, 0)\right]\phi'(\xi, 0)$
T	temperature	K
T	dimensionless temperature	(–)
T_f	wall temperature	K
T_∞	ambient temperature	K
u, v	velocity components	m/s
u_∞	free stream velocity	m/s
x	axial coordinate	m
y	normal coordinate	m

Greek Symbols

α	thermal diffusivity	m^2/s
α_1	nonlinear density-temperature (NDT) parameter	$\frac{\beta_1(T_f-T_\infty)}{\beta_0}$
α_2	nonlinear density-concentration (NDC) parameter	$\frac{\beta_3(C_w-C_\infty)}{\beta_2}$
β_0, β_1	coefficients of thermal expansion of first and second orders	(1/K, 1/K)
β_2, β_3	coefficients of solutal expansion of first and the second order	(m^3/kgmol, m^3/kgmol)
η	similarity variable	(–)
λ	mixed convection parameter	$\frac{Ra}{Pe}$
Ω	inclination of angle	(–)
μ^*	fluid consistency of the power-law fluid	kg/(ms)
ν	kinematic viscosity	m^2/s
ρ	density of the fluid	kg/m^3
ψ	stream function	(–)
ξ	stream-wise coordinate	(–)

Subscripts

w	conditions at the wall	(–)
∞	conditions at the ambient medium	

Acknowledgements This work was supported by of Council of Scientific and Industrial Research (CSIR), New Delhi, India (Project No 25 (0246)/15 /EMR-II).

References

1. Shenoy A (1993) Darcy-Forchheimer natural, forced and mixed convection heat transfer in non-Newtonian power-law fluid saturated porous media. Transp Porous Media 11(3):219–241
2. Shenoy A (1994) Non-Newtonian fluid heat transfer in porous media. Adv Heat Transf 24:102–191
3. Gorla RSR, Kumari M (1999) Nonsimilar solutions for mixed convection in non-Newtonian fluids along a wedge with variable surface temperature in a porous medium. Int J Numerical Methods Heat Fluid Flow 9(5):601–611
4. Ibrahim F, Abdel Gaid S, Gorla RSR (2000) Non-Darcy mixed convection flow along a vertical plate embedded in a non-Newtonian fluid saturated porous medium with surface mass transfer. Int J Numerical Methods Heat Fluid Flow 10(4):397–408
5. Kumari M, Nath G (2004) Non-Darcy mixed convection in power-law fluids along a non-isothermal horizontal surface in a porous medium. Int J Eng Sci 42(3–4):353–369
6. Cheng CY (2008) Double-diffusive natural convection along a vertical wavy truncated cone in non-Newtonian fluid saturated porous media with thermal and mass stratification. Int Commun Heat Mass Transf 35(8):985–990
7. Kairi R, RamReddy C (2014) Solutal dispersion and viscous dissipation effects on non-Darcy free convection over a cone in power-law fluids. Heat Transf-Asian Res 43(5):476–488
8. Nield DA, Bejan A et al (2006) Convection in porous media, vol 3. Springer, Heidelberg
9. Murthy P (2000) Effect of double dispersion on mixed convection heat and mass transfer in non-Darcy porous medium. J Heat Transf 122(3):476–484
10. El Amin MF (2004) Double dispersion effects on natural convection heat and mass transfer in non-Darcy porous medium. Appl Math Comput 156(1):1–17

11. Kairi R, Murthy P (2010) Effect of double dispersion on mixed convection heat and mass transfer in a non-Newtonian fluid saturated non-Darcy porous medium. J Porous Media 13(8):749–757
12. RamReddy C (2013) Effect of double dispersion on convective flow over a cone. Int J Nonlinear Sci 15(4):309–321
13. Bouaziz A, Hanini S (2016) Double dispersion for double diffusive boundary layer in non-Darcy saturated porous medium filled by a nanofluid. J Mech 32(4):441–451
14. Barrow H, Sitharamarao T (1971) Effect of variation in volumetric expansion coefficient on free convection heat transfer. Br Chem Eng 16(8):704–709
15. Vajravelu K, Sastri K (1977) Fully developed laminar free convection flow between two parallel vertical walls-I. Int J Heat Mass Transf 20(6):655–660
16. Partha M (2010) Nonlinear convection in a non-Darcy porous medium. Appl Math Mech 31(5):565–574
17. Kameswaran P, Sibanda P, Partha M, Murthy P (2014) Thermophoretic and nonlinear convection in non-Darcy porous medium. J Heat Transf 136(4)
18. Munir AF, Tasawar H, Bashir A (2014) Peristaltic flow in an asymmetric channel with convective boundary conditions and Joule heating. J Central South Univ 21(4):1411–1416
19. Hayat T, Hussain M, Shehzad S, Alsaedi A (2016) Flow of a power-law nanofluid past a vertical stretching sheet with a convective boundary condition. J Appl Mech Tech Phys 57(1):173–179
20. RamReddy C, Naveen P (2019) Analysis of activation energy in quadratic convective flow of a micropolar fluid with chemical reaction and suction/injection effects. Multidisc Modeling Mater Struct 16(1):169–190
21. Murthy P, Singh P (1999) Heat and mass transfer by natural convection in a non-Darcy porous medium. Acta Mechanica 138(3–4):243–254
22. Chen HT et al (1988) Free convection flow of non-Newtonian fluids along a vertical plate embedded in a porous medium. J Heat Transf (Trans ASME (Am Soc Mech Eng) Ser C) (United States) 110(1)
23. Huang M, Huang J, Chou Y, Chen C (1989) Effects of Prandtl number on free convection heat transfer from a vertical plate to a non-Newtonian fluid. J Heat Transf (Trans ASME (Am Soc Mech Eng) Ser C) (United States) 111(1)
24. Chamkha AJ, Ahmed SE, Aloraier AS (2010) Melting and radiation effects on mixed convection from a vertical surface embedded in a non-Newtonian fluid saturated non-Darcy porous medium for aiding and opposing eternal flows. Int J Phys Sci 5(7):1212–1224
25. Prasad J, Hemalatha K, Prasad B (2014) Mixed convection flow from vertical plate embedded in non-Newtonian fluid saturated non-Darcy porous medium with thermal dispersion-radiation and melting effects. J Appl Fluid Mech 7(3):385–394
26. Sparrow E, Yu H (1971) Local non-similarity thermal boundary-layer solutions. J Heat Transf 93(4):328–334
27. Awad F, Sibanda P, Motsa SS, Makinde OD (2011) Convection from an inverted cone in a porous medium with cross-diffusion effects. Comput Math Appl 61(5):1431–1441
28. Khidir AA, Narayana M, Sibanda P, Murthy P (2015) Natural convection from a vertical plate immersed in a power-law fluid saturated non-Darcy porous medium with viscous dissipation and Soret effects. Afrika Matematika 26(7):1495–1518
29. RamReddy C, Naveen P (2019) Analysis of activation energy and thermal radiation on convective flow of a power-law fluid under convective heating and chemical reaction. Heat Transf-Asian Res 48(6):2122–2154
30. Canuto C, Hussaini MY, Quarteroni A, Thomas Jr A et al (2012) Spectral methods in fluid dynamics. Springer Science & Business Media
31. Chaoyang W, Chuanjing T, Xiaofen Z (1990) Mixed convection of non-Newtonian fluids from a vertical plate embedded in a porous medium. Acta Mechanica Sinica 6(3):214–220
32. Chen C (2004) Heat and mass transfer in MHD flow by natural convection from a permeable, inclined surface with variable wall temperature and concentration. Acta Mechanica 172(3):219–235

33. Aziz A (2009) A similarity solution for laminar thermal boundary layer over a flat plate with a convective surface boundary condition. Commun Nonlinear Sci Numerical Simul 14(4):1064–1068
34. Makinde O (2010) On MHD heat and mass transfer over a moving vertical plate with a convective surface boundary condition. Canadian J Chem Eng 88(6):983–990
35. Pal D, Chatterjee S (2013) Soret and dufour effects on MHD convective heat and mass transfer of a power-law fluid over an inclined plate with variable thermal conductivity in a porous medium. Appl Math Comput 219(14):7556–7574
36. Sui J, Zheng L, Zhang X, Chen G (2015) Mixed convection heat transfer in power-law fluids over a moving conveyor along an inclined plate. Int J Heat Mass Transf 85:1023–1033

Micropolar Fluid Flow over a Frustum of Cone Subjected to Convective Boundary Condition: Darcy–Forchheimer Model

T. Pradeepa and Ch. RamReddy

Abstract This paper emphasizes the Soret and viscous dissipation effects on mixed convective flow of an incompressible micropolar fluid over a vertical frustum of a cone embedded in a non-Darcy porous medium subject to convective boundary condition. The similarity solution does not attain for this complicated fluid flow problem. Using non-similarity transformations, the governing boundary layer equations are converted into a set of non-dimensional partial differential equations. Prior to being these non-similarity equations are linearized by quasilinearization method and solved by the Chebyshev spectral collocation method. Several features emerging from these parameters, namely micropolar, viscous dissipation, Biot, and Soret numbers on physical quantities of the flow, are explored in detail.

Keywords Convective boundary condition · Truncated cone · Micropolar fluid · Non-Darcy porous medium · Spectral quasi-linearization method

1 Introduction

The convective heat and mass transfer analysis in Darcy and/or non-Darcy porous medium have received significant attention from theoretical as well as practical point of view, owing to its applications mentioned in many areas such as geothermal and petroleum resources, enhanced oil recovery, drying of porous solids, cooling of nuclear reactors, thermal insulation, solid matrix heat exchanges, and other practical interesting designs. The non-Darcy (Forchheimer model) is a modification of classical Darcy model by incorporating the inertial effects (i.e., addition of a squared term

T. Pradeepa (✉)
Department of Mathematics, Telangana Social Welfare Residential Degree College, Mahabubabad, Telangana State, India
e-mail: pradeepa.23@gmail.com

Ch. RamReddy
Department of Mathematics, National Institute of Technology Warangal, Warangal 506004, India
e-mail: chramreddy@nitw.ac.in; chittetiram@gmail.com

S. S. Ray et al. (eds.), *Applied Analysis, Computation and Mathematical Modelling in Engineering*, Lecture Notes in Electrical Engineering 897,
https://doi.org/10.1007/978-981-19-1824-7_9

of velocity) in the momentum equation. The literary work on the convective flow due to buoyancy and external forces in a non-Darcy (Forchheimer model) porous medium has been provided by [1–3] (for more details, see the references cited therein).

Micropolar fluids are the subclass of micro-fluids initiated by Eringen [4]. Compared to the classical Newtonian fluids, the flow motion of micropolar fluids distinguishes by two supplementary variables, i.e., the spin vector, responsible for the micro-rotations, and the micro-inertia tensor describes the distribution of atoms and molecules inside the fluid elements in addition to the velocity vector. Thus, micropolar fluids are able to delineate the rheological behavior of animal blood, drug suspension in pharmacology, liquid crystal, colloidal fluids, plasma, etc. The comprehensive review of micropolar fluid mechanics has been reported by [5–8]. The locally produced thermal energy due to viscous stress mechanism, commonly known as viscous dissipation, influences forced, mixed and free convective flows for fluid saturated porous medium and clear viscous fluids. It has unavoidable role in the convective transport mechanism when the fluid flow field is at low temperature or in high gravitational force field or of extreme size. Extensive research can be found in the literature to study the viscous dissipation effect on micropolar fluid flow over different geometries. To mention a few [9–12].

From literature survey, it is found that this type of flow study over truncated cone is applicable in polymer industry, processing of edible items or slurries, melted plastics at industrial level due to involvement of cone-shaped bodies in these areas. However, no literature is observed regarding the mixed convective transport in a Darcy/non-Darcy porous medium saturated by Newtonian/non-Newtonian fluids with truncated cone as a geometry. Yih [13] examined the numerical solution for the natural convection flow from the vertical truncated cone through saturated porous medium using Keller box method. The buoyancy-driven convective flow of a nanofluid from the frustum of a cone embedded in a porous medium by taking thermophoresis and Brownian motion effects has been elaborated by Cheng [14]. Postelnicu [15] considered the local non-similarity solution for the micropolar fluid flow due to buoyancy forces subjected to flux condition. Patrulescu et al. [16] discussed the mixed convection boundary layer flow of a fluid with three nanoparticles through a truncated cone and observed that existence of dual solution for flow reversal.

The main intention of this study is to understand the mixed convective transport over a convectively heated truncated cone embedded in a non-Darcy porous medium saturated by an incompressible micropolar fluid. In addition, viscous dissipation and thermal diffusion effects are taken into consideration. The similarity solution does not obtain for the intricate flow situation, and hence, the non-similarity solution is attained by using spectral quasilinearization method and the usefulness of pertinent parameters discussed through graphical representations.

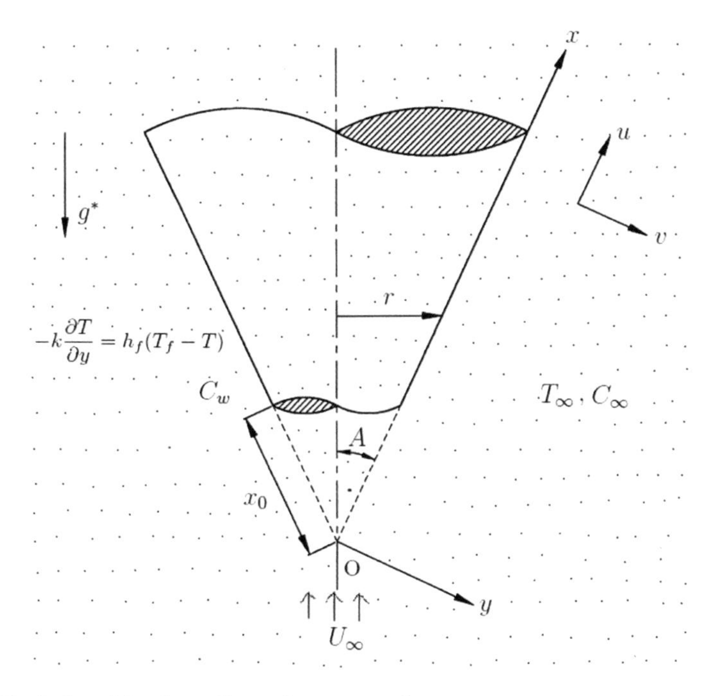

Fig. 1 Physical model and coordinates for a truncated cone

2 Mathematical Analysis

The steady, 2D, laminar mixed convective flow of a micropolar fluid over a truncated cone in a non-Darcy porous medium is considered. The porous medium is considered to be isotropic and homogeneous. The solutal concentration is taken to be constant and is given by C_w. The outer flow velocity is in the form of U_∞. At the ambient media, the temperature and concentration are T_∞ and C_∞, respectively. By convection, the surface of the vertical frustum of a cone is either cooled or heated from a fluid of temperature T_f with $T_f < T_\infty$ relating to a cooled surface and $T_f > T_\infty$ relating to a heated surface. The physical geometry of the problem is shown in Fig. (1). The coordinate system is such that the x-coordinate is taken along the vertical frustum of a cone and y-coordinate is measured normal to it.

By employing Boussinesq approximations, the governing equations for incompressible micropolar fluid using Darcy–Forchheimer model [17] are given by

$$\frac{\partial (u\, r)}{\partial x} + \frac{\partial (v\, r)}{\partial y} = 0 \tag{1}$$

$$\frac{\rho}{\epsilon^2}\left(u\,\frac{\partial u}{\partial x} + v\,\frac{\partial u}{\partial y}\right) = \left(\frac{\mu + \kappa}{\epsilon}\right)\frac{\partial^2 u}{\partial y^2} + \kappa\,\frac{\partial \omega}{\partial y} - \frac{\mu}{K_p}\,(u - U_\infty) - \frac{\rho\, b}{K_p}\,(u^2 - U_\infty^2) \tag{2}$$

$$+ \rho g^* \left(\beta_T (T - T_\infty) + \beta_C (C - C_\infty)\right) \cos A$$

$$\frac{\rho j}{\epsilon} \left(u \frac{\partial \omega}{\partial x} + v \frac{\partial \omega}{\partial y} \right) = \gamma \frac{\partial^2 \omega}{\partial y^2} - \kappa \left(2\omega + \frac{1}{\epsilon} \frac{\partial u}{\partial y} \right) \tag{3}$$

$$u \frac{\partial T}{\partial x} + v \frac{\partial T}{\partial y} = \alpha \frac{\partial^2 T}{\partial y^2} + \left(\frac{\mu + \kappa}{\rho C_p} \right) \left(\frac{\partial u}{\partial y} \right)^2 \tag{4}$$

$$u \frac{\partial C}{\partial x} + v \frac{\partial C}{\partial y} = D \frac{\partial^2 C}{\partial y^2} + \frac{D K_T}{T_m} \frac{\partial^2 T}{\partial y^2} \tag{5}$$

where u and v denote the components of velocity in x and y directions, respectively, T represents the temperature, ω is the microrotation component, b indicates the Forchheimer constant, g^* represents the acceleration due to gravity, γ specifies spin-gradient viscosity, r is the truncated cone radius, ρ is the fluid density, C is the concentration, x_0 represents the frustum of a cone leading edge distance measured from the origin, C_p is the specific heat, j represents the micro-inertia density, ϵ indicates porosity, μ represents dynamic coefficient of viscosity, K_T specifies thermal diffusion ratio, β_T represents coefficient of thermal expansion, D represents the solutal diffusivity, κ represents the vortex viscosity, K_p indicates permeability, β_C is the coefficient of solutal expansion, T_m is mean fluid temperature, and α denotes thermal diffusivity. Further, followed the work of several authors by presuming that $\gamma = \left(\mu + \frac{\kappa}{2}\right) j$ [18].

The boundary conditions are

$$u = 0, \quad v = 0, \quad \omega = -n \frac{\partial u}{\partial y}, \quad -k \frac{\partial T}{\partial y} = h_f (T_f - T), \quad C = C_w \quad \text{at} \quad y = 0 \tag{6a}$$

$$u = U_\infty, \quad \omega = 0, \quad T = T_\infty, \quad C = C_\infty \quad \text{as} \quad y \to \infty \tag{6b}$$

here, the subscripts w and ∞ represent the conditions at the wall and boundary layer outer edge, respectively, n is material constant parameter, k is thermal conductivity of fluid, and h_f indicates the convective heat transfer coefficient.

Now, introduce a stream function ψ as

$$u = \frac{1}{r} \frac{\partial \psi}{\partial y}, \quad v = -\frac{1}{r} \frac{\partial \psi}{\partial x} \tag{7}$$

When the thickness of boundary layer is sufficiently less relative to the local radius of a truncated cone, then the local radius to a point in the boundary layer can be approximated by truncated cone radius:

$$r = xsinA \tag{8}$$

Here, defining the non-similarity transformations in the following form

$$\eta = \frac{y}{\overline{x}} Re_x^{1/2},\ \xi = \frac{\overline{x}}{x_0} = \frac{x - x_0}{x_0},\ \omega = \frac{\nu\, Re_x^{3/2}}{\overline{x}^2}\, g\left(\xi, \eta\right), \tag{9}$$

$$\psi = r\nu Re_x^{1/2} f\left(\xi, \eta\right),\ \theta\left(\xi, \eta\right) = \frac{T - T_\infty}{T_f - T_\infty},\ \phi\left(\xi, \eta\right) = \frac{C - C_\infty}{C_w - C_\infty}$$

where $\overline{x} = x - x_0$ and $Re_x = \dfrac{U_\infty \overline{x}}{\nu}$ is the local Reynolds number.
Substitute (7) - (9) into (2)- (5), the set of equations reduces to the following form

$$\frac{1}{\epsilon}\left(\frac{1}{1-N}\right) f''' + \frac{1}{\epsilon^2}\left(R + \frac{1}{2}\right) f f'' + \frac{\xi}{Da\, Re_{x_0}}(1 - f') + \xi \frac{Fs}{Da}\left(1 - (f')^2\right) \tag{10}$$

$$+ \left(\frac{N}{1-N}\right) g' + \xi\, \lambda\left(\theta + \mathcal{B}\phi\right) = \frac{\xi}{\epsilon^2}\left(f' \frac{\partial f'}{\partial \xi} - f'' \frac{\partial f}{\partial \xi}\right)$$

$$\left(\frac{2-N}{2-2N}\right) g'' + \frac{1}{\epsilon}\left(R + \frac{1}{2}\right) f\, g' + \frac{1}{2\epsilon} f'\, g - \xi\left(\frac{N}{1-N}\right)\left(2g + \frac{1}{\epsilon} f''\right) = \frac{\xi}{\epsilon}\left(f' \frac{\partial g}{\partial \xi} - g' \frac{\partial f}{\partial \xi}\right) \tag{11}$$

$$\frac{1}{Pr}\theta'' + \left(R + \frac{1}{2}\right) f\, \theta' + \left(\frac{1}{1-N}\right) \varepsilon\, (f'')^2 = \xi\left(f' \frac{\partial \theta}{\partial \xi} - \theta' \frac{\partial f}{\partial \xi}\right) \tag{12}$$

$$\frac{1}{Sc}\phi'' + \left(R + \frac{1}{2}\right) f\, \phi' + Sr\, \theta'' = \xi\left(f' \frac{\partial \phi}{\partial \xi} - \phi' \frac{\partial f}{\partial \xi}\right) \tag{13}$$

where the primes denote partial derivative with respect to η alone. $N = \left(\dfrac{\kappa}{\mu + \kappa}\right)$ represents the coupling number [19], $Gr = \dfrac{g^* \beta_T (T_f - T_\infty)\overline{x}^3 cosA}{\nu^2}$ denotes the thermal Grashof number, $\varepsilon = \dfrac{U_\infty^2}{C_p(T_f - T_\infty)}$ is the viscous dissipation parameter, $Gc = \dfrac{g^* \beta_C (C_w - C_\infty)\overline{x}^3 cosA}{\nu^2}$ represents the solutal Grashof number, $Fs = \dfrac{b}{x_0}$ represents the Forchheimer number, $Pr = \dfrac{\nu}{\alpha}$ indicates the Prandtl number, $Da = \dfrac{K_p}{x_0^2}$ is the Darcy parameter, $Sr = \dfrac{DK_T(T_f - T_\infty)}{\nu T_m (C_w - C_\infty)}$ is the Soret number, $Sc = \dfrac{\nu}{D}$ is the Schmidt number, $\mathcal{B} = \dfrac{Gc}{Gr}$ is the buoyancy ratio, $\lambda = \dfrac{Gr_{x_0}}{Re_{x_0}^2}$ indicates the mixed convection parameter and $Re_{x_0} = \dfrac{U_\infty x_0}{\nu}$ is the Reynolds number. Also, notice that $\lambda < 0$, $\lambda = 0$ and $\lambda > 0$ correspond to opposing flow, forced convection flow, and assisting flow, respectively.

The boundary conditions are

$$\left(R + \frac{1}{2}\right) f(\xi, \eta) + \xi \frac{\partial f}{\partial \xi} = 0,\ f'(\xi, \eta) = 0,\ g(\xi, \eta) = -nf''(\xi, \eta), \tag{14a}$$

$$\theta'(\xi, \eta) = -\xi^{1/2} Bi(1 - \theta(\xi, \eta)),\ \phi(\xi, \eta) = 1 \quad at\ \eta = 0,$$

$$f'(\xi, \eta) = 1,\ g(\xi, \eta) = 0,\ \theta(\xi, \eta) = 0,\ \phi(\xi, \eta) = 0 \quad as\ \eta \to \infty. \tag{14b}$$

where $R = \dfrac{\xi}{1 + \xi}$, $Bi = \dfrac{h_f x_0}{k Re_{x_0}^{1/2}}$ represents the Biot number. R becomes zero when $\xi = 0$; hence, the present problem diminishes to mixed convective flow along a vertical plate in a micropolar fluid. As $\xi \to \infty$, $R \to 1$, since $\xi = (x - x_0)/x_0$, ξ becoming large means x is far downstream or cross section of truncated cone radius leading edge is very small.

3 Skin Friction, Wall Couple Stress, Heat and Mass Transfer Coefficients

The wall shear stress and wall couple stress are:

$$\tau_w = \left[(\mu + \kappa)\frac{\partial u}{\partial y} + \kappa\omega\right]_{y=0},\ m_w = \gamma \left[\frac{\partial \omega}{\partial y}\right]_{y=0},$$

The heat transfer and mass transfer rates:

$$q_w = -k \left[\frac{\partial T}{\partial y}\right]_{y=0},\ q_m = -D \left[\frac{\partial C}{\partial y}\right]_{y=0}$$

The non-dimensional skin friction $C_f = \dfrac{2\tau_w}{\rho U_\infty^2}$, wall couple stress $M_w = \dfrac{m_w}{\rho U_\infty^2 x_0}$, the local Nusselt number $Nu_x = \dfrac{q_w \bar{x}}{k(T_f - T_\infty)}$ and local Sherwood number $Sh_x = \dfrac{q_m \bar{x}}{D(C_w - C_\infty)}$ are given by

$$\left.\begin{aligned} &C_f Re_x^{1/2} = 2\left(\frac{1-nN}{1-N}\right)f''(\xi,0), \qquad M_w Re_x = \left(\frac{2-N}{2-2N}\right)g'(\xi,0), \\ &\frac{Nu_x}{Re_x^{1/2}} = -\theta'(\xi,0), \qquad \frac{Sh_x}{Re_x^{1/2}} = -\phi'(\xi,0). \end{aligned}\right\} \tag{15}$$

where $Re_x = \dfrac{U_\infty \bar{x}}{\nu}$ is the local Reynold's number.

4 Solution of the Problem

The non-homogeneous and nonlinear coupled partial differential equations(PDE's) (10)–(13) along with boundary conditions (14) have been solved numerically by spectral quasi-linearization method (SQLM) [20, 21]. Essentially, quasilinearization technique is the generalized Newton–Raphson method initiated by Bellman and Kalaba [22] for solving the functional equations. By applying quasilinearization procedure to Eqs. (10)–(13), the resultant equations are:

$$\frac{1}{\epsilon}\left(\frac{1}{1-N}\right)f'''_{r+1} + a_{1,r}f''_{r+1} + \left(\frac{N}{1-N}\right)g'_{r+1} + a_{2,r}f'_{r+1} + a_{3,r}f_{r+1} + a_{4,r} \tag{16}$$

$$+\xi\,\lambda\,\theta_{r+1} + \xi\,\lambda\,\mathcal{B}\,\phi_{r+1} - a_{5,r}\frac{\partial f'_{r+1}}{\partial\xi} - a_{6,r}\frac{\partial f_{r+1}}{\partial\xi} = 0,$$

$$\left(\frac{2-N}{2-2N}\right)g''_{r+1} - \frac{\xi}{\epsilon}\left(\frac{N}{1-N}\right)f''_{r+1} + b_{1,r}\,g'_{r+1} + b_{2,r}\,g_{r+1} + b_{3,r}f'_{r+1} \tag{17}$$

$$+b_{4,r}f_{r+1} + b_{5,r} - b_{6,r}\frac{\partial g_{r+1}}{\partial\xi} - b_{7,r}\frac{\partial f_{r+1}}{\partial\xi} = 0,$$

$$\frac{1}{Pr}\theta''_{r+1} + c_{1,r}\,\theta'_{r+1} + c_{2,r}f''_{r+1} + c_{3,r}f'_{r+1} + c_{4,r}f_{r+1} + c_{5,r} - c_{6,r}\frac{\partial\theta_{r+1}}{\partial\xi} - c_{7,r}\frac{\partial f_{r+1}}{\partial\xi} = 0, \tag{18}$$

$$\frac{1}{Sc}\phi''_{r+1} + d_{1,r}\,\phi'_{r+1} + Sr\,\theta''_{r+1} + d_{2,r}f'_{r+1} + d_{3,r}f_{r+1} + d_{4,r} - d_{5,r}\frac{\partial\phi_{r+1}}{\partial\xi} - d_{6,r}\frac{\partial f_{r+1}}{\partial\xi} = 0, \tag{19}$$

where

$$a_{1,r} = \frac{1}{\epsilon^2}\left(R+\frac{1}{2}\right)f_r + \frac{\xi}{\epsilon^2}\frac{\partial f_r}{\partial\xi};\ a_{2,r} = \frac{-\xi}{Da\,Re_{x_0}} - \xi\frac{2Fs}{Da}f'_r - \frac{\xi}{\epsilon^2}\frac{\partial f'_r}{\partial\xi};\ a_{3,r} = \frac{1}{\epsilon^2}\left(R+\frac{1}{2}\right)f''_r;$$

$$a_{4,r} = \frac{-1}{\epsilon^2}\left(R+\frac{1}{2}\right)f_r f''_r + \frac{\xi}{Da\,Re_{x_0}} + \xi\frac{Fs}{Da} + \xi\frac{Fs}{Da}f'^2_r + \frac{\xi}{\epsilon^2}f'_r\frac{\partial f'_r}{\partial\xi} - \frac{\xi}{\epsilon^2}f''_r\frac{\partial f_r}{\partial\xi};\ a_{5,r} = \frac{\xi}{\epsilon^2}f'_r;\ a_{6,r} = \frac{-\xi}{\epsilon^2}f''_r;$$

$$b_{1,r} = \frac{1}{\epsilon}\left(R + \frac{1}{2}\right) f_r + \frac{\xi}{\epsilon}\frac{\partial f_r}{\partial \xi};\ b_{2,r} = \frac{1}{2\epsilon} f_r' - 2\xi\left(\frac{N}{1-N}\right);\ b_{3,r} = \frac{1}{2\epsilon} g_r - \frac{\xi}{\epsilon}\frac{\partial g_r}{\partial \xi};$$

$$b_{4,r} = \frac{1}{\epsilon}\left(R + \frac{1}{2}\right) g_r';\ b_{5,r} = \frac{-1}{\epsilon}\left(R + \frac{1}{2}\right) f_r\, g_r' - \frac{1}{2\epsilon} f_r'\, g_r + \frac{\xi}{\epsilon} f_r' \frac{\partial g_r}{\partial \xi} - \frac{\xi}{\epsilon} g_r' \frac{\partial f_r}{\partial \xi};\ b_{6,r} = \frac{\xi}{\epsilon} f_r';\ b_{7,r} = \frac{-\xi}{\epsilon} g_r';$$

$$c_{1,r} = \left(R + \frac{1}{2}\right) f_r + \xi\,\frac{\partial f_r}{\partial \xi};\ c_{2,r} = 2\varepsilon\left(\frac{1}{1-N}\right) f_r'';\ c_{3,r} = -\xi\,\frac{\partial \theta_r}{\partial \xi};\ c_{4,r} = \left(R + \frac{1}{2}\right)\theta_r';\ c_{7,r} = -\xi\theta_r';$$

$$c_{5,r} = -\left(R + \frac{1}{2}\right) f_r\,\theta_r' - \varepsilon\left(\frac{1}{1-N}\right) f_r''^2 + \xi f_r' \frac{\partial \theta_r}{\partial \xi} - \xi\theta_r' \frac{\partial f_r}{\partial \xi};\ c_{6,r} = \xi f_r';$$

$$d_{1,r} = \left(R + \frac{1}{2}\right) f_r + \xi\,\frac{\partial f_r}{\partial \xi};\ d_{2,r} = -\xi\,\frac{\partial \phi_r}{\partial \xi};\ d_{3,r} = \left(R + \frac{1}{2}\right)\phi_r';\ d_{6,r} = -\xi\phi_r';$$

$$d_{4,r} = -\left(R + \frac{1}{2}\right) f_r\,\phi_r' + \xi f_r' \frac{\partial \phi_r}{\partial \xi} - \xi\phi_r' \frac{\partial f_r}{\partial \xi};\ d_{5,r} = \xi f_r';$$

Discretize Eqs. (16) to (19) using the spectral collocation method (i.e., Chebyshev) [23, 24] in the direction of η, and the implicit finite difference method is applied in ξ direction. The collocation points on (η, ξ) are interpreted as

$$\tau_j = \cos\left(\frac{\pi j}{N_x}\right),\quad \xi^n = n\Delta\xi \qquad j = 0, 1, 2, \ldots, N_x,\ n = 0, 1, 2, \ldots, N_t \tag{20}$$

where N_x+1 indicates the number of collocation points in η direction, $\Delta\xi$ is the spacing, and $N_t + 1$ is total number of collocation points in the direction of ξ.

The primitive concept beyond this method is the representation of a derivative matrix D, used to approximate the derivative coefficients $f(\eta)$ of unknown variables at the grid points as the matrix vector product:

$$\frac{df}{d\eta} = \sum_{k=0}^{N_x} D_{lk} f(\tau_k) = \mathbf{DF},\ l = 0, 1, \ldots, N_x, \tag{21}$$

Here, the vector function at the collocation(grid) point is represented by $\mathbf{F} = [f(\tau_0), f(\tau_1), f(\tau_2)\ldots, f(\tau_{N_x})]^T$; similarly, the vector functions corresponding to ϕ, g, and θ are termed as Φ, $\mathbf{G}$, and Θ, respectively. The derivative is scaled as $\mathbf{D} = \dfrac{2\mathcal{D}}{L}$; here, D represents derivative with respect to η. The derivatives of higher order are expressed as exponents of $\mathbf{D}$,

$$f^{(m)} = \mathbf{D}^m\mathbf{F},\ g^{(m)} = \mathbf{D}^m\mathbf{G},\ \theta^{(m)} = \mathbf{D}^m\Theta,\ \phi^{(m)} = \mathbf{D}^m\Phi. \tag{22}$$

where $\mathbf{D}$ indicates the Chebyshev derivative matrix of size $(N_x + 1) \times (N_x + 1)$, and m is the order of derivative. With centering about a midpoint halfway between ξ^n and ξ^{n+1} finite difference scheme is imposed, this midpoint is elucidated as $\xi^{n+\frac{1}{2}} = \left(\xi^{n+1} + \xi^n\right)/2$. Applying $\xi^{n+\frac{1}{2}}$ to any other function, for instance $f(\xi, \eta)$ and its related derivative obtained as

$$f(\xi^{n+\frac{1}{2}}, \eta_j) = f_j^{n+\frac{1}{2}} = \frac{f_j^{n+1} + f_j^n}{2} \tag{23}$$

$$\left(\frac{\partial f}{\partial \xi}\right)^{n+\frac{1}{2}} = \frac{f_j^{n+1} - f_j^n}{\Delta\xi} \tag{24}$$

Applying finite difference in ξ and spectral methods on Eqs. (16)–(19) gives

$$\begin{bmatrix} A_{11} & A_{12} & A_{13} & A_{14} \\ A_{21} & A_{22} & A_{23} & A_{24} \\ A_{31} & A_{32} & A_{33} & A_{34} \\ A_{41} & A_{42} & A_{43} & A_{44} \end{bmatrix} \begin{bmatrix} F_{r+1}^{n+1} \\ G_{r+1}^{n+1} \\ \Theta_{r+1}^{n+1} \\ \Phi_{r+1}^{n+1} \end{bmatrix} = \begin{bmatrix} B_{11} & B_{12} & B_{13} & B_{14} \\ B_{21} & B_{22} & B_{23} & B_{24} \\ B_{31} & B_{32} & B_{33} & B_{34} \\ B_{41} & B_{42} & B_{43} & B_{44} \end{bmatrix} \begin{bmatrix} F_{r+1}^{n} \\ G_{r+1}^{n} \\ \Theta_{r+1}^{n} \\ \Phi_{r+1}^{n} \end{bmatrix} + \begin{bmatrix} K_1 \\ K_2 \\ K_3 \\ K_4 \end{bmatrix}$$

where A_{ij}, B_{ij}, $(i, j = 1, 2, 3, 4)$ are $(N_x + 1) \times (N_x + 1)$ matrices and K_i, $(i = 1, 2, 3, 4)$ is $(N_x + 1) \times 1$ vectors defined as

$$A_{11} = \frac{1}{2}\left[\frac{1}{\epsilon}\left(\frac{1}{1-N}\right)\mathbf{D}^3 + a_{1,r}^{n+\frac{1}{2}}\mathbf{D}^2 + a_{2,r}^{n+\frac{1}{2}}\mathbf{D} + a_{3,r}^{n+\frac{1}{2}}\right] - \frac{a_{5,r}^{n+\frac{1}{2}}\mathbf{D}}{\Delta\xi} - \frac{a_{6,r}^{n+\frac{1}{2}}}{\Delta\xi};$$

$$A_{12} = \frac{1}{2}\left[\left(\frac{N}{1-N}\right)\mathbf{D}\right];\; A_{13} = \frac{1}{2}\xi\,\lambda\mathbf{I};\; A_{14} = \frac{1}{2}\xi\,\lambda\mathcal{B}\,\mathbf{I};$$

$$A_{21} = \frac{1}{2}\left[\frac{\xi}{\epsilon}\left(\frac{-N}{1-N}\right)\mathbf{D}^2 + b_{3,r}^{n+\frac{1}{2}}\mathbf{D} + b_{4,r}^{n+\frac{1}{2}}\right] - \frac{b_{7,r}^{n+\frac{1}{2}}}{\Delta\xi};\; A_{24} = \mathbf{0};$$

$$A_{22} = \frac{1}{2}\left[\left(\frac{2-N}{2-2N}\right)\mathbf{D}^2 + b_{1,r}^{n+\frac{1}{2}}\mathbf{D} + b_{2,r}^{n+\frac{1}{2}}\right] - \frac{b_{6,r}^{n+\frac{1}{2}}}{\Delta\xi};\; A_{23} = \mathbf{0};$$

$$A_{31} = \frac{1}{2}\left[c_{2,r}^{n+\frac{1}{2}}\mathbf{D}^2 + c_{3,r}^{n+\frac{1}{2}}\mathbf{D} + c_{4,r}^{n+\frac{1}{2}}\right] - \frac{c_{7,r}^{n+\frac{1}{2}}}{\Delta\xi};\; A_{32} = \mathbf{0};$$

$$A_{33} = \frac{1}{2}\left[\frac{1}{Pr}\mathbf{D}^2 + c_{1,r}^{n+\frac{1}{2}}\mathbf{D}\right] - \frac{c_{6,r}^{n+\frac{1}{2}}}{\Delta\xi};\; A_{34} = \mathbf{0};$$

$$A_{41} = \frac{1}{2}\left[d_{2,r}^{n+\frac{1}{2}}\mathbf{D} + d_{3,r}^{n+\frac{1}{2}}\right] - \frac{d_{6,r}^{n+\frac{1}{2}}}{\Delta\xi};\ A_{42} = \mathbf{0};$$

$$A_{43} = \frac{1}{2}Sr\mathbf{D}^2;\ A_{44} = \frac{1}{2}\left[\frac{1}{Sc}\mathbf{D}^2 + d_{1,r}^{n+\frac{1}{2}}\mathbf{D}\right] - \frac{d_{5,r}^{n+\frac{1}{2}}}{\Delta\xi};$$

$$B_{11} = -\frac{1}{2}\left[\frac{1}{\epsilon}\left(\frac{1}{1-N}\right)\mathbf{D}^3 + a_{1,r}^{n+\frac{1}{2}}\mathbf{D}^2 + a_{2,r}^{n+\frac{1}{2}}\mathbf{D} + a_{3,r}^{n+\frac{1}{2}}\right] - \frac{a_{5,r}^{n+\frac{1}{2}}\mathbf{D}}{\Delta\xi} - \frac{a_{6,r}^{n+\frac{1}{2}}}{\Delta\xi};$$

$$B_{12} = -\frac{1}{2}\left[\left(\frac{N}{1-N}\right)\mathbf{D}\right];\ B_{13} = -\frac{1}{2}\xi\lambda\mathbf{I};\ B_{14} = -\frac{1}{2}\lambda\xi\mathcal{B}\mathbf{I}$$

$$B_{21} = -\frac{1}{2}\left[\frac{\xi}{\epsilon}\left(\frac{-N}{1-N}\right)\mathbf{D}^2 + b_{3,r}^{n+\frac{1}{2}}\mathbf{D} + b_{4,r}^{n+\frac{1}{2}}\right] - \frac{b_{7,r}^{n+\frac{1}{2}}}{\Delta\xi};$$

$$B_{22} = -\frac{1}{2}\left[\left(\frac{2-N}{2-2N}\right)\mathbf{D}^2 + b_{1,r}^{n+\frac{1}{2}}\mathbf{D} + b_{2,r}^{n+\frac{1}{2}}\right] - \frac{b_{6,r}^{n+\frac{1}{2}}}{\Delta\xi};\ B_{23} = \mathbf{0};\ B_{24} = \mathbf{0};$$

$$B_{31} = -\frac{1}{2}\left[c_{2,r}^{n+\frac{1}{2}}\mathbf{D}^2 + c_{3,r}^{n+\frac{1}{2}}\mathbf{D} + c_{4,r}^{n+\frac{1}{2}}\right] - \frac{c_{7,r}^{n+\frac{1}{2}}}{\Delta\xi};\ B_{32} = \mathbf{0};$$

$$B_{33} = -\frac{1}{2}\left[\frac{1}{Pr}\mathbf{D}^2 + c_{1,r}^{n+\frac{1}{2}}\mathbf{D}\right] - \frac{c_{6,r}^{n+\frac{1}{2}}}{\Delta\xi};\ B_{34} = \mathbf{0};$$

$$B_{41} = -\frac{1}{2}\left[d_{2,r}^{n+\frac{1}{2}}\mathbf{D} + d_{3,r}^{n+\frac{1}{2}}\right] - \frac{d_{6,r}^{n+\frac{1}{2}}}{\Delta\xi};\ B_{42} = \mathbf{0};$$

$$B_{43} = -\frac{1}{2}Sr\mathbf{D}^2;\ B_{44} = -\frac{1}{2}\left[\frac{1}{Sc}\mathbf{D}^2 + d_{1,r}^{n+\frac{1}{2}}\mathbf{D}\right] - \frac{d_{5,r}^{n+\frac{1}{2}}}{\Delta\xi};$$

$$K_1 = -a_{4,r}^{n+\frac{1}{2}};\ K_2 = -b_{5,r}^{n+\frac{1}{2}};\ K_3 = -c_{5,r}^{n+\frac{1}{2}};\ K_4 = -d_{4,r}^{n+\frac{1}{2}}.$$

The collocation points $N_x = 100$ are used in this study for all cases. Noticed that the SQLM depends on the value of any quantity computation, suppose F_{r+1}^{n+1} at each time step. This is attained by iterating using the QLM. The iteration calculations are carried until some appropriate tolerance level, ϵ_1, is attained. The tolerance level is stated as the maximum values of the infinity norm of the difference between the values of the calculated quantities, that is

$$max\{\|f{\prime}_{r+1}^{1^{n+1}} - f{\prime}_{r+1}^{1^{n}}\|_{\infty},\ \|g_{r+1}^{n+1} - g_{r+1}^{n}\|_{\infty},\ \|\theta_{r+1}^{n+1} - \theta_{r+1}^{n}\|_{\infty},\ \|\phi_{r+1}^{n+1} - \phi_{r+1}^{n}\|_{\infty}\} < \epsilon_1 \quad (25)$$

An adequately small step size $\Delta\xi$ is considered to ensure the results accuracy.

5 Results and Discussion

The resulting nonlinear, non-homogeneous coupled partial differential equations (10)–(13) together with the boundary conditions (14) have been solved numerically using spectral quasilinearisation method. To analyze the effects of Soret, coupling number, viscous dissipation and Biot number the computations are executed for $Re_{x_0} = 200$, $Da = 0.5$, $Pr = 0.72$, $\epsilon = 0.6$, $Sc = 0.22$, and $n = 0$. In order to test the validity of code generated the existing problem numerical scheme at $\xi = 0$, has been compared with the results attained by Lloyd and Sparrow [25], Ramreddy and Pradeepa [26] for $Bi \to \infty$, $Sr = 0$, $\mathcal{B} = 0$, $N = 0$, $Da \to \infty$, $\varepsilon = 0$, $Fs = 0$ and $\lambda = 0$. The results are shown in Table 1, and the agreement is good.

Figure 2 illustrates the variation of Forchheimer number and viscous dissipation on dimensionless skin friction$\left(C_f Re_x^{1/2}\right)$, wall couple stress$(M_w Re_x)$, Nusselt number $\left(Nu_x/Re_x^{1/2}\right)$ and Sherwood number $\left(Sh_x/Re_x^{1/2}\right)$ against streamwise coordinate (ξ). Figure 2 exhibits that as viscous dissipation parameter rises the skin friction and mass transfer rate enhances, wall couple stress and heat transfer rate diminish for both Darcian ($Fs = 0.0$) and non-Darcian flows ($Fs = 0.5$). Moreover, it is observed that the skin friction, heat, and mass transfer rates are lower but wall couple stress is higher for Darcian flow compared with that of non-Darcian flow. For both Darcian and non-Darcian flow, there is no significant effect on the skin friction, wall couple stress, and heat transfer rate but mass transfer rate increases slightly in the case of vertical plate ($\xi = 0$), whereas in truncated cone ($\xi > 0$) skin friction and mass transfer rate

Table 1 Comparison analysis of $-\theta'(0, 0)$ with the proposed method(SQLM) and that of results obtained by Lloyd and Sparrow [25]

Pr		$-\theta'(0, 0)$	
Pr	Lloyd and Sparrow [25]	Ramreddy and Pradeepa [26]	Present
0.003	0.02937	0.03967	0.03967
0.01	0.0515	0.05382	0.05382
0.03	0.08439	0.08443	0.08443
0.72	0.2956	0.29564	0.29564
10	0.7281	0.72814	0.72814
100	1.572	1.57184	1.57184

Ramreddy and Pradeepa [26] when $n = 0.0$, $\varepsilon = 0$, $\mathcal{B} = 0$, $Da \to \infty$, $N = 0$, $\epsilon = 1$, $Da \to \infty$, $Bi \to \infty$, $\xi = 0$, $Sr = 0$, $Fs = 0$ and $\lambda = 0$

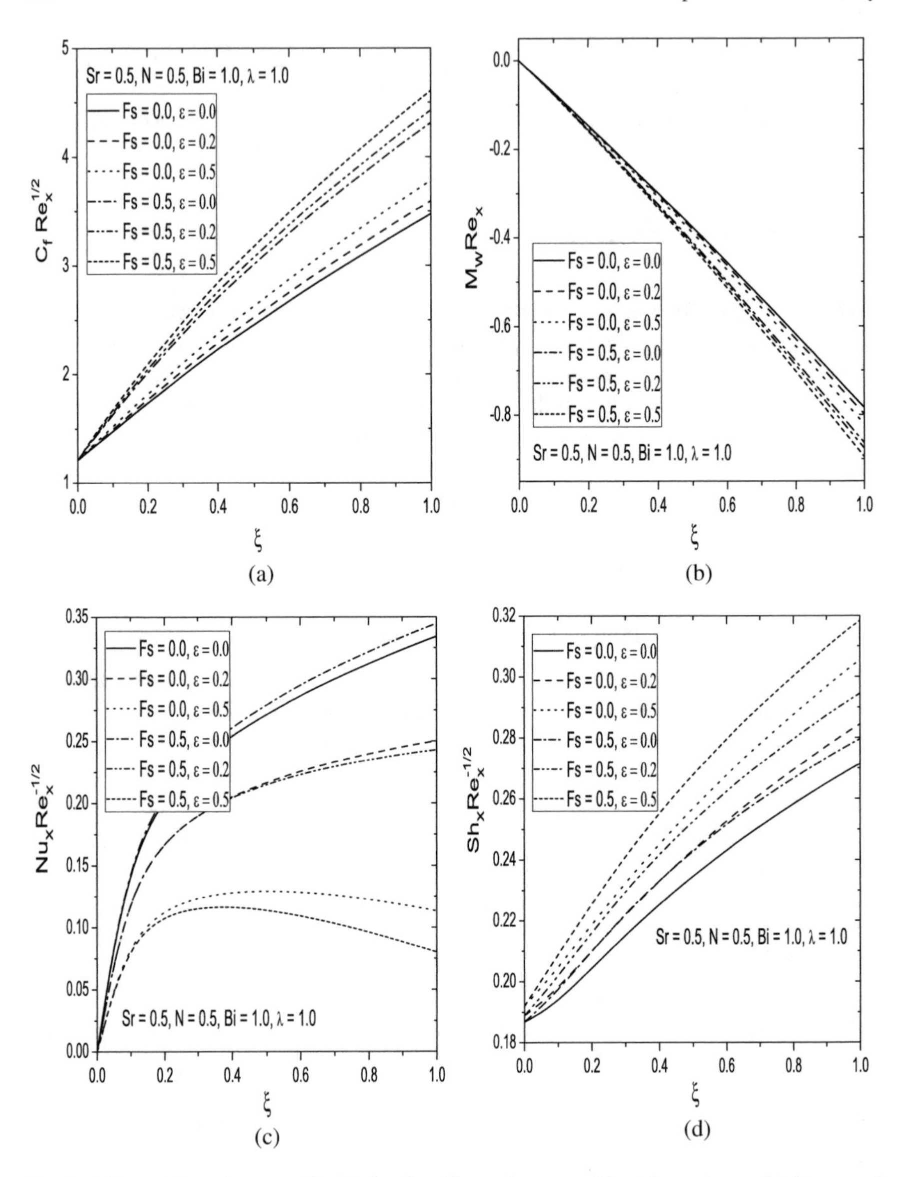

Fig. 2 Effect of Fs and ε on **a** skin friction **b** wall couple stress **c** Nusselt number and **d** Sherwood number

raises, but wall couple stress and Nusselt number diminishes with increase of viscous dissipation parameter.

The influence of Biot and Soret numbers on dimensionless skin friction $\left(C_f Re_x^{1/2}\right)$, wall couple stress$(M_w Re_x)$, Nusselt number $\left(Nu_x/Re_x^{1/2}\right)$, and Sherwood number $\left(Sh_x/Re_x^{1/2}\right)$ is depicted in Fig. 3. The diffusion of mass due to the temperature

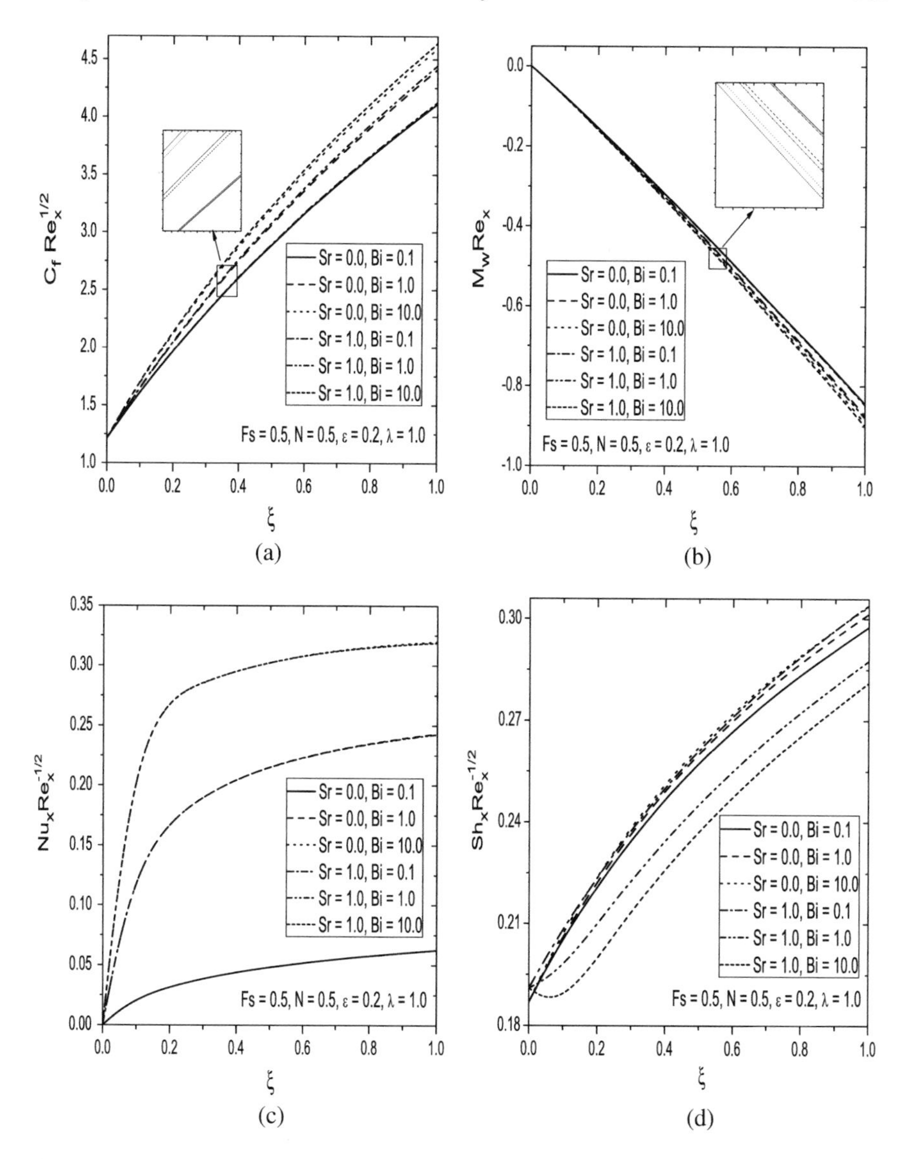

Fig. 3 Effect of *Sr* and *Bi* on **a** skin friction **b** wall couple stress **c** Nusselt number and **d** Sherwood number

gradients is delineated as Soret number *Sr*. Figures 3a, b display that for both absence($Sr = 0.0$) and presence($Sr = 1.0$) of Soret number, the wall couple stress diminishes, and skin friction enhances with the raise of Biot number. Figure 3c displays that for both the presence and absence of Soret number, the $\left(Nu_x/Re_x^{1/2}\right)$ augments nonlinearly with an increase of Biot number, but there is no immense effect with rise in Soret number. While raising the Biot number the $\left(Sh_x/Re_x^{1/2}\right)$ enhances in the absence of Soret number $Sr = 0.0$, and with an existence of Soret number $Sr = 1.0$, it diminishes which is shown in Fig. 3d.

The dimensionless skin friction $\left(C_f Re_x^{1/2}\right)$, wall couple stress $(M_w Re_x)$, Nusselt number $\left(Nu_x/Re_x^{1/2}\right)$, and Sherwood number $\left(Sh_x/Re_x^{1/2}\right)$ with mixed convection parameter variation for both micropolar $(N = 0.5)$ viscous fluids $(N = 0)$ is shown in Fig. 4. The coupling number(N) characterizes the rotational and linear motion of fluid particles. Figure 4a illustrates that for both $(N = 0)$ and $(N = 0.5)$ the skin friction in opposing flow case is low and in aiding flow is more as compared with forced convection flow. Further noticed that the skin friction is high in case of micropolar fluid than compared with that of viscous fluid case because micropolar fluid offers high resistance which emerge from fluid particles motion. Figure 4b portrays that for micropolar fluid the wall couple stress is less in case of aiding flow than that of opposing flow case. Figures 4c, d represent that the heat and mass transfer rates of micropolar fluid are lower than that of viscous fluid. Moreover, for both fluids $(N = 0.0 \text{ and } N = 0.5)$, the $\left(Nu_x/Re_x^{1/2}\right)$ and $\left(Sh_x/Re_x^{1/2}\right)$ are less in case of opposing flow as compared with forced convection and aiding flow.

Figure 5 portrays the effect of material constant parameter and porosity on dimensionless skin friction$\left(C_f Re_x^{1/2}\right)$, wall couple stress$(M_w Re_x)$, Nusselt number $\left(Nu_x/Re_x^{1/2}\right)$, and Sherwood number $\left(Sh_x/Re_x^{1/2}\right)$ against streamwise coordinate (ξ). Figure 5a exhibits that the skin friction decreases and then increases with increase of porosity. The skin friction decreases with increase of material constant parameter. The wall couple stress enhances with the enhancement of both ϵ and n displayed in Fig. 5b. Figure 5c shows that Nusselt number increases and decreases with increase of ϵ and n, respectively. As the porosity increases the Sherwood number decreases first and then increases which is clearly observed in Fig. 5d. Moreover, Sherwood number increases with increase of material constant parameter.

6 Conclusions

A mathematical model of steady mixed convection incompressible micropolar fluid flow over a truncated cone embedded in a saturated porous medium with viscous dissipation and thermal diffusion effects is investigated in this paper. In addition, Forchheimer porous medium with convective boundary condition is incorporated. The resulting non-similarity equations are solved using spectral quasilinearization method. Based on the analysis carried out, the main conclusions are drawn

- For both presence $(Fs = 0.5)$ and absence $(Fs = 0.0)$ of Forchheimer number, the Sherwood number and skin friction increases, whereas wall couple stress decreases with raise in viscous dissipation.

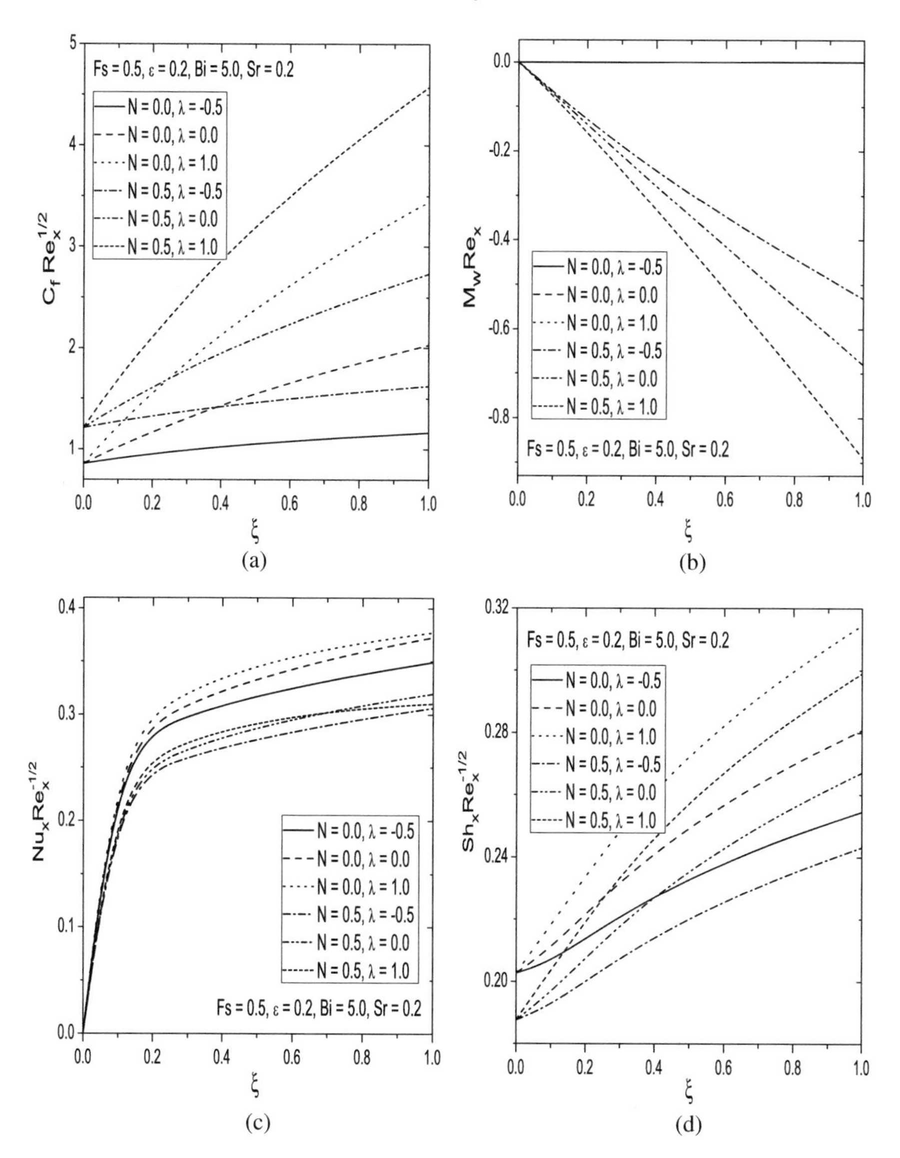

Fig. 4 Effect of N and λ on **a** skin friction **b** wall couple stress **c** Nusselt number and **d** Sherwood number

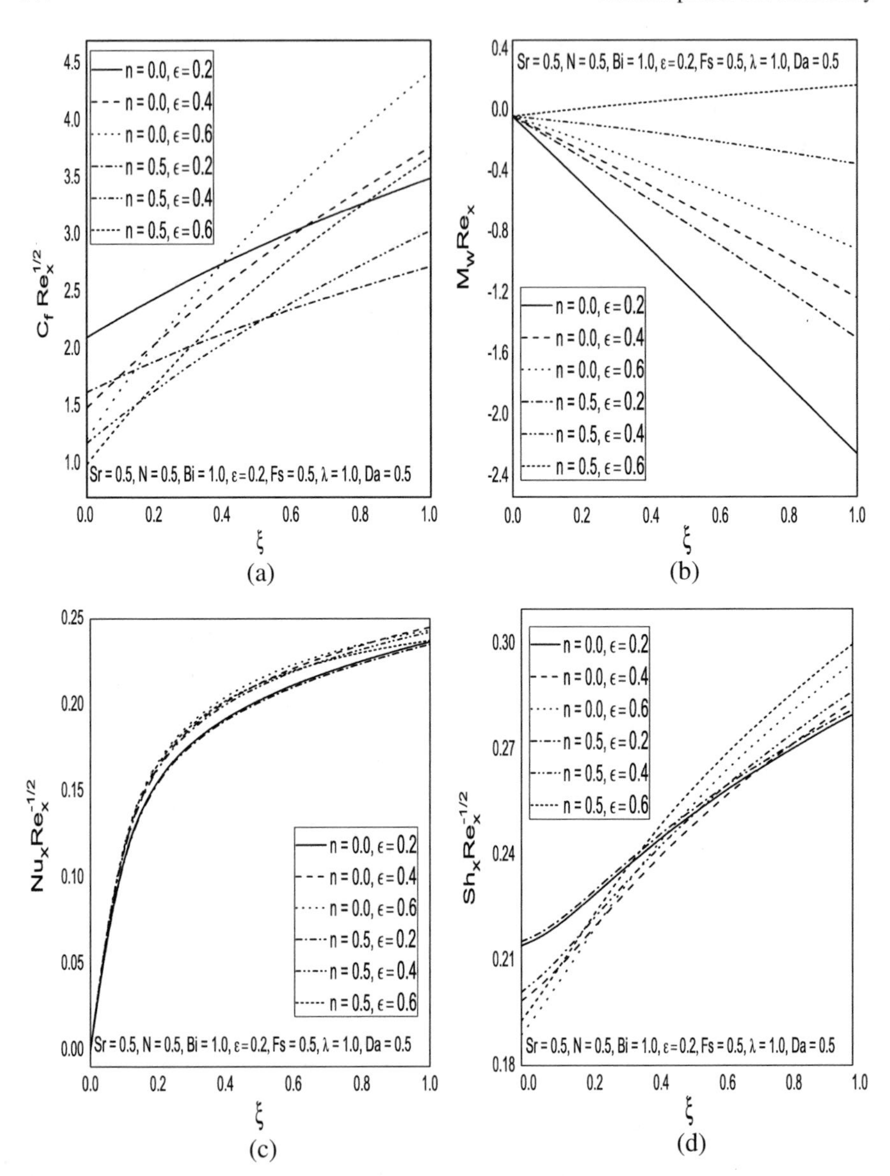

Fig. 5 Effect of n and ϵ on **a** skin friction **b** wall couple stress **c** Nusselt number and **d** Sherwood number

- The raise in Biot number results to diminish the wall couple stress, but raises Nusselt number, skin friction for both the cases, i.e., presence and absence of Soret number. Meanwhile, as the Biot number increases, the Sherwood number shows the opposite behavior for the cases of $Sr = 0.0$ and $Sr = 1.0$.
- In the opposing flow, the Sherwood number and skin friction are more for both viscous and micropolar fluids. The Nusselt number is less for opposing flow as compared with forced convection and aiding flow.
- The skin friction and Sherwood number show similar behavior with increase of porosity, but they show opposite behavior with increase of material constant parameter. The wall couple stress and Nusselt number increases with increase of porosity; however, they show opposite behavior with increase of material constant parameter.

References

1. Ingham DB, Pop I (2005) Transport phenomena in porous media. Elsevier, Oxford
2. Nield DA, Bejan A (2013) Convection in porous media, 4th edn. Springer, New York
3. Vafai K (2015) Handbook of porous media, 3rd edn. CRC Press, Taylor and Francis Group
4. Eringen AC (1966) Theory of micropolar fluids. J Math Mech 16:1–18
5. Ariman T, Turk MA, Sylvester ND (1973) Microcontinuum fluid mechanics—a review. Int J Eng Sci 11(8):905–930
6. Ariman T, Turk MA, Sylvester ND (1974) Applications of microcontinuum fluid mechanics. Int J Eng Sci 12(4):273–293
7. Lukaszewicz G (1999) Micropolar fluids—theory and applications. Birkhauser, Basel
8. Eremeyev V, Lebedev L, Altenbach H (2013) Foundations of micropolar mechanics. Springer, New York, Heidelberg
9. Gebhart B (1962) Effect of viscous dissipation in natural convection. J Fluid Mech 14:225–235
10. El-Amin MF, Mohammadein AA (2005) Effects of viscous dissipation and Joule heating on magnetohydrodynamic Hiemenz flow of a micropolar fluid. Heat Transf Eng 26(6):75–81
11. El-Aziz MA (2009) Viscous dissipation effect on mixed convection flow of a micropolar fluid over an exponentially stretching sheet. Canadian J Phys 87(4):359–368
12. Ahmad K, Ishak A, Nazar R (2013) Micropolar fluid flow and heat transfer over a nonlinearly stretching plate with viscous dissipation. Math Problems Eng 2013:1–5
13. Yih KA (1999) Coupled heat and mass transfer by free convection over a truncated cone in porous media: VWT/VWC or VHF/VMF. Acta Mechanica 137(1–2):83–97
14. Cheng CY (2012) Natural convection boundary layer flow over a truncated cone in a porous medium saturated by a nanofluid. Int Commun Heat Mass Transf 39:231–235
15. Postelnicu A (2012) Free convection from a truncated cone subject to constant wall heat flux in a micropolar fluid. Meccanica 47:1349–1357
16. Patrulescu FO, Groşan T, Pop I (2014) Mixed convection boundary layer flow from a vertical truncated cone in a nanofluid. Int J Numerical Methods Heat Fluid Flow 24(5):1175–1190
17. Srinivasacharya D, Ramreddy Ch (2012) Mixed convection in a doubly stratified micropolar fluid saturated non-Darcy porous medium. Canadian J Chem Eng 90(5):1311–1322
18. Ahmadi G (1976) Self-similar solution of incompressible micropolar boundary layer flow over a semi-infinite plate. Int J Eng Sci 14:639–646
19. Cowin SC (1968) Polar fluids. Phys Fluids 11:1919–1927
20. Srinivasacharya D, Motsa SS, Surender O (2015) Numerical study of free convection in a doubly stratified non-Darcy porous medium using spectral quasilinearization method. Int J Nonlinear Sci Numerical Simul 16:173–183

21. RamReddy C, Pradeepa T (2016) Non-similarity solution of micropolar fluid flow over a truncated cone with Soret and viscous dissipation effects using spectral quasilinearization method. Int J Appl Comput Math 1–15. https://doi.org/10.1007/s40819-016-0227-y
22. Bellman RE, Kalaba RE (1965) Quasilinearisation and non-linear boundary-value problems. Elsevier, New York
23. Trefethen LN (2000) Spectral methods in MATLAB. SIAM
24. Canuto C, Hussaini MY, Quarteroni A, Zang TA (2006) Spectral methods fundamentals in single domains. Springer, New York
25. Lloyd JR, Sparrow EM (1970) Combined free and forced convective flow on vertical surfaces. Int J Heat Mass Transfer 13:434–438
26. Ramreddy C, Pradeepa T (2017) Non-similarity solutions for viscous dissipation and Soret effects in micropolar fluid over a truncated cone with convective boundary condition: spectral quasilinearization approach. Int J Nonlinear Sci Numerical Simul 18(5):327–342

Convergence and Comparison Theorems for Three-Step Alternating Iteration Method for Rectangular Linear System

Smrutilekha Das, Debadutta Mohanty, and Chinmay Kumar Giri

Abstract The three-step alternating iteration method was introduced by Nandi et al. (J. Appl. Math. Comput. 60:485–515, 2019) for solving a rectangular linear system $Ax = b$ by using proper weak regular splittings of type I. In this article, we expand the theory of such alternating iterations by using proper weak regular splittings of type II.

Keywords Moore–Penrose inverse · Proper splitting · Convergence theorems · Comparison theorems · Three-step alternating iteration method

Mathematics Subject Classification (2010) 15A09 · 65F10 · 65F20

1 Introduction

Let us consider a system of linear equations of the form

$$Ax = b, \quad A \in \mathbb{R}^{m\times n}, \ x \in \mathbb{R}^n, \ b \in \mathbb{R}^m. \tag{1.1}$$

The original version of this chapter was revised: The second author's affiliation has been changed to "Department of Mathematics, Seemanta Mahavidyalaya, Jharpokharia, Odisha, 757086, India". The correction to this chapter is available at https://doi.org/10.1007/978-981-19-1824-7_25.

S. Das
Department of Mathematics, Maharaja Sriram Chandra Bhanja Deo University, Takhatpur, Baripada, Odisha 757003, India

D. Mohanty
Department of Mathematics, Seemanta Mahavidyalaya, Jharpokharia, Odisha 757086, India

C. K. Giri (✉)
Department of Mathematics, National Institute of Science and Technology, Berhampur, Odisha 761008, India
e-mail: ckg2357@gmail.com

S. S. Ray et al. (eds.), *Applied Analysis, Computation and Mathematical Modelling in Engineering*, Lecture Notes in Electrical Engineering 897,
https://doi.org/10.1007/978-981-19-1824-7_10

When the coefficient matrix A is very large and sparse, iterative methods become more efficient. In this direction, by using the notion of proper splitting Berman and Plemmons [4] proposed the following iterative method:

$$x^{k+1} = U^{\dagger}Vx^{k} + U^{\dagger}b, \quad k = 0, 1, 2, \ldots, \tag{1.2}$$

where $U^{\dagger}V$ is the iteration matrix and $U^{\dagger}$ is the Moore–Penrose inverse of U. The same authors also proved that the above iterative method converges to $A^{\dagger}b$ for any initial guess x^{0} if and only if the spectral radius of the iteration matrix $U^{\dagger}V$ is less than 1 (see Corollary 1, [4] for instance). Therefore, the rate of convergence of the iterative method (1.2) depends on $\rho(U^{\dagger}V)$ and so, the spectral radius of the iteration matrix plays an important role in the comparison of the rate of convergence of different iterative methods of the same system. Many authors such as Berman and Plemmons [4], Climent et al. [7], Climent and Perea [8], Jena et al. [13], Mishra [14, 18], Mishra and Mishra [17], Baliarsingh and Mishra [1], Giri and Mishra [9–12], Shekhar et al. [21], and others have introduced several convergence and comparison results for different subclasses of a proper splitting.

In particular, Mishra [14] proposed the concept of alternating iteration method for rectangular matrices by extending the work of Benzi and Szyld [2]. Mishra [14] considered two proper splittings of $A \in \mathbb{R}^{m\times n}$, namely $A = M - N = U - V$, and proposed the following iterative method

$$x^{k+1} = U^{\dagger}VM^{\dagger}Nx^{k} + U^{\dagger}(VM^{\dagger} + I)b, \quad k = 0, 1, 2, \ldots$$

to solve (1.1). Recently, Nandi et al. [19] introduced the three-step alternating iteration method by extending the work of Mishra [14]. Now we recall the same. Let $A = M - N = P - Q = U - V$ be proper splittings of $A \in \mathbb{R}^{m\times n}$. The same authors considered

$$x^{k+1/3} = M^{\dagger}Nx^{k} + M^{\dagger}b, \tag{1.3}$$

$$x^{k+1/2} = P^{\dagger}Qx^{k+1/3} + P^{\dagger}b, \tag{1.4}$$

$$x^{k+1} = U^{\dagger}Vx^{k+1/2} + U^{\dagger}b, \quad k = 0, 1, 2, \ldots. \tag{1.5}$$

By simplifying (1.3), (1.4), and (1.5), one can formulate the following iterative method known as *the three-step alternating iterative method for rectangular matrices*

$$x^{k+1} = U^{\dagger}VP^{\dagger}QM^{\dagger}Nx^{k} + U^{\dagger}(VP^{\dagger}QM^{\dagger} + VP^{\dagger} + I)b, \quad k = 0, 1, 2, \ldots, \tag{1.6}$$

where $H = U^{\dagger}VP^{\dagger}QM^{\dagger}N$ is the iteration matrix of the iterative method (1.6). The same authors [19] then studied the convergence criteria for the above iterative method by assuming the splittings $A = M - N = P - Q = U - V$ are proper weak regular of type I (see Theorem 3.1). The convergence of three-step alternating iterations for a singular linear system using the *Group Inverse* (see [3] for the definition) for type II matrix splitting is studied in [22]. However, convergence of three-step alternating for a rectangular linear system in the case of type II splittings is not yet considered in the literature.

The main objective of this paper is to study the convergence of three-step alternating iteration method by considering the splittings are proper weak regular of type II. To fulfill this objective, we organize the contents of the paper as follows. In Sect. 2, we introduce notations, definitions, and some preliminary results that are frequently used while proving our main results. In Sect. 3, we prove our main results. Here, we derived convergence and comparison results for the three-step alternating iteration method by considering splittings $A = M - N = P - Q = U - V$ are proper weak regular of type II. The findings are verified through numerical examples in Sect. 4. Finally, we concluded this article in Sect. 5.

2 Preliminaries

Throughout the article, all the matrices are considered as real of order $m \times n$, unless stated otherwise. The symbol $\mathbb{R}^{m\times n}$ denotes the set of all real matrices of order $m \times n$ and by $\mathbb{R}^n$ we mean an n-dimensional Euclidean space. The transpose, the range space and the null space of a matrix A are denoted by A^{T}, $\mathcal{R}(A)$ and $\mathcal{N}(A)$, respectively. Let L and M be complementary subspaces of $\mathbb{R}^n$, and $P_{L,M}$ be a projector onto L along M. Then $P_{L,M}A = A$ if and only if $\mathcal{R}(A) \subseteq L$ and $AP_{L,M} = A$ if and only if $\mathcal{N}(A) \supseteq M$. If $L \perp M$, then we denote $P_{L,M}$ by P_L. Let $\lambda_1, \lambda_2, \ldots, \lambda_n$ be the eigenvalues of $A \in \mathbb{R}^{n\times n}$. Then the spectral radius of $A \in \mathbb{R}^{n\times n}$ is denoted by $\rho(A)$ and is defined by $\rho(A) = \max\{|\lambda_1|, |\lambda_2|, \ldots, |\lambda_n|\}$, whereas $\sigma\{A\}$ denotes the set of all eigenvalues of A.

2.1 Nonnegative Matrices

$A \in \mathbb{R}^{m\times n}$ is called nonnegative (positive) if each entry of A is nonnegative (positive) and is denoted by $A \geq 0$ $(A > 0)$. For $A,\ B \in \mathbb{R}^{m\times n}$, $A \geq B$ means $A - B \geq 0$. The same notation and nomenclature are also used for vectors. The next results deal with the nonnegativity of a matrix and its spectral radius.

Theorem 2.1 (Theorem 2.20, [23])
Let $B \in \mathbb{R}^{n\times n}$ and $B \geq 0$. Then

(i) B has a nonnegative real eigenvalue equal to its spectral radius.
(ii) To $\rho(B)$, there corresponds an eigenvector $x \geq 0$.

Theorem 2.2 (Theorem 2.1.11, [6])
Let $B \in \mathbb{R}^{n\times n}$, $B \geq 0$, $x \geq 0$ $(x \neq 0)$, and α is a positive scalar.

(i) If $\alpha x \leq Bx$, then $\alpha \leq \rho(B)$.

(ii) If $Bx \leq \alpha x$, $x > 0$, *then* $\rho(B) \leq \alpha$.

Theorem 2.3 (Theorem 3.15, [23])
Let $B \in \mathbb{R}^{n\times n}$ *and* $B \geq 0$. *Then* $\rho(B) < 1$ *if and only if* $(I - B)^{-1}$ *exists and* $(I - B)^{-1} = \sum_{n=0}^{\infty} B^n \geq 0$.

2.2 The Moore–Penrose Inverse of a Matrix

For $A \in \mathbb{R}^{m\times n}$, the unique matrix $X \in \mathbb{R}^{n\times m}$ satisfying the following four equations known as Penrose equations:

$$AXA = A, \quad XAX = X, \quad (AX)^{\mathrm{T}} = AX \text{ and } (XA)^{\mathrm{T}} = XA$$

is called the Moore–Penrose inverse of A. It always exists and is denoted by $A^{\dagger}$ (see [20]). Next, we collect some well-known properties of the Moore–Penrose inverse of $A \in \mathbb{R}^{m\times n}$ which will be used frequently in this article, namely: $\mathcal{R}(A^{\dagger}) = \mathcal{R}(A^{\mathrm{T}})$; $\mathcal{N}(A^{\dagger}) = \mathcal{N}(A^{\mathrm{T}})$; $AA^{\dagger} = P_{\mathcal{R}(A^{\dagger})}$; $A^{\dagger}A = P_{\mathcal{R}(A^{\mathrm{T}})}$. In particular, if $x \in \mathcal{R}(A)$, then $x = A^{\dagger}Ax$ (for more details, see [3]). $A \in \mathbb{R}^{m\times n}$ is called semimonotone, if $A^{\dagger} \geq 0$.

2.3 Proper Splittings

A splitting $A = U - V$ of $A \in \mathbb{R}^{m\times n}$ is called a proper splitting if $\mathcal{R}(U) = \mathcal{R}(A)$ and $\mathcal{N}(U) = \mathcal{N}(A)$. A few properties of a proper splitting are summarized below.

Theorem 2.4 (Theorem 1, [4])
Let $A = U - V$ *be a proper splitting of* $A \in \mathbb{R}^{m\times n}$. *Then*

(a) $A = U(I - U^{\dagger}V)$,
(b) $(I - U^{\dagger}V)$ *is nonsingular,*
(c) $A^{\dagger} = (I - U^{\dagger}V)^{-1}U^{\dagger}$.

Theorem 2.5 (Theorem 1, [7])
Let $A = U - V$ *be a proper splitting of* $A \in \mathbb{R}^{m\times n}$. *Then*

(a) $A = (I - VU^{\dagger})U$,
(b) $(I - VU^{\dagger})$ *is nonsingular,*
(c) $A^{\dagger} = U^{\dagger}(I - VU^{\dagger})^{-1}$.

Theorem 2.6 (Theorem 1, [15])
Let $A = U - V$ *be a proper splitting of* $A \in \mathbb{R}^{m\times n}$. *Then*

(a) $AA^{\dagger} = UU^{\dagger}$ *and* $A^{\dagger}A = U^{\dagger}U$,

(b) $U^\dagger VA^\dagger = A^\dagger VU^\dagger$,
(c) $U^\dagger VA^\dagger V = A^\dagger VU^\dagger V$,
(d) $VU^\dagger VA^\dagger = VA^\dagger VU^\dagger$.

We refer [5, 16, 17, 21] for methods of construction of a proper splitting for a given $A \in \mathbb{R}^{m\times n}$. Different subclasses of a proper splitting are recalled next.

Definition 2.7 A proper splitting $A = U - V$ of $A \in \mathbb{R}^{m\times n}$ is called a

(i) proper regular splitting [13], if $U^\dagger \geq 0$ and $V \geq 0$.
(ii) proper weak regular splitting of type I [13], if $U^\dagger \geq 0$ and $U^\dagger V \geq 0$.
(iii) proper weak regular splitting of type II [12], if $U^\dagger \geq 0$ and $VU^\dagger \geq 0$.

For the above class of proper splitting, we have the following convergence result.

Theorem 2.8 (Theorem 2.4, [17])
Let $A = U - V$ be any of the above class of splittings of $A \in \mathbb{R}^{m\times n}$. Then $A^\dagger \geq 0$ if and only if $\rho(VU^\dagger) = \rho(U^\dagger V) < 1$.

It is well known that matrix splittings having a smaller radius of iteration matrix gives a faster rate of convergence for (1.2). Therefore, we have the following comparison result for proper weak regular splittings of different types.

Theorem 2.9 (Theorem 3.3, [9])
Let $A = U_1 - V_1 = U_2 - V_2$ be two proper weak regular splitting of different types a semimonotone matrix $A \in \mathbb{R}^{m\times n}$. If $U_1^\dagger \geq U_2^\dagger$, then $\rho(U_1^\dagger V_1) \leq \rho(U_2^\dagger V_2) < 1$.

3 Classical Three-Step Alternating Iteration Method

This section deals with the convergence of the three-step alternating iteration method, when the splittings $A = M - N = P - Q = U - V$ are proper weak regular of type II. Before proving our main results, we recall the following results.

Theorem 3.1 (Theorem 15, [19])
Let $A = M - N = P - Q = U - V$ be three proper weak regular splittings of type I of a semimonotone matrix $A \in \mathbb{R}^{m\times n}$. Then $\rho(H) = \rho(U^\dagger VP^\dagger QM^\dagger N) < 1$.

Theorem 3.2 (Theorem 16, [19])
Let $A = M - N = P - Q = U - V$ be three proper weak regular splittings of type I of a semimonotone matrix $A \in \mathbb{R}^{m\times n}$. Then the unique splitting $A = B - C$ induced by H with $B = M(M + U - A + VP^\dagger N)^\dagger U$ is a proper weak regular splitting of type I if $\mathcal{R}(M + U - A + VP^\dagger N) = \mathcal{R}(A)$ and $\mathcal{N}(M + U - A + VP^\dagger N) = \mathcal{N}(A)$.

A few properties of the matrix B mentioned in the above theorem are obtained next.

Lemma 3.3 *Let $A = M - N = P - Q = U - V$ be three proper splittings of $A \in \mathbb{R}^{m\times n}$. If $B = M(M + U - A + VP^{\dagger}N)^{\dagger}U$, $\mathcal{R}(M + U - A + VP^{\dagger}N) = \mathcal{R}(A)$ and $\mathcal{N}(M + U - A + VP^{\dagger}N) = \mathcal{N}(A)$, then $B^{\dagger} = U^{\dagger}(M + U - A + VP^{\dagger}N)M^{\dagger}$, $\mathcal{R}(B) = \mathcal{R}(A)$ and $\mathcal{N}(B) = \mathcal{N}(A)$.*

Proof Let $X = U^{\dagger}(M + U - A + VP^{\dagger}N)M^{\dagger}$. Since $\mathcal{R}(M + U - A + VP^{\dagger}N) = \mathcal{R}(A)$, $\mathcal{N}(M + U - A + VP^{\dagger}N) = \mathcal{N}(A)$ and $A = M - N = P - Q = U - V$ are proper splittings, we have $(M + U - A + VP^{\dagger}N)(M + U - A + VP^{\dagger}N)^{\dagger} = AA^{\dagger} = MM^{\dagger} = PP^{\dagger} = UU^{\dagger}$. Hence,

$$
\begin{aligned}
XB &= U^{\dagger}(M + U - A + VP^{\dagger}N)M^{\dagger}M(M + U - A + VP^{\dagger}N)^{\dagger}U \\
&= U^{\dagger}(M + U - A + VP^{\dagger}N)(M + U - A + VP^{\dagger}N)^{\dagger}U \\
&= U^{\dagger}U \\
&= (U^{\dagger}U)^{\mathrm{T}} \\
&= (XB)^{\mathrm{T}},
\end{aligned}
$$

and

$$
\begin{aligned}
BX &= M(M + U - A + VP^{\dagger}N)^{\dagger}UU^{\dagger}(M + U - A + VP^{\dagger}N)M^{\dagger} \\
&= M(M + U - A + VP^{\dagger}N)^{\dagger}(M + U - A + VP^{\dagger}N)M^{\dagger} \\
&= MM^{\dagger} \\
&= (MM^{\dagger})^{\mathrm{T}} \\
&= (BX)^{\mathrm{T}}.
\end{aligned}
$$

Therefore, $XBX = U^{\dagger}UU^{\dagger}(M + U - A + VP^{\dagger}N)M^{\dagger} = U^{\dagger}(M + U - A + VP^{\dagger}N) M^{\dagger} = X$ and $BXB = MM^{\dagger}M(M + U - A + VP^{\dagger}N)^{\dagger}U = M(M + U - A + VP^{\dagger} N)^{\dagger}U = B$. Hence, $X = B^{\dagger}$.

Next we will show that $\mathcal{R}(B) = \mathcal{R}(A)$ and $\mathcal{N}(B) = \mathcal{N}(A)$. To do this, we will first prove that $\mathcal{N}(U) = \mathcal{N}(B)$. Clearly, $\mathcal{N}(U) \subseteq \mathcal{N}(B)$. Let $x \in \mathcal{N}(B)$, i.e., $Bx = 0$. By pre-multiplying $M^{\dagger}$ to $Bx = 0$ and using the fact that $M^{\dagger}M = P_{\mathcal{R}((M+U-A+VP^{\dagger}N)^{\mathrm{T}})} = P_{\mathcal{R}(A^{\mathrm{T}})}$, we get $M^{\dagger}M(M + U - A + VP^{\dagger}N)^{\dagger}Ux = (M + U - A + VP^{\dagger}N)^{\dagger}Ux = 0$. Again, pre-multiplying $(M + U - A + VP^{\dagger}N)$ and using the fact $(M + U - A + VP^{\dagger}N)(M + U - A + VP^{\dagger}N)^{\dagger} = P_{\mathcal{R}(M+U-A+VP^{\dagger}N)} = P_{\mathcal{R}(A)} = P_{\mathcal{R}(U)}$, we get $(M + U - A + VP^{\dagger}N)(M + U - A + VP^{\dagger}N)^{\dagger}Ux = Ux = 0$, i.e., $x \in \mathcal{N}(U)$. Hence, $\mathcal{N}(B) \subseteq \mathcal{N}(U)$. Next, we will show that $\mathcal{R}(B) = \mathcal{R}(A)$ which is equivalent to prove $\mathcal{N}(B^{\mathrm{T}}) = \mathcal{N}(A^{\mathrm{T}})$. But $B = M(M + U - A + VP^{\dagger}N)^{\dagger}U$ implies $\mathcal{N}(M^{\mathrm{T}}) \subseteq \mathcal{N}(B^{\mathrm{T}})$. So, we need to show the other way, i.e., $\mathcal{N}(B^{\mathrm{T}}) \subseteq \mathcal{N}(M^{\mathrm{T}})$. Let $x \in \mathcal{N}(B^{\mathrm{T}})$, then $(M(M + U - A + VP^{\dagger}N)^{\dagger}U)^{\mathrm{T}}x = 0$. By pre-multiplying $(U^{\dagger})^{\mathrm{T}}$, we get $(UU^{\dagger})^{\mathrm{T}}((M + U - A + VP^{\dagger}N)^{\dagger})^{\mathrm{T}}M^{\mathrm{T}}x = ((M + U - A + VP^{\dagger}N)^{\dagger}UU^{\dagger})^{\mathrm{T}}M^{\mathrm{T}} x = 0$ and using the fact $UU^{\dagger} = P_{\mathcal{R}(A)} = P_{\mathcal{R}(M+U-A+VP^{\dagger}N)} = (M + U - A + VP^{\dagger} N)(M + U - A + VP^{\dagger}N)^{\dagger}$, we get $((M + U - A + VP^{\dagger}N)^{\dagger})^{\mathrm{T}}M^{\mathrm{T}}x = 0$. Again, pre-multiplying $(M + U - A + VP^{\dagger}N)^{\mathrm{T}}$ and using the fact $(M+U - A + VP^{\dagger}N)^{\dagger}(M +$

$U - A + VP^{\dagger}N) = P_{\mathcal{R}((M+U-A+VP^{\dagger}N)^{\mathrm{T}})} = P_{\mathcal{R}(A^{\mathrm{T}})} = M^{\dagger}M$, we obtain $(M + U - A + VP^{\dagger}N)^{\mathrm{T}}((M + U - A + VP^{\dagger}N)^{\dagger})^{\mathrm{T}}M^{\mathrm{T}}x = (M^{\dagger}M)^{\mathrm{T}}M^{\mathrm{T}}x = M^{\mathrm{T}}x = 0$. Hence, $\mathcal{N}(B^{\mathrm{T}}) \subseteq \mathcal{N}(M^{\mathrm{T}}) = \mathcal{N}(A^{\mathrm{T}})$. □

We have the following two expressions for $B^{\dagger}$,

$$\begin{aligned} B^{\dagger} &= U^{\dagger}(M + U - A + VP^{\dagger}N)M^{\dagger} \\ &= U^{\dagger}MM^{\dagger} + U^{\dagger}UM^{\dagger} - U^{\dagger}AM^{\dagger} + U^{\dagger}VP^{\dagger}NM^{\dagger} \\ &= U^{\dagger} + U^{\dagger}VM^{\dagger} + U^{\dagger}VP^{\dagger}NM^{\dagger} \end{aligned}$$

and

$$\begin{aligned} B^{\dagger} &= U^{\dagger}(M + U - A + VP^{\dagger}N)M^{\dagger} \\ &= U^{\dagger}MM^{\dagger} + U^{\dagger}UM^{\dagger} - U^{\dagger}AM^{\dagger} + U^{\dagger}VP^{\dagger}NM^{\dagger} \\ &= M^{\dagger} + U^{\dagger}NM^{\dagger} + U^{\dagger}VP^{\dagger}NM^{\dagger}. \end{aligned}$$

The example given below shows that the alternating scheme is convergent even though $A = M - N = P - Q = U - V$ are not proper weak regular splittings of type I.

Example 3.4 Let $A = \begin{bmatrix} 5 & -3 & 5 \\ -3 & 5 & -3 \end{bmatrix}$. Then, $A^{\dagger} = \begin{bmatrix} 0.1563 & 0.0938 \\ 0.1875 & 0.3125 \\ 0.1563 & 0.0938 \end{bmatrix} \geq 0$. Consider $A = \begin{bmatrix} 15 & -9 & 15 \\ -3 & 10 & -3 \end{bmatrix} - \begin{bmatrix} 10 & -6 & 10 \\ 0 & 5 & 0 \end{bmatrix} = M - N$. Then $\mathcal{R}(M) = \mathcal{R}(A)$, $\mathcal{N}(M) = \mathcal{N}(A)$, $M^{\dagger} = \begin{bmatrix} 0.0407 & 0.0366 \\ 0.0244 & 0.1220 \\ 0.0407 & 0.0366 \end{bmatrix} \geq 0$, $M^{\dagger}N = \begin{bmatrix} 0.4065 & -0.0610 & 0.4065 \\ 0.2439 & 0.4634 & 0.2439 \\ 0.4065 & -0.0610 & 0.4065 \end{bmatrix} \ngeq 0$ and $NM^{\dagger} = \begin{bmatrix} 0.6667 & 0 \\ 0.1220 & 0.6098 \end{bmatrix} \geq 0$. Hence, $A = M - N$ is a proper weak regular splitting of type II but not I.

Further consider $A = \begin{bmatrix} 20 & -12 & 20 \\ -3 & 10 & -3 \end{bmatrix} - \begin{bmatrix} 15 & -9 & 15 \\ 0 & 5 & 0 \end{bmatrix} = P - Q$. Then $\mathcal{R}(P) = \mathcal{R}(A)$, $\mathcal{N}(P) = \mathcal{N}(A)$, $P^{\dagger} = \begin{bmatrix} 0.0305 & 0.0366 \\ 0.0183 & 0.1220 \\ 0.0305 & 0.0366 \end{bmatrix} \geq 0$, $P^{\dagger}Q = \begin{bmatrix} 0.4573 & -0.0915 & 0.4573 \\ 0.2744 & 0.4451 & 0.2744 \\ 0.4573 & -0.0915 & 0.4573 \end{bmatrix} \ngeq 0$ and $QP^{\dagger} = \begin{bmatrix} 0.7500 & 0 \\ 0.0915 & 0.6098 \end{bmatrix} \geq 0$. Hence, $A = P - Q$ is a proper weak regular splitting of type II but not I.

Again consider $A = \begin{bmatrix} 25 & -15 & 25 \\ -3 & 10 & -3 \end{bmatrix} - \begin{bmatrix} 20 & -12 & 20 \\ 0 & 5 & 0 \end{bmatrix} = U - V$. Then $\mathcal{R}(U) = \mathcal{R}(A)$, $\mathcal{N}(U) = \mathcal{N}(A)$, $U^{\dagger} = \begin{bmatrix} 0.0244 & 0.0366 \\ 0.0146 & 0.1220 \\ 0.0244 & 0.0366 \end{bmatrix} \geq 0$,

$U^{\dagger}V = \begin{bmatrix} 0.4878 & -0.1098 & 0.4878 \\ 0.2927 & 0.4341 & 0.2927 \\ 0.4878 & -0.1098 & 0.4878 \end{bmatrix} \ngeq 0$ and $VU^{\dagger} = \begin{bmatrix} 0.8000 & 0 \\ 0.0732 & 0.6098 \end{bmatrix} \geq 0$. Hence, $A = U - V$ is a proper weak regular splitting of type II but not I. But $\rho(H) = \rho(U^{\dagger}VP^{\dagger}QM^{\dagger}N) = 0.4 < 1$, and A is semimonotone.

In the above example, we can see that $H = U^{\dagger}VP^{\dagger}QM^{\dagger}N$
$= \begin{bmatrix} 0.3046 & -0.1147 & 0.3046 \\ 0.3486 & 0.0176 & 0.3486 \\ 0.3046 & -0.1147 & 0.3046 \end{bmatrix} \ngeq 0$ but $S = VU^{\dagger}QP^{\dagger}NM^{\dagger} =$
$\begin{bmatrix} 0.4 & 0 \\ 0.1191 & 0.2267 \end{bmatrix} \geq 0.$

Motivated by the above example, we will now introduce a similar result as Theorem 3.1 by considering the given splittings are proper weak regular of type II. With this objective, we first derived the following properties of H and S which are useful to prove further results.

Theorem 3.5 *Let $A = M - N = P - Q = U - V$ be three proper splittings of $A \in \mathbb{R}^{m\times n}$ and $S = VU^{\dagger}QP^{\dagger}NM^{\dagger}$. Then,*

(i) $AA^{\dagger}S = S = SAA^{\dagger}$ and $A^{\dagger}AH = H = HA^{\dagger}A$, where H is the iteration matrix of the iterative method (1.6).
(ii) $S = AHA^{\dagger}$, $H = A^{\dagger}SA$ and $\rho(S) = \rho(H)$.
(iii) $I - S$ and $I - H$ are invertible if $\mathcal{R}(M + U - A + VP^{\dagger}N) = \mathcal{R}(A)$ and $\mathcal{N}(M + U - A + VP^{\dagger}N) = \mathcal{N}(A)$.

Proof (i) $AA^{\dagger}S = AA^{\dagger}VU^{\dagger}QP^{\dagger}NM^{\dagger} = S$ using the fact that $\mathcal{R}(S) \subseteq \mathcal{R}(V) \subseteq \mathcal{R}(A)$. Again $SAA^{\dagger} = VU^{\dagger}QP^{\dagger}NM^{\dagger}AA^{\dagger} = VU^{\dagger}QP^{\dagger}NM^{\dagger}MM^{\dagger} = S$. Similar argument yields the other equality.

(ii) By Theorem 2.6, we have $U^{\dagger}VA^{\dagger} = A^{\dagger}VU^{\dagger}$, $P^{\dagger}QA^{\dagger} = A^{\dagger}QP^{\dagger}$ and $M^{\dagger}NA^{\dagger} = A^{\dagger}NM^{\dagger}$. So

$$\begin{aligned} S &= AA^{\dagger}S \\ &= AA^{\dagger}VU^{\dagger}QP^{\dagger}NM^{\dagger} \\ &= AU^{\dagger}VA^{\dagger}QP^{\dagger}NM^{\dagger} \\ &= AU^{\dagger}VP^{\dagger}QA^{\dagger}NM^{\dagger} \\ &= AU^{\dagger}VP^{\dagger}QM^{\dagger}NA^{\dagger} \\ &= AHA^{\dagger}. \end{aligned}$$

We then have $A^{\dagger}S = A^{\dagger}AHA^{\dagger} = HA^{\dagger}$. Again, post-multiplying A, we get $H = A^{\dagger}SA$.

Consider $Sx = \lambda x$, where λ is an eigenvalue of S, and x is its corresponding eigenvector. Then $x \in \mathcal{R}(S) \subseteq \mathcal{R}(A)$. Now $\lambda x = Sx = AHA^{\dagger}x$. Pre-multiplying $A^{\dagger}$, we get $\lambda y = Hy$, where $y = A^{\dagger}x$. Therefore, λ is an eigenvalue of H if, $y \neq 0$. Suppose that $y = A^{\dagger}x = 0$. Then $x \in \mathcal{N}(A^{\dagger}) = \mathcal{N}(A^{T})$. So, we have $x \in \mathcal{R}(A) \cap \mathcal{N}(A^{T}) =$

$\{0\}$ a contradiction. Hence, $y \neq 0$ and so $\sigma\{S\} \subseteq \sigma\{H\}$. For the other way, consider $Hy = \mu y$, where μ is an eigenvalue of H, and y is its corresponding eigenvector. Then $y \in \mathcal{R}(H) \subseteq \mathcal{R}(A^{\mathrm{T}})$. Also $\mu y = A^{\dagger}SAy$. Pre-multiplying A, we get $\mu z = Sz$, where $z = Ay$. Suppose that $z = 0$. Then $y \in \mathcal{N}(A)$. So $y \in \mathcal{R}(A^{\mathrm{T}}) \cap \mathcal{N}(A)$ which yields $y = 0$, a contradiction. Hence, $\sigma\{H\} \subseteq \sigma\{S\}$. Therefore, $\rho(S) = \rho(H)$.

(iii) We will prove this by the method of contradiction. Suppose that $I - S$ is not invertible. Then 1 is an eigenvalue of S. Therefore, $x = VU^{\dagger}QP^{\dagger}NM^{\dagger}x \in \mathcal{R}(VU^{\dagger}QP^{\dagger}NM^{\dagger}) \subseteq \mathcal{R}(V) \subseteq \mathcal{R}(U) = \mathcal{R}(P) = \mathcal{R}(M) = \mathcal{R}(A)$. So, $x = UU^{\dagger}x = PP^{\dagger}x = MM^{\dagger}x$ and hence

$$
\begin{aligned}
x &= VU^{\dagger}QP^{\dagger}NM^{\dagger}x \\
&= (U - A)U^{\dagger}(P - A)P^{\dagger}(M - A)M^{\dagger}x \\
&= (UU^{\dagger} - AU^{\dagger})(PP^{\dagger} - AP^{\dagger})(MM^{\dagger} - AM^{\dagger})x \\
&= (AA^{\dagger} - AU^{\dagger})(AA^{\dagger} - AP^{\dagger})(AA^{\dagger} - AM^{\dagger})x \\
&= (AA^{\dagger} - AU^{\dagger})(AA^{\dagger} - AP^{\dagger})(AA^{\dagger}x - AM^{\dagger}x) \\
&= (AA^{\dagger} - AU^{\dagger})(AA^{\dagger} - AP^{\dagger})(x - AM^{\dagger}x) \\
&= (AA^{\dagger} - AU^{\dagger})(AA^{\dagger}x - AA^{\dagger}AM^{\dagger}x - AP^{\dagger}x + AP^{\dagger}AM^{\dagger}x) \\
&= (AA^{\dagger} - AU^{\dagger})(x - AM^{\dagger}x - AP^{\dagger}x + AP^{\dagger}AM^{\dagger}x) \\
&= (x - AM^{\dagger}x - AP^{\dagger}x + AP^{\dagger}AM^{\dagger}x - AU^{\dagger}x + AU^{\dagger}AM^{\dagger}x \\
&\quad + AU^{\dagger}AP^{\dagger}x - AU^{\dagger}AP^{\dagger}AM^{\dagger}x) \\
&= x - A(M^{\dagger} + P^{\dagger} - P^{\dagger}AM^{\dagger} + U^{\dagger} - U^{\dagger}AM^{\dagger} - U^{\dagger}AP^{\dagger} \\
&\quad + U^{\dagger}AP^{\dagger}AM^{\dagger})x \\
&= x - A(M^{\dagger}MM^{\dagger} + P^{\dagger}PP^{\dagger} + U^{\dagger} - P^{\dagger}AM^{\dagger} - U^{\dagger}AM^{\dagger} - U^{\dagger}AP^{\dagger} \\
&\quad + U^{\dagger}AP^{\dagger}AM^{\dagger})x \\
&= x - A(U^{\dagger}UM^{\dagger} + U^{\dagger}UP^{\dagger} + U^{\dagger} - P^{\dagger}AM^{\dagger} - U^{\dagger}AM^{\dagger} - U^{\dagger}AP^{\dagger} \\
&\quad + U^{\dagger}AP^{\dagger}AM^{\dagger})x \\
&= x - A(U^{\dagger}UM^{\dagger} + U^{\dagger}UP^{\dagger}PP^{\dagger} + U^{\dagger}UU^{\dagger} - P^{\dagger}PP^{\dagger}AM^{\dagger} - U^{\dagger}AM^{\dagger} \\
&\quad - U^{\dagger}AP^{\dagger}PP^{\dagger} + U^{\dagger}AP^{\dagger}AM^{\dagger})x \\
&= x - A(U^{\dagger}UM^{\dagger} + U^{\dagger}UP^{\dagger}MM^{\dagger} + U^{\dagger}MM^{\dagger} - U^{\dagger}UP^{\dagger}AM^{\dagger} - U^{\dagger}AM^{\dagger} \\
&\quad - U^{\dagger}AP^{\dagger}M^{\dagger} + U^{\dagger}AP^{\dagger}AM^{\dagger})x \\
&= x - A(U^{\dagger}(U + M - A + UP^{\dagger}M - UP^{\dagger}A - AP^{\dagger}M + AP^{\dagger}A)M^{\dagger})x \\
&= x - A(U^{\dagger}(U + M - A + VP^{\dagger}N)M^{\dagger})x \\
&= x - AB^{\dagger}x.
\end{aligned}
$$

Then $AB^{\dagger}x = 0$. Thus, $B^{\dagger}x \in \mathcal{N}(A) = \mathcal{N}(B)$ and so $BB^{\dagger}x = 0$. But $x \in \mathcal{R}(A) = \mathcal{R}(B)$. (We have used the facts that $\mathcal{N}(B) = \mathcal{N}(A)$ and $\mathcal{R}(B) = \mathcal{R}(A)$ which follows from Lemma 3.3.) Therefore, $x = BB^{\dagger}x = 0$, a contradiction. Thus, $I - S$ is invertible.

The other proof is explained next. Let $(I - H)x = (I - U^{\dagger}VP^{\dagger}QM^{\dagger}N)x = 0$. So, $x \in \mathcal{R}(U^{\dagger}) = \mathcal{R}(U^{\mathrm{T}}) = \mathcal{R}(A^{\mathrm{T}})$. Substituting $V = U - A$, $Q = P - A$ and $N = M - A$ in $(I - H)x = 0$ and then simplifying, we get $B^{\dagger}Ax = 0$. Pre-multiplying B and using the fact $\mathcal{R}(B) = \mathcal{R}(A)$, we get $x \in \mathcal{N}(A)$. Hence, $x = 0$ yielding a contradiction. Thus, $I - H$ is invertible. □

Theorem 3.6 *Let $A = M - N = P - Q = U - V$ be three proper weak regular splittings of type II of a semimonotone matrix $A \in \mathbb{R}^{m\times n}$ and $S = VU^{\dagger}QP^{\dagger}NM^{\dagger}$. Then $\rho(H) = \rho(U^{\dagger}VP^{\dagger}QM^{\dagger}N) < 1$.*

Proof We have

$$\begin{aligned}
0 \le S &= VU^{\dagger}QP^{\dagger}NM^{\dagger}\\
&= (U - A)U^{\dagger}(P - A)P^{\dagger}(M - A)M^{\dagger}\\
&= (UU^{\dagger} - AU^{\dagger})(PP^{\dagger} - AP^{\dagger})(MM^{\dagger} - AM^{\dagger})\\
&= (AA^{\dagger} - AU^{\dagger})(AA^{\dagger} - AP^{\dagger})(AA^{\dagger} - AM^{\dagger})\\
&= AA^{\dagger} - AM^{\dagger} - AP^{\dagger} - AU^{\dagger} + AP^{\dagger}AM^{\dagger} + AU^{\dagger}AM^{\dagger}\\
&\quad + AU^{\dagger}AP^{\dagger} - AU^{\dagger}AP^{\dagger}AM^{\dagger}
\end{aligned}$$

since $A = M - N = P - Q = U - V$ are proper weak regular splittings of type II. Now

$$\begin{aligned}
A^{\dagger}(I - S) &= A^{\dagger} - A^{\dagger}S\\
&= A^{\dagger}AA^{\dagger} - A^{\dagger}S\\
&= A^{\dagger}(AA^{\dagger} - S)\\
&= A^{\dagger}(AA^{\dagger} - AA^{\dagger} + AM^{\dagger} + AP^{\dagger} + AU^{\dagger} - AP^{\dagger}AM^{\dagger} - AU^{\dagger}AM^{\dagger}\\
&\quad - AU^{\dagger}AP^{\dagger} + AU^{\dagger}AP^{\dagger}AM^{\dagger})\\
&= A^{\dagger}AM^{\dagger} + A^{\dagger}AP^{\dagger} + A^{\dagger}AU^{\dagger} - A^{\dagger}AP^{\dagger}AM^{\dagger} - A^{\dagger}AU^{\dagger}AM^{\dagger}\\
&\quad - A^{\dagger}AU^{\dagger}AP^{\dagger} + A^{\dagger}AU^{\dagger}AP^{\dagger}AM^{\dagger}\\
&= M^{\dagger} + P^{\dagger} + U^{\dagger} - P^{\dagger}AM^{\dagger} - U^{\dagger}AM^{\dagger} - U^{\dagger}AP^{\dagger} + U^{\dagger}AP^{\dagger}AM^{\dagger}\\
&= M^{\dagger} + P^{\dagger}(M - A)M^{\dagger} + U^{\dagger}(P - A)P^{\dagger} + U^{\dagger}(AP^{\dagger}A - A)M^{\dagger}\\
&= M^{\dagger} + P^{\dagger}NM^{\dagger} + U^{\dagger}QP^{\dagger} + U^{\dagger}(AP^{\dagger}A - PP^{\dagger}A)M^{\dagger}\\
&= M^{\dagger} + P^{\dagger}NM^{\dagger} + U^{\dagger}QP^{\dagger} - U^{\dagger}(P - A)P^{\dagger}AM^{\dagger}\\
&= M^{\dagger} + P^{\dagger}NM^{\dagger} + U^{\dagger}QP^{\dagger} - U^{\dagger}QP^{\dagger}(M - N)M^{\dagger}\\
&= M^{\dagger} + P^{\dagger}NM^{\dagger} + U^{\dagger}QP^{\dagger}NM^{\dagger} \ge 0.
\end{aligned}$$

Hence,

$$\begin{aligned}
0 &\le A^{\dagger}(I - S)(I + S + S^2 + S^3 + \cdots + S^m)\\
&= A^{\dagger}(I - S^{m+1})\\
&\le A^{\dagger}
\end{aligned}$$

for each $m \in \mathbb{N}$. So, the partial sums of the series $\sum_{m=0}^{\infty} S^m$ is uniformly bounded. Hence, $\rho(S) < 1$. Thus $\rho(H) = \rho(S) < 1$ by Theorem 3.5 (ii). □

The next result showing that the matrix B in the splitting $A = B - C$ induced by H can also be expressed as the product of A and $(I - H)^{-1}$.

Lemma 3.7 *Let $A = M - N = P - Q = U - V$ be three proper splittings of $A \in \mathbb{R}^{m\times n}$ such that $\rho(H) < 1$, $\mathcal{R}(M + P - A + VP^{\dagger}N) = \mathcal{R}(A)$ and $\mathcal{N}(M + P - A + VP^{\dagger}N) = \mathcal{N}(A)$. Then the unique splitting $A = B - C$ induced by H is a proper splitting such that $H = B^{\dagger}C$, where $B = A(I - H)^{-1}$.*

Proof Let $B = A(I - H)^{-1}$ and $C = B - A$. Then

$$\begin{aligned} B^{\dagger}C &= B^{\dagger}(B - A) \\ &= B^{\dagger}B - B^{\dagger}A \\ &= B^{\dagger}B - (I - H)A^{\dagger}A \\ &= A^{\dagger}A - A^{\dagger}A + HA^{\dagger}A \\ &= H \end{aligned}$$

using the fact that $A = B - C$ is a proper splitting which is shown next.

Let $Z = (I - H)A^{\dagger}$. Then $BZ = AA^{\dagger}$ which is symmetric and $ZBZ = (I - H)A^{\dagger}A A^{\dagger} = (I - H)A^{\dagger} = Z$. Again

$$\begin{aligned} ZB &= (I - H)A^{\dagger}A(I - H)^{-1} \\ &= (A^{\dagger}A - HA^{\dagger}A)(I - H)^{-1} \\ &= (A^{\dagger}A - A^{\dagger}AH)(I - H)^{-1} \\ &= A^{\dagger}A \end{aligned}$$

which is symmetric and $BZB = AA^{\dagger}A(I - H)^{-1} = A(I - H)^{-1} = B$. So, we have

$$\begin{aligned} B^{\dagger} &= (I - H)A^{\dagger} \\ &= A^{\dagger} - U^{\dagger}VP^{\dagger}QM^{\dagger}NA^{\dagger} \\ &= A^{\dagger} - (U^{\dagger}(U - A)P^{\dagger}(P - A)M^{\dagger}(M - A)A^{\dagger}) \\ &= A^{\dagger} - (P^{\dagger} - U^{\dagger}AP^{\dagger})(PM^{\dagger} - AM^{\dagger})(MA^{\dagger} - AA^{\dagger}) \\ &= A^{\dagger} - (M^{\dagger} - P^{\dagger}AM^{\dagger} - U^{\dagger}AM^{\dagger} + U^{\dagger}AP^{\dagger}AM^{\dagger})(MA^{\dagger} - AA^{\dagger}) \\ &= M^{\dagger} + P^{\dagger} - P^{\dagger}AM^{\dagger} + U^{\dagger} - U^{\dagger}AM^{\dagger} - U^{\dagger}AP^{\dagger} + U^{\dagger}AP^{\dagger}AM^{\dagger} \\ &= M^{\dagger} + U^{\dagger} - U^{\dagger}AM^{\dagger} + P^{\dagger} - P^{\dagger}(P - Q)M^{\dagger} - U^{\dagger}(U - V)P^{\dagger} \\ &\quad + U^{\dagger}(U - V)P^{\dagger}(M - N)M^{\dagger} \\ &= M^{\dagger} + U^{\dagger} - U^{\dagger}AM^{\dagger} + U^{\dagger}VP^{\dagger}NM^{\dagger}. \end{aligned}$$

Hence, $A = B - C$ is a proper splitting, by Lemma 3.3.

For uniqueness, suppose that there exists another splitting $A = \bar{B} - \bar{C}$ such that $H = \bar{B}^{\dagger}\bar{C}$. Then $\bar{B}H = \bar{B}\bar{B}^{\dagger}\bar{C} = \bar{C} = \bar{B} - A$. So $\bar{B}(I - H) = A$. Hence $\bar{B} = A(I - H)^{-1} = B$. □

Theorem 3.8 *Let $A = M - N = P - Q = U - V$ be three proper weak regular splittings of type II of a semimonotone matrix $A \in \mathbb{R}^{m\times n}$ and $S = VU^{\dagger}QP^{\dagger}NM^{\dagger}$. Then, H and S induce the same proper splitting $A = B - C$ if $\mathcal{R}(M + U - A + VP^{\dagger}N) = \mathcal{R}(A)$ and $\mathcal{N}(M + U - A + VP^{\dagger}N) = \mathcal{N}(A)$. Furthermore, the unique proper splitting $A = X - Y$ induced by matrix S is also a proper weak regular splitting of type II.*

Proof By Lemma 3.7, we have $B = A(I - H)^{-1}$. Let us consider $X = (I - S)^{-1}A$ and $Y = X - A$. We will show that the matrices H and S induce the same proper splitting $A = B - C$. Since $H = HA^{\dagger}A$ and $S = AHA^{\dagger}$, so $S^k = AH^kA^{\dagger}$, for any nonnegative integer k. By Theorem 3.6, we have $\rho(S) < 1$. Also, $S \geq 0$. Therefore, Theorem 2.3 yields

$$\begin{aligned} X &= (I - S)^{-1}A \\ &= \sum_{k=o}^{\infty} S^k A \\ &= \sum_{k=o}^{\infty} AH^kA^{\dagger}A \\ &= \sum_{k=o}^{\infty} AH^k \\ &= A(I - H)^{-1} \\ &= B. \end{aligned}$$

Then $\mathcal{R}(X) = \mathcal{R}(B) = \mathcal{R}(A)$ and $\mathcal{N}(X) = \mathcal{N}(B) = \mathcal{N}(A)$ which in turn yields $A = X - Y$ is a proper splitting.

To prove $A = X - Y$ is a proper weak regular splitting of type II, consider $Z = A^{\dagger}(I - S)$. Then $ZX = A^{\dagger}(I - S)(I - S)^{-1}A = A^{\dagger}A$. Hence, ZX is symmetric and $ZXZ = A^{\dagger}AA^{\dagger}(I - S) = A^{\dagger}(I - S) = Z$. Using the property $AA^{\dagger}S = S = SAA^{\dagger}$, we obtain

$$\begin{aligned} XZ &= (I - S)^{-1}AA^{\dagger}(I - S) \\ &= (I - S)^{-1}(AA^{\dagger} - AA^{\dagger}S) \\ &= (I - S)^{-1}(AA^{\dagger} - SAA^{\dagger}) \\ &= AA^{\dagger}. \end{aligned}$$

So, XZ is symmetric and

$$\begin{aligned} XZX &= AA^{\dagger}(I-S)^{-1}A \\ &= AA^{\dagger}\sum_{k=0}^{\infty} S^k A \\ &= \sum_{k=0}^{\infty} S^k A \\ &= (I-S)^{-1}A \\ &= X. \end{aligned}$$

Hence, $X^{\dagger} = A^{\dagger}(I-S) \geq 0$ (see the proof of Theorem 3.6 for $A^{\dagger}(I-S) \geq 0$). Therefore,

$$\begin{aligned} YX^{\dagger} &= (X-A)X^{\dagger} \\ &= XX^{\dagger} - AX^{\dagger} \\ &= AA^{\dagger} - AA^{\dagger}(I-S) \\ &= S \geq 0. \end{aligned}$$

Thus $A = X - Y$ is a proper weak regular splitting of type II induced by S. Let $A = X_1 - Y_1$ be another splitting induced by S such that $S = Y_1 X_1^{\dagger}$. Then $SX_1 = Y_1 X_1^{\dagger} X_1 = Y_1 = X_1 - A$. So $A = X_1 - SX_1 = (I-S)X_1$ which implies $X_1 = (I-S)^{-1}A = X$. Hence, $A = X - Y$ is the unique proper weak regular splitting of type II induced by S. □

The next results confirm that the proposed alternating iterative scheme converges faster than (1.2) under suitable assumptions.

Theorem 3.9 *Let $A = M - N = P - Q = U - V$ be three proper regular splittings of a semimonotone matrix $A \in \mathbb{R}^{m\times n}$ with $\mathcal{R}(M + U - A + VP^{\dagger}N) = \mathcal{R}(A)$ and $\mathcal{N}(M + U - A + VP^{\dagger}N) = \mathcal{N}(A)$. Then $\rho(H) \leq \min\{\rho(M^{\dagger}N),\ \rho(P^{\dagger}Q),\ \rho(U^{\dagger}V)\} < 1$.*

Proof By Theorem 3.8, $A = B - C$ is a proper weak regular splitting of type II induced by H, and from (1.6), we have

$$B^{\dagger} = U^{\dagger}(VP^{\dagger}QM^{\dagger} + VP^{\dagger} + I) = U^{\dagger}VP^{\dagger}QM^{\dagger} + U^{\dagger}VP^{\dagger} + U^{\dagger} \geq U^{\dagger}.$$

and

$$\begin{aligned} B^{\dagger} &= U^{\dagger}VP^{\dagger}QM^{\dagger} + U^{\dagger}VP^{\dagger} + U^{\dagger} \\ &= U^{\dagger}VP^{\dagger}QM^{\dagger} + U^{\dagger}UP^{\dagger} - U^{\dagger}AP^{\dagger} + U^{\dagger}MM^{\dagger} \\ &= U^{\dagger}VP^{\dagger}QM^{\dagger} + P^{\dagger} + U^{\dagger}(M-A)M^{\dagger} \\ &= U^{\dagger}VP^{\dagger}QM^{\dagger} + P^{\dagger} + U^{\dagger}NM^{\dagger} \geq P^{\dagger}. \end{aligned}$$

Also,

$$\begin{aligned} B^{\dagger} &= U^{\dagger}(M + U - A + VP^{\dagger}N)M^{\dagger} \\ &= U^{\dagger}MM^{\dagger} + U^{\dagger}UM^{\dagger} - U^{\dagger}(M - N)M^{\dagger} + U^{\dagger}VP^{\dagger}NM^{\dagger} \\ &= U^{\dagger} + M^{\dagger} - U^{\dagger}MM^{\dagger} + U^{\dagger}NM^{\dagger} + U^{\dagger}VP^{\dagger}NM^{\dagger} \\ &= U^{\dagger} + M^{\dagger} - U^{\dagger} + U^{\dagger}NM^{\dagger} + U^{\dagger}VP^{\dagger}NM^{\dagger} \\ &= M^{\dagger} + U^{\dagger}NM^{\dagger} + U^{\dagger}VP^{\dagger}NM^{\dagger} \geq M^{\dagger}. \end{aligned}$$

Hence, by applying Theorem 2.9 to the pair of the splittings $A = B - C$ and $A = P - Q$, $A = B - C$ and $A = U - V$, and $A = B - C$ and $A = B - C$, we have $\rho(H) \leq \rho(P^{\dagger}Q) < 1$, $\rho(H) \leq \rho(M^{\dagger}N) < 1$ and $\rho(H) \leq \rho(U^{\dagger}V) < 1$, respectively. Therefore, $\rho(H) \leq \min\{\rho(P^{\dagger}Q),\ \rho(M^{\dagger}N),\ \rho(U^{\dagger}V)\} < 1$. □

Theorem 3.10 *Let $A = P - Q = M - N = U - V$ be three proper weak regular splittings of type II of a semimonotone matrix $A \in \mathbb{R}^{n\times n}$ with $\mathcal{R}(M + U - A + VP^{\dagger}N) = \mathcal{R}(A)$ and $\mathcal{N}(M + U - A + VP^{\dagger}N) = \mathcal{N}(A)$. Let $A = B - C$ be the proper weak regular splitting of type II induced by H (or S). If $PB^{\dagger} \geq I$, $MB^{\dagger} \geq I$ and $UB^{\dagger} \geq I$, then $\rho(H) \leq \min\{\rho(P^{\dagger}Q),\ \rho(M^{\dagger}N),\ \rho(U^{\dagger}V)\} < 1$.*

Proof Consider the pair of splittings $A = B - C$ and $A = P - Q$. By Theorem 2.5, we have

$$B^{\dagger}(I - CB^{\dagger})^{-1} = P^{\dagger}(I - QP^{\dagger})^{-1}. \tag{3.1}$$

Pre-multiplying (3.1) by P, we obtain

$$PB^{\dagger}(I - CB^{\dagger})^{-1} = PP^{\dagger}(I - QP^{\dagger})^{-1} = (I - QP^{\dagger})^{-1}. \tag{3.2}$$

As $CB^{\dagger} \geq 0$, there exists an eigenvector $x \geq 0$ such that $CB^{\dagger}x = \rho(CB^{\dagger})x$ by Theorem 2.1 . Post-multiplying (3.2) by x, we get $PB^{\dagger}(I - CB^{\dagger})^{-1}x = (I - QP^{\dagger})^{-1}x$, i.e., $\frac{PB^{\dagger}x}{1-\rho(B^{\dagger}C)} = (I - QP^{\dagger})^{-1}x$. Using the fact $PB^{\dagger} \geq I$, we have

$$\frac{x}{1 - \rho(CB^{\dagger})} \leq \frac{PB^{\dagger}x}{1 - \rho(CB^{\dagger})} = (I - QP^{\dagger})^{-1}x.$$

Thus, $\rho(B^{\dagger}C) \leq \rho(P^{\dagger}Q)$ by Theorem 2.2 (i). Similarly, for the pair of splittings $A = B - C$ and $A = M - N$, and $A = B - C$ and $A = U - V$, we have $\rho(H) \leq \rho(M^{\dagger}N) < 1$ and $\rho(H) \leq \rho(U^{\dagger}V) < 1$, respectively. Hence, $\rho(H) \leq \min\{\rho(P^{\dagger}Q),\ \rho(M^{\dagger}N),\ \rho(U^{\dagger}V)\} < 1$. □

Theorem 3.11 *Let $A = M - N = P - Q = U - V$ be three proper regular splittings of a semimonotone matrix $A \in \mathbb{R}^{m\times n}$ with $\mathcal{R}(M + U - A + VP^{\dagger}N) = \mathcal{R}(A)$ and $\mathcal{N}(M + U - A + VP^{\dagger}N) = \mathcal{N}(A)$. Then $\rho(H) \leq \min\{\rho(P^{\dagger}QM^{\dagger}N),\ \rho(U^{\dagger}VP^{\dagger}Q),\ \rho(U^{\dagger}VM^{\dagger}N)\} < 1$.*

Proof Suppose $A = B - C$ be the splitting induced by the matrix H. Let $A = B_1 - C_1$, $A = B_2 - C_2$, and $A = B_3 - C_3$ be the splitting induced by the matrices $M^{\dagger}NP^{\dagger}Q$, $M^{\dagger}NU^{\dagger}V$, and $P^{\dagger}QU^{\dagger}V$, respectively. Then

$$\begin{aligned} B^{\dagger} &= M^{\dagger} + U^{\dagger} - U^{\dagger}AM^{\dagger} + U^{\dagger}VP^{\dagger}NM^{\dagger} \\ &\geq M^{\dagger} + U^{\dagger} - U^{\dagger}AM^{\dagger} \\ &= U^{\dagger}(M + U - A)M^{\dagger} = B_2^{\dagger} \end{aligned}$$

and by Theorems 3.8 and 2.9, we have $\rho(B^{\dagger}C) \leq \rho(B_2^{\dagger}C_2)$, i.e., $\rho(H) \leq \rho(U^{\dagger}V M^{\dagger}N)$. Again,

$$\begin{aligned} B^{\dagger} &= M^{\dagger} + P^{\dagger} + U^{\dagger} - P^{\dagger}AM^{\dagger} - U^{\dagger}AP^{\dagger} + U^{\dagger}AP^{\dagger}AM^{\dagger} \\ &= M^{\dagger} + P^{\dagger} - P^{\dagger}AM^{\dagger} + U^{\dagger} - U^{\dagger}AM^{\dagger} - U^{\dagger}AP^{\dagger} + U^{\dagger}AP^{\dagger}AM^{\dagger} \\ &= M^{\dagger} + P^{\dagger} - P^{\dagger}AM^{\dagger} + U^{\dagger} - U^{\dagger}(M - N)M^{\dagger} - U^{\dagger}(P - Q)P^{\dagger} \\ &\quad + U^{\dagger}(P - Q)P^{\dagger}(M - N)M^{\dagger} \\ &= M^{\dagger} + P^{\dagger} - P^{\dagger}AM^{\dagger} + U^{\dagger}NM^{\dagger} - U^{\dagger} + U^{\dagger}QP^{\dagger} \\ &\quad + U^{\dagger}(PP^{\dagger} - QP^{\dagger})(MM^{\dagger} - NM^{\dagger}) \\ &= M^{\dagger} + P^{\dagger} - P^{\dagger}AM^{\dagger} + U^{\dagger}QP^{\dagger}NM^{\dagger} \\ &\geq M^{\dagger} + P^{\dagger} - P^{\dagger}AM^{\dagger} \\ &= P^{\dagger}(M + P - A)M^{\dagger} = B_1^{\dagger} \end{aligned}$$

and by Theorems 3.8 and 2.9, we have $\rho(B^{\dagger}C) \leq \rho(B_1^{\dagger}C_1)$, i.e., $\rho(H) \leq \rho(P^{\dagger}Q M^{\dagger}N)$.

Similarly, we can obtain

$$\begin{aligned} B^{\dagger} &= U^{\dagger} + P^{\dagger} - U^{\dagger}AP^{\dagger} + U^{\dagger}VP^{\dagger}QM^{\dagger} \\ &\geq U^{\dagger} + P^{\dagger} - U^{\dagger}AP^{\dagger} \\ &= B_3^{\dagger} \end{aligned}$$

and $\rho(H) \leq \rho(U^{\dagger}VP^{\dagger}Q)$. Hence $\rho(H) \leq \min\{\rho(P^{\dagger}QM^{\dagger}N), \rho(U^{\dagger}VP^{\dagger}Q), \rho(U^{\dagger}VM^{\dagger}N)\} < 1$. □

The following example demonstrates the above results.

Example 3.12 Consider the matrices in Example 3.4. Then $S = VU^{\dagger}QP^{\dagger}NM^{\dagger}$ $= \begin{bmatrix} 0.4000 & 0 \\ 0.1191 & 0.2267 \end{bmatrix}$ and the splitting induced by S is $A = X - Y$ $= \begin{bmatrix} 8.3333 & -5 & 8.3333 \\ -2.5960 & 5.6957 & -2.5960 \end{bmatrix} - \begin{bmatrix} 3.3333 & -2 & 3.3333 \\ 0.4040 & 0.6957 & 0.4040 \end{bmatrix}$. Here, $\mathcal{R}(A) = \mathcal{R}(X)$,

$\mathcal{N}(A) = \mathcal{N}(X)$, $X^\dagger = \begin{bmatrix} 0.0826 & 0.0725 \\ 0.0753 & 0.2417 \\ 0.0826 & 0.0725 \end{bmatrix} \geq 0$ and $YX^\dagger = \begin{bmatrix} 0.4000 & 0 \\ 0.1191 & 0.2267 \end{bmatrix} \geq 0.$

Hence, the induced splitting $A = X - Y$ is a proper weak regular splitting of type II. Also, $PX^\dagger = \begin{bmatrix} 2.4000 & 0 \\ 0.2573 & 1.9816 \end{bmatrix} \geq I$, $MX^\dagger = \begin{bmatrix} 1.8000 & 0 \\ 0.2573 & 1.9816 \end{bmatrix} \geq I$, $UX^\dagger = \begin{bmatrix} 3 & 0 \\ 0.2573 & 1.9816 \end{bmatrix} \geq I$, respectively. Further $\rho(H) = 0.4 < \min\{\rho(M^\dagger N),\ \rho(P^\dagger Q),\ \rho(U^\dagger V)\} = \min\{0.6666,\ 0.75,\ 0.8\} < 1$ and $\rho(H) = 0.4 < \min\{\rho(P^\dagger Q M^\dagger N),\ \rho(U^\dagger V P^\dagger Q),\ \rho(U^\dagger V M^\dagger N)\} = \min\{0.5,\ 0.6,\ 0.5333\} < 1$.

4 Numerical Computation

In this section, we demonstrate a few numerical examples to validate the proposed theory. The estimation of error bounds, time in seconds, the number of iterations (IT), and the spectral radius of the corresponding iteration matrix are evaluated using MATLAB. We use MATLAB R2015a for the numerical computations in a Windows operating system with configurations: Intel(R) Xenon(R) E-2224 with 16 GB RAM. To terminate the iterative process, we use the stopping criteria $||x_k - x_{k-1}|| < \epsilon$, where $\epsilon = 10^{-7}$.

To show that the three-step iterative scheme converges faster than the two-step and single-step iterative schemes, we consider two different inconsistent linear system and we consider three proper splittings of the coefficient matrices.

Example 4.1 Let us consider a linear system of the form (1.1) with $A = \begin{bmatrix} 1 & 1 & 0 \\ 0 & 0 & 8 \\ 11 & 11 & 0 \\ 0 & 0 & 2 \end{bmatrix}$ and $b = [1, 1, 1, 1]^{\mathrm{T}}$. We have

$A^\dagger = \begin{bmatrix} 0.0041 & 0 & 0.0451 & 0 \\ 0.0041 & 0 & 0.0451 & 0 \\ 0 & 0.1176 & 0 & 0.0294 \end{bmatrix} \geq 0.$ Setting $M = \begin{bmatrix} 1 & 1 & 0 \\ 0 & 0 & 12 \\ 11 & 11 & 0 \\ 0 & 0 & 3 \end{bmatrix}$,

$P = \begin{bmatrix} 1 & 1 & 0 \\ 0 & 0 & 28 \\ 11 & 11 & 0 \\ 0 & 0 & 7 \end{bmatrix}$ and $U = \begin{bmatrix} 1 & 1 & 0 \\ 0 & 0 & 16 \\ 11 & 11 & 0 \\ 0 & 0 & 16 \end{bmatrix}$, we get three proper regular splittings $A = M - N = P - Q = U - V$ of A. The comparison of rate of convergence of single-step, two-step, and three-step is given in Table 1.

Table 1 Comparison table for Example 4.1

Alternate iterations	Splittings	IT	$\|\|A^{\dagger}b - x_n\|\|_2$	Time	ρ
Three-step	$A = M - N = P - Q = U - V$	7	4.9835e−08	0.000261	0.1190
Two-step	$A = M - N = P - Q$	10	8.6099e−08	0.000369	0.2381
Two-step	$A = M - N = U - V$	8	8.7555e−08	0.000349	0.1667
Two-step	$A = P - Q = U - V$	14	8.0775e−08	0.000512	0.3571
Single-step	$A = M - N$	13	9.2239e−08	0.000469	0.3333
Single-step	$A = P - Q$	43	7.6557e−08	0.001516	0.7143
Single-step	$A = U - V$	21	7.0123e−08	0.000753	0.5

Example 4.2 Let us consider another linear system of the form (1.1) with $A = \begin{bmatrix} 1.41 & 0 & 1.41 \\ 0 & 1.41 & 0 \\ 1.41 & 0 & 1.41 \\ 2.82 & 0 & 2.82 \end{bmatrix}$ and $b = [2, 4, 1, 1]^T$. We have

$A^{\dagger} = \begin{bmatrix} 0.0591 & 0 & 0.0591 & 0.1182 \\ 0.0000 & 0.7092 & 0 & 0 \\ 0.0591 & 0 & 0.0591 & 0.1182 \end{bmatrix} \geq 0$. Setting $M = \begin{bmatrix} 2.115 & 0 & 2.115 \\ 0 & 2.115 & 0 \\ 2.115 & 0 & 2.115 \\ 4.23 & 0 & 4.23 \end{bmatrix}$, $P = \begin{bmatrix} 4.512 & 0 & 4.512 \\ 0 & 4.5120 & 0 \\ 4.512 & 0 & 4.512 \\ 9.024 & 0 & 9.024 \end{bmatrix}$ and $U = \begin{bmatrix} 3.525 & 0 & 3.525 \\ 0 & 3.525 & 0 \\ 3.525 & 0 & 3.525 \\ 7.05 & 0 & 7.05 \end{bmatrix}$, we get three proper regular splittings $A = M - N = P - Q = U - V$ of A. The comparison of rate of convergence of single-step, two-step, and three-step is given in Table 2.

From Tables 1 and 2, we observe that the spectral radius of the iteration matrix of three-step alternating iteration schemes is significantly less which ultimately leads to faster converges of these schemes. One may note that for a small size linear system 4×3, the iteration number (IT) and time in seconds is quite less in case of three-step. So, for large inconsistent linear system, one can expect a significantly less iterations and time consumption which eventually leads to a faster convergence and thus computationally more feasible.

Table 2 Comparison table for Example 4.2

Alternate iterations	Splittings	IT	$\|\|A^{\dagger}b - x_n\|\|_2$	Time	ρ
Three-step	$A = M - N = P - Q = U - V$	9	5.0376e−08	0.000332	0.1375
Two-step	$A = M - N = P - Q$	12	6.0162e−08	0.000434	0.2292
Two-step	$A = M - N = U - V$	11	5.8726e−08	0.000408	0.2
Two-step	$A = P - Q = U - V$	14	5.8342e−08	0.000730	0.4125
Single-step	$A = M - N$	16	8.2157e−08	0.001231	0.6
Single-step	$A = P - Q$	46	9.3766e−08	0.001643	0.6875
Single-step	$A = U - V$	34	8.2157e−08	0.001231	0.6

5 Conclusion

In this paper, we have settled the problem of finding the least-squares solution of minimum norm of a given inconsistent linear system using iterative method. In this direction, we have derived some suitable sufficient conditions for the convergence of the three-step alternating iterative scheme in case of proper weak regular splittings of type II (Theorem 3.6) and then we have shown that the three-step alternating iteration scheme converges faster than the usual iteration scheme and the two-step alternating iteration scheme (Theorems 3.9, 3.10 and 3.11). Finally, we have validated our theoretical results by performing numerical computations in Sect. 4. Our results further expand the existing theory and give a faster numerical solution.

References

1. Baliarsingh AK, Mishra D (2017) Comparison results for proper nonnegative splittings of matrices. Results Math 71:93–109
2. Benzi M, Szyld DB (1997) Existence and uniqueness of splittings for stationary iterative methods with applications to alternating methods. Numer Math 76:309–321
3. Ben-Israel A, Greville TNE (2003) Generalized inverses. Theory and applications. Springer, New York
4. Berman A, Plemmons RJ (1974) Cones and iterative methods for best square least squares solutions of linear systems. SIAM J Numer Anal 11:145–154
5. Berman A, Neumann M (1976) Proper splittings of rectangular matrices. SIAM J Appl Math 31:307–312
6. Berman A, Plemmons RJ (1994) Nonnegative matrices in the mathematical sciences. SIAM, Philadelphia

7. Climent J-J, Devesa A, Perea C (2000) Convergence results for proper splittings. In: Recent advances in applied and theoretical mathematics. World Scientific and Engineering Society Press, Singapore, pp 39–44
8. Climent J-J, Perea C (2003) Iterative methods for least square problems based on proper splittings. J Comput Appl Math 158:43–48
9. Giri CK, Mishra D (2018) More on convergence theory of proper multisplittings. Khayyam J Math 4:144–154
10. Giri CK, Mishra D (2017) Comparison results for proper multisplittings of rectangular matrices. Adv Oper Theory 2:334–352
11. Giri CK, Mishra D (2017) Some comparison theorems for proper weak splittings of type II. J Anal 25:267–279
12. Giri CK, Mishra D (2017) Additional results on convergence of alternating iterations involving rectangular matrices. Numer Funct Anal Optim 38:160–180
13. Jena L, Mishra D, Pani S (2014) Convergence and comparison theorems for single and double decomposition of rectangular matrices. Calcolo 51:141–149
14. Mishra D (2018) Proper weak regular splitting and its applications to convergence of alternating methods. Filomat 32:6563–6573
15. Mishra D, Sivakumar KC (2012) Comparison theorems for subclass of proper splittings of matrices. Appl Math Lett 25:2339–2343
16. Mishra D, Sivakumar KC (2012) On splitting of matrices and nonnegative generalized inverses. Oper Matrices 6:85–95
17. Mishra N, Mishra D (2018) Two-stage iterations based on composite splittings for rectangular linear systems. Comput Math Appl 75:2746–2756
18. Mishra D (2014) Nonnegative splittings for rectangular matrices. Comput Math Appl 67:136–144
19. Nandi AK, Sahoo JK, Ghosh P (2019) Three-step alternating and preconditioned scheme for rectangular matrices. J Appl Math Comput 60:485–515
20. Penrose R (1955) A generalized inverse for matrices. Proc Cambridge Philos Soc 51:406–413
21. Shekhar V, Giri CK, Mishra D (2020) Convergence theory of iterative methods based on proper splittings and proper multisplittings for rectangular linear systems. Filomat 34:1835–1851
22. Shekhar V, Mishra D, More on convergence theory of three-step alternating iteration scheme, preprint
23. Varga RS (2000) Matrix iterative analysis. Springer, Berlin

Thermal Hydraulic Performance of Helical Baffle Shell and Tube Heat Exchanger Using RSM Method

Ravi Gugulothu, Narsimhulu Sanke, Sahith Nagadesi, and Ratna Kumari Jilugu

Abstract Owing to the world-wide industrial utilization of shell and tube heat exchangers (STHEs), minimization of design parameters is prime target to design engineers as well as utilize. Traditionally helical and segmental baffle STHEs are designed by pioneer works, based on pioneer's research work 40° helix angle helical baffle STHE is chosen to optimize the design parameters. The response surface methodology (RSM) has been applied successfully to model and analyze the process and also to determine the optimum operating conditions. The optimization command in design expert software was used to get the best combination among design variables. All the solutions of the examined model had significantly influenced the parameters which indicate that the designed statistical model is fitted well with the efficiency of the model.

Keywords Heat exchanger · Helix angle · Reynolds number · Nusselt number and RSM

1 Introduction

Nowadays, the most commonly used equipment in industrial application is shell and tube heat exchanger, because of certain advantages like ease of manufacturing,

R. Gugulothu (✉) · N. Sanke
Department of Mechanical Engineering, University College of Engineering, Osmania University, Hyderabad, India
e-mail: ravi.gugulothu@gmail.com

N. Sanke
e-mail: nsanke@osmania.ac.in

S. Nagadesi
Department of Mechanical Engineering, JNTUH College of Engineering Hyderabad, Hyderabad, Telangana Stats, India

R. K. Jilugu
Quality Control, 02-Electrical Machines, BHEL Ramachandrapuram, Hyderabad, Telangana State, India

S. S. Ray et al. (eds.), *Applied Analysis, Computation and Mathematical Modelling in Engineering*, Lecture Notes in Electrical Engineering 897,
https://doi.org/10.1007/978-981-19-1824-7_11

maintenance and ease of adaptability at different operating conditions as temperatures and pressures [1]. Ravi et al. [2, 3] conducted a review on STHEs using baffles like segmental, helical, ring shaped, circular and tri-sectional baffles. It was noticed that there is a scope to design a STHE having helical baffles of 40° helix angle to enhance the performance of STHE. Wang et al. [4] studied the performance of the STHEs using segmental baffle and circumferential overlap trisection helical baffles of helix angles 12°, 16°, 20°, 24° and 28°. They concluded that the smaller inclined helix angle gives better performance when compared with helical baffles and segmental baffles with different helix angles. Halle et al. [5] experimentally studied pressure drop on the shell side for different segmental baffled configurations like square, triangular, rotated square and rotated triangular with 1.25 tube pitch to diameter ratio. They found that the higher pressure drop in the 10 inch nozzle was compared with 14 inch nozzles, more pressure drop in 8 cross pass than the 6 cross pass and 90° tube pattern layout is lesser than the 30° tube pattern layout. Gugulothu et al. [6] conducted numerical studies on STHEs using segmental baffle and evaluated the impact of local hydrodynamic parameters to improve the efficiency of it. Stehlik et al. [7] have been studied pressure drop and heat transfer correction factors for segmental baffle, helical baffle based on Bell Delaware method. They noticed that the pressure drop decreases a factor ranging from 0.26 to 0.6 as the angle of inclination increases 5°–25°. They concluded that helical baffle tubular heat exchanger has lesser heat transfer and lower pumping costs when comparing with segmental baffle heat exchanger. Reddy et al. [8] numerically studied helical coil tube heat exchanger by varying mass flow rate and concluded that by enhancing heat transfer coefficient by 10%. Gugulothu et al. [9] reviewed the enhancement of heat transfer using passive techniques to save the energy to improve the popularity.

2 Experimental Methods

The STHE has been designed by Ravi et al. 2018 19 and considered the 102.10 mm as outer shell diameter, 4 number of tubes, rotating (45°) square as tube layouts and 3 mm as baffle thickness. Figure 1 shows the STHE of helical baffle, which passes hot fluid in tube side and cold fluid in the shell side at different mass flow rate. In this research paper, tube outer diameter, tube length, tube side fluid temperature and

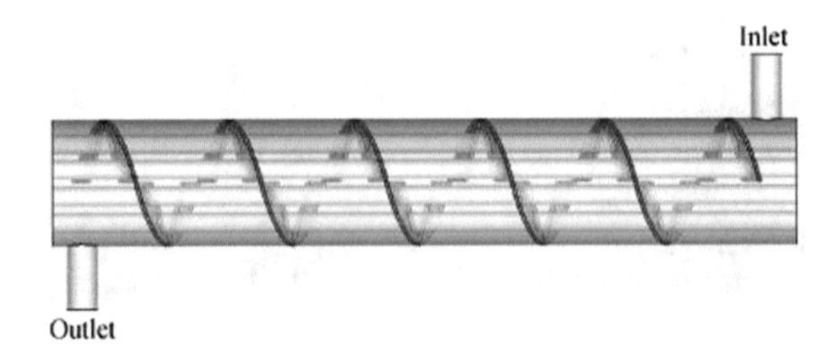

Fig. 1 STHE with a helical baffle [1]

helix angles are considering as varying parameters. In this research work, cold fluid is flowing in shell side at room temperature, and hot fluid is flowing in tube side at different temperatures (42, 57 and 72 °C).

$$\text{Shell - side Nusselt number}: \mathrm{Nu}_s = \frac{h_s \times D_e}{k_{\text{fluid}}} \tag{1}$$

$$\text{Shell side heat transfer coefficient}\, h_s = 0.023\left(\frac{k_{\text{fluid}}}{D_e}\right)\mathrm{Re}_s^{0.8}\mathrm{Pr}_s^{0.3} \tag{2}$$

$$\text{Effective diameter in mm}\, D_e = \frac{4\left(P_t^2 - \left(0.25\pi - d_0^2\right)\right)}{\pi d_0} \tag{3}$$

$$\text{Reynolds number}\, \mathrm{Re}_s = \frac{V_s \times D_e}{v} \tag{4}$$

$$\text{Prandtl number}\, \mathrm{Pr}_s = \frac{\mu C_p}{k_{\text{fluid}}} \tag{5}$$

$$\text{Shell side pressure drop}\, \Delta P_s = \frac{\rho L f_s V_s^2}{2B} \tag{6}$$

$$\text{Friction factor}\, f_s = \frac{0.184}{\mathrm{Re}_s^2} \tag{7}$$

$$\text{Helix baffle spacing in mm}\, B = \pi D_i \tan(\beta) \tag{8}$$

$$\text{Tube side Nusselt number}\, \mathrm{Nu}_t = \frac{h_t \times d_i}{k_{\text{fluid}}} \tag{9}$$

$$\text{Tube side heat transfer coefficient}\, h_t = 0.023\left(\frac{k_{\text{fluid}}}{d_i}\right)\mathrm{Re}_t^{0.8}\mathrm{Pr}_t^{0.4} \tag{10}$$

$$\text{Reynolds number}\, \mathrm{Re}_t = \frac{V_t \times d_i}{v} \tag{11}$$

$$\text{Prandtl number}\, \mathrm{Pr}_t = \frac{\mu C_p}{k_{\text{fluid}}} \tag{12}$$

$$\text{Shell side pressure drop}\, \Delta P_t = \frac{\rho L f_t V_t^2}{2d_i} \tag{13}$$

$$\text{Friction factor}\, f_t = \frac{0.184}{\mathrm{Re}_t^2} \tag{14}$$

$$\text{Tube pitch in mm}\, P_t = 1.25 d_0 \tag{15}$$

$$\text{Number of tubes } N_t = 4 \tag{16}$$

3 Response Surface Methodology (RSM)

A tool named response surface methodology (RSM) from design expert software is used in the mathematical modeling, statistical analysis and optimization of model. It determines correlation between the independent input variables and the dependent output variables using accumulated data. The output responses are accumulated with experiments based on the input variable combinations provided by RSM and are analyzed for selecting the best model providing the relationship between the input and output factors. For each of the factors, three levels were considered for the study which are based on Box–Behnken design in RSM at low (−1), medium (0) and high (+1) to get an overall response at one center point by considering the three levels. The higher R-square value close to 1 is an indication that the obtained responses have high relevance to the experimental values, whereas the value closer to zero is an indication that the obtained responses were not having much relevance to the experimental study. Tube outer diameter is $22.2 < d_0 < 31.8$ mm, tube length is $1000 < L < 2000$ mm, tube side temperature is $42 < T_t < 77$ °C, helix angle in degree $15 < \beta < 45$.

4 Results and Discussion

Tables 4, 5, 6 and 7 show the results of ANOVA test, which were used to investigate the statistical significance of quadratic model with variables. Statistical significance of the proposed model and interactions between the four variables was assessed by the probability value, i.e., *P*-value. *P*-valves are lesser than 0.0001 which indicate the regression models developed are significant (Tables 1, 2 and 3).

The **Predicted R^2** is 0.9832 which has reasonably agreed with the **Adjusted R^2** which is 0.9920; i.e., the difference between the values is less than 0.2.

$$\begin{aligned} \text{Nu} = {} & 20684 + 78.88C + 86.55D + 21.99CD \\ & + 3.17\left(A^2 + B^2\right) + 30.18C^2 + 80.82D^2 \end{aligned} \tag{17}$$

The **Predicted R^2** value which is 0.9990 has reasonably agreed with the **Adjusted R^2** value which is 0.9995; i.e., the difference is lesser than 0.2.

The **Predicted R^2** value which is 0.9740 has reasonably agreed with the **Adjusted R^2** value which is 0.9875; i.e., the difference is lesser than 0.2.

The **Predicted R^2** value which is 0.9815 has reasonably agreed with the **Adjusted R^2** value which is 0.9922; i.e., the difference is lesser than 0.2.

Table 1 Thermal properties of fluids

Shell side fluid properties	
Kinematic viscosity (νs)	8.2103E−07
Dynamic viscosity (μs)	0.0012
Specific heat (Cps)	4178
Thermal conductivity shell (ks)	0.15
Thermal conductivity (kshellfluid)	0.6138
Density of shell side fluid (ρs)	997.35
Tube side fluid properties	
Kinematic viscosity (νt)	7.4774E−07
Dynamic viscosity (μt)	0.0012
Specific heat (Cpt)	4178
Thermal conductivity tube (kt)	0.15
Thermal conductivity (ktubefluid)	0.6201
Density of tube side fluid (ρt)	996.3

Table 2 Design parameters

Parameters	−1	0	1
Length (L_s) in mm	1000	1500	2000
Helix angle (β)	15	30	45
Tube outer diameter (d_0) in mm	22.2	25.4	31.8
Tube side temperature in K	300	350	400

The normal probability plots considering residuals of Nusselt number, friction factor, Reynolds number and the performance evaluation are shown in Fig. 2. It shows the residuals falling near a straight line suggesting that errors are normally distributed under normal distribution and are minimal in value.

Figure 3 indicates that the perceived points on the above plot demonstrate that actual values are distributed fairly closer to the straight line and exhibit a significant correlation. The normal plot of residuals suggests that the percentage probability is fitted well with the externally studentized residuals. The responses correspond to the straight lines suggesting an indication for design space.

Figure 4 demonstrates the analyzed data suggesting the effect of parameters individually and in combinations for the adopted model.

Figure 5 indicates the plots for predicted values versus actual values, which shows the fitness of the model with the values being scattered along the straight line. In the above figure, all the points are closed to fitted line.

Figure 6 shows the plots of observed values on the predicted value, which is measured with DFFITS, whose limits are between +2.32379 and −2.32379. Depending on the above criteria and by analyzing the diagnostic case statistics of the data, our model is well fitted to optimize all the independent variables and adequate.

Table 3 Factor and its responses

S. No	L_s	β	d_0	T	Nu	f	Re	PEC
1.	−1	−1	1	1	515.196	0.012	777,559.011	0.291
2.	−1	−1	−1	−1	177.226	0.016	204,845.436	0.325
3.	−1	1	−1	1	308.568	0.014	409,690.873	0.208
4.	−1	1	1	−1	295.903	0.014	388,779.505	0.453
5.	−1	1	−1	−1	177.226	0.016	204,845.436	0.325
6.	−1	1	1	1	515.196	0.012	777,559.011	0.291
7.	−1	−1	1	−1	295.903	0.014	388,779.505	0.453
8.	−1	0	0	0	209.555	0.015	252,574.269	0.251
9.	−1	−1	−1	1	308.568	0.014	409,690.873	0.208
10.	0	−1	0	0	209.555	0.015	252,574.269	0.357
11.	0	0	0	0	209.555	0.015	252,574.269	0.357
12.	0	0	0	1	364.856	0.013	505,148.537	0.325
13.	0	1	0	0	209.555	0.015	252,574.269	0.357
14.	0	0	0	−1	209.555	0.015	252,574.269	0.507
15.	0	0	1	0	295.903	0.014	388,779.505	0.453
16.	0	0	−1	0	177.226	0.016	204,845.436	0.325
17.	1	1	−1	−1	177.226	0.016	204,845.436	0.592
18.	1	1	1	1	515.196	0.012	777,559.011	0.530
19.	1	−1	1	−1	295.903	0.014	388,779.505	0.826
20.	1	−1	−1	1	308.568	0.014	409,690.873	0.380
21.	1	1	1	−1	295.903	0.014	388,779.505	0.826
22.	1	−1	1	1	515.196	0.012	777,559.011	0.530
23.	1	1	−1	1	308.568	0.014	409,690.873	0.380
24.	1	0	0	0	209.555	0.015	252,574.269	0.458
25.	1	−1	−1	−1	177.226	0.016	204,845.436	0.592

5 Conclusion

In this work, thermal hydraulic performance of a counter flow shell and tube heat exchanger with helical baffle was numerically carried out using RSM method. The second-order response surface model is utilized to establish the relation between design parameters and response variables. The tube outer diameter, tube length, tube side fluid temperature and helix angle are considering as the effecting parameters of the geometry to study the thermal hydraulic performance, Nusselt number, friction factor, Reynolds number and performance evaluation are the three objective functions. We can conclude that the RSM is employed successfully for obtaining the optimized combination of the process parameters/variables to design STHE for process industries.

Table 4 ANOVA test for Nusselt number

Source	Sum of squares	DF	Mean square	*F*-value	*p*-value
Model	0.00003113	14	22,238.51	212.48	<0.0001
A-A	0.0000	1	0.0000	0.0000	1.0000
B-B	0.0000	1	0.0000	0.0000	1.0000
C-C	0.00001120	1	0.00001120	1070.16	<0.0001
D-D	0.00001348	1	0.00001348	1288.20	<0.0001
AB	0.0000	1	0.0000	0.0000	1.0000
AC	0.0000	1	0.0000	0.0000	1.0000
AD	0.0000	1	0.0000	0.0000	1.0000
BC	0.0000	1	0.0000	0.0000	1.0000
BD	0.0000	1	0.0000	0.0000	1.0000
CD	7735.46	1	7735.46	73.91	<0.0001
A^2	25.54	1	25.54	0.2440	0.6320
B^2	25.54	1	25.54	0.2440	0.6320
C^2	2319.07	1	2319.07	22.16	0.0008
D^2	16,633.94	1	16,633.94	158.93	<0.0001
Residual	1046.63	10	104.66		
Cor total	0.00003124	24			

Fit—statistics

Std. dev	10.23	R^2	0.9966
Mean	291.32	Adj. R^2	0.9920
C.V. %	3.51	Pred. R^2	0.9832
		Adeq Prec.	44.6024

Table 5 ANOVA test for friction factor

Source	Sum of squares	DF	Mean square	F-value	p-value
Model	0.0000	14	0.000002769	3657.26	<0.0001
A-A	0.0000	1	0.0000	0.0000	1.0000
B-B	0.0000	1	0.0000	0.0000	1.0000
C-C	0.0000	1	0.0000	19,407.42	<0.0001
D-D	0.0000	1	0.0000	22,456.29	<0.0001
AB	0.0000	1	0.0000	0.0000	1.0000
AC	0.0000	1	0.0000	0.0000	1.0000
AD	0.0000	1	0.0000	0.0000	1.0000
BC	0.0000	1	0.0000	0.0000	1.0000
BD	0.0000	1	0.0000	0.0000	1.0000
CD	0.00000006161	1	0.00000006161	81.36	<0.0001
A^2	0.00000000009917	1	0.00000000009917	0.1310	0.7250
B^2	0.00000000009917	1	0.00000000009917	0.1310	0.7250
C^2	0.0000002269	1	0.0000002269	299.65	<0.0001
D^2	0.000002462	1	0.000002462	3251.52	<0.0001
Residual	0.000000007572	10	0.000000007572		
Cor total	0.0000	24			

Fit—statistics

Std. dev	0.0000	R^2	0.9998
Mean	0.0144	Adj. R^2	0.9995
C.V. %	0.1915	Pred. R^2	0.9990
		Adeq Prec.	175.9801

Table 6 ANOVA test for Reynolds number

Source	Sum of squares	DF	Mean square	F-value	p-value
Model	8.957	14	6.398	136.90	<0.0001
A-A	0.0000	1	0.0000	0.0000	1.0000
B-B	0.0001	1	0.0001	2.612E−13	1.0000
C-C	3.176	1	3.176	679.73	<0.0001
D-D	3.834	1	3.834	820.48	<0.0001
AB	0.0001	1	0.0001	2.612E−13	1.0000
AC	0.0001	1	0.0001	2.612E−13	1.0000
AD	0.0001	1	0.0001	2.612E−13	1.0000
BC	0.0000	1	0.0000	0.0000	1.0000
BD	0.0001	1	0.0001	2.612E−13	1.0000
CD	3.383	1	3.383	72.40	<0.0001
A^2	1.247	1	1.247	0.2669	0.6167
B^2	1.247	1	1.247	0.2669	0.6167
C^2	6.686	1	6.686	14.31	0.0036
D^2	4.524	1	4.524	96.82	<0.0001
Residual	4.673	10	4.673		
Cor total	9.003	24			

Fit—statistics

Std. dev	21,617.34	R^2	0.9948
Mean	0.00003895	Adj. R^2	0.9875
C.V. %	5.55	Pred. R^2	0.9740
		Adeq Prec.	36.1246

Table 7 ANOVA test for performance evaluation

Source	Sum of squares	DF	Mean square	*F*-value	*p*-value
Model	0.6265	14	0.0448	218.15	<0.0001
A-A	0.2962	1	0.2962	1443.91	<0.0001
B-B	0.0000	1	0.0000	0.0000	1.0000
C-C	0.0969	1	0.0969	472.27	<0.0001
D-D	0.1712	1	0.1712	834.68	<0.0001
AB	0.0000	1	0.0000	0.0000	1.0000
AC	0.0076	1	0.0076	36.82	0.0001
AD	0.0132	1	0.0132	64.20	<0.0001
BC	0.0000	1	0.0000	0.0000	1.0000
BD	0.0000	1	0.0000	0.0000	1.0000
CD	0.0042	1	0.0042	20.64	0.0011
A^2	0.000001827	1	0.000001827	0.0089	0.9267
B^2	0.000005700	1	0.000005700	0.0278	0.8709
C^2	0.0029	1	0.0029	14.13	0.0037
D^2	0.0094	1	0.0094	45.94	<0.0001
Residual	0.0021	10	0.0002		
Cor total	0.6286	24			

Fit—statistics

Std. dev	0.0143	R^2	0.9967
Mean	0.4239	Adj. R^2	0.9922
C.V. %	3.38	Pred. R^2	0.9815
		Adeq Prec.	53.9333

Fig. 2 Normal probability plot residuals

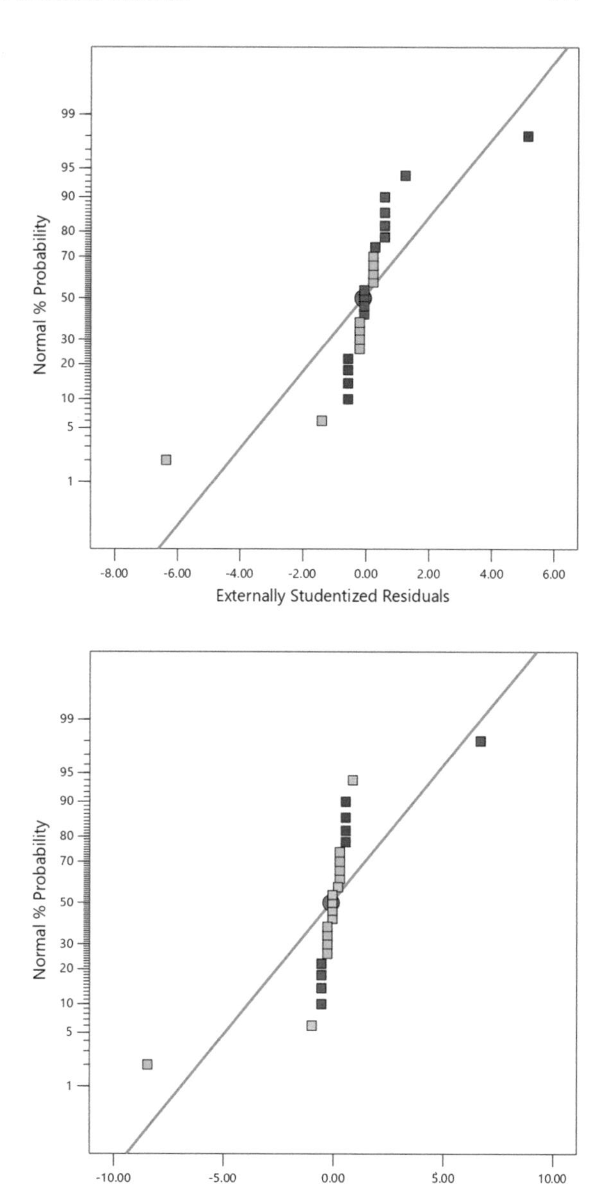

Fig. 2 (continued)

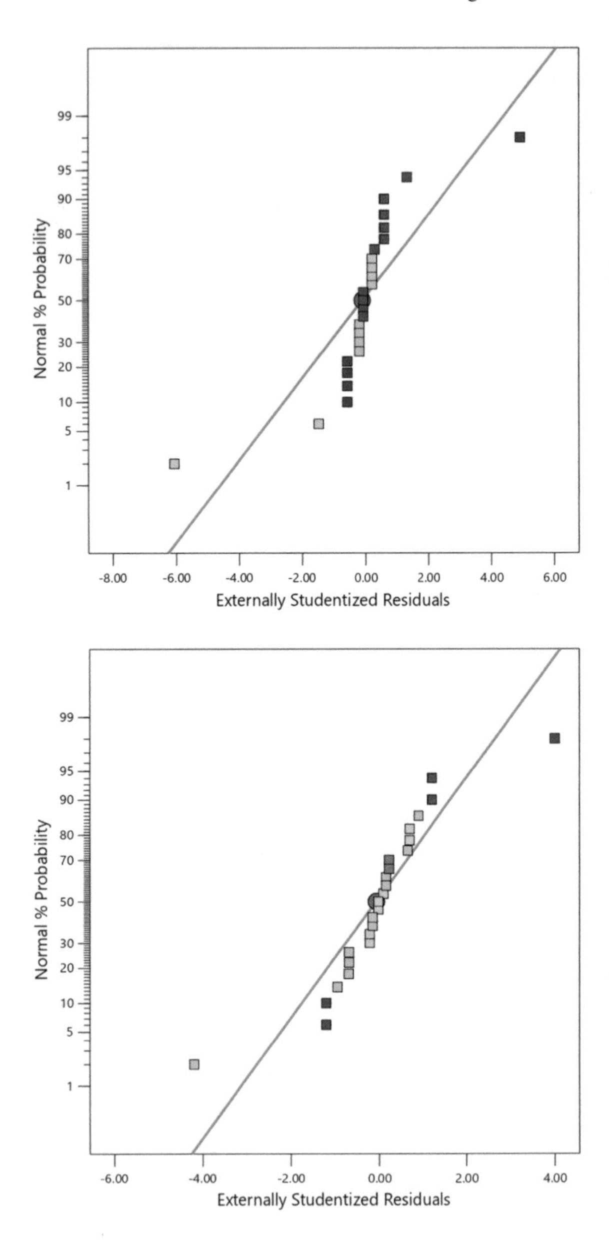

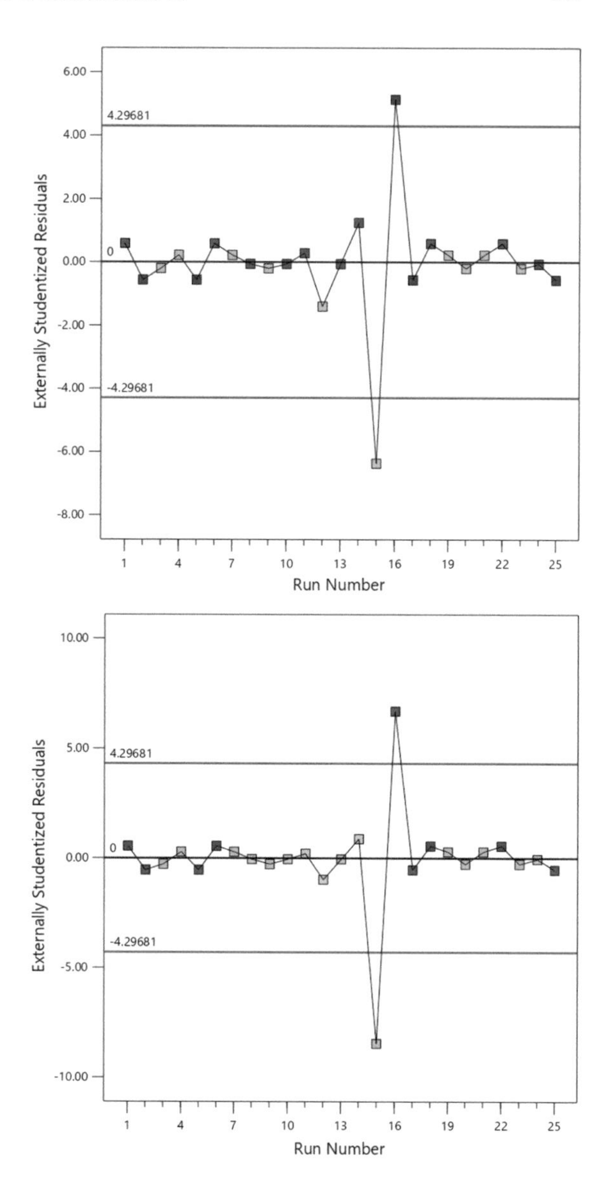

Fig. 3 Second-order RSM values

Fig. 3 (continued)

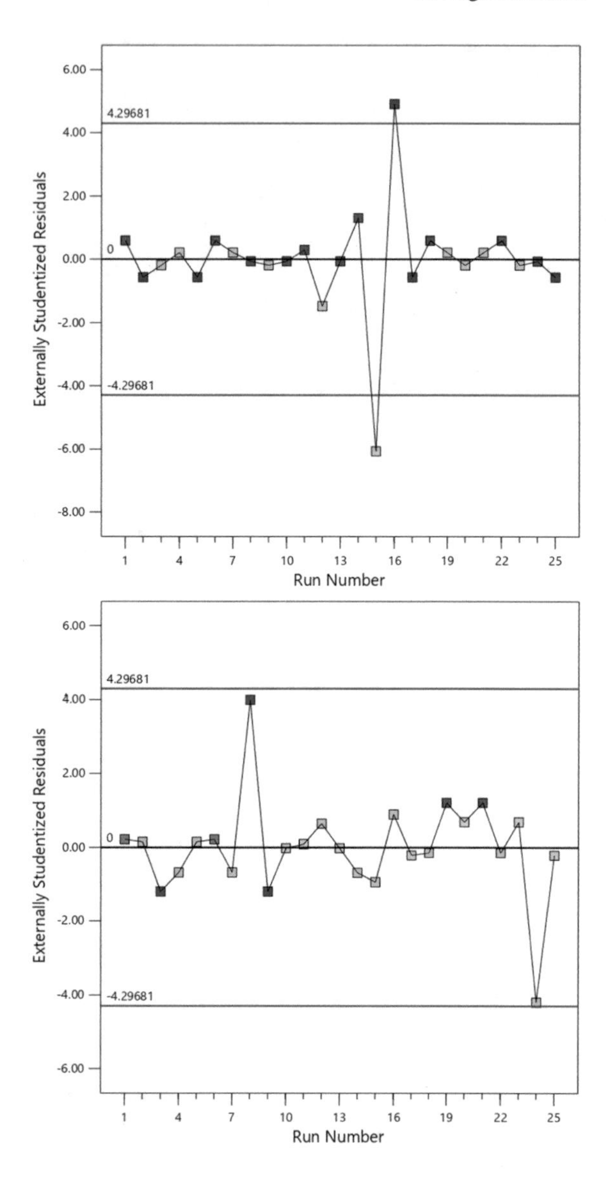

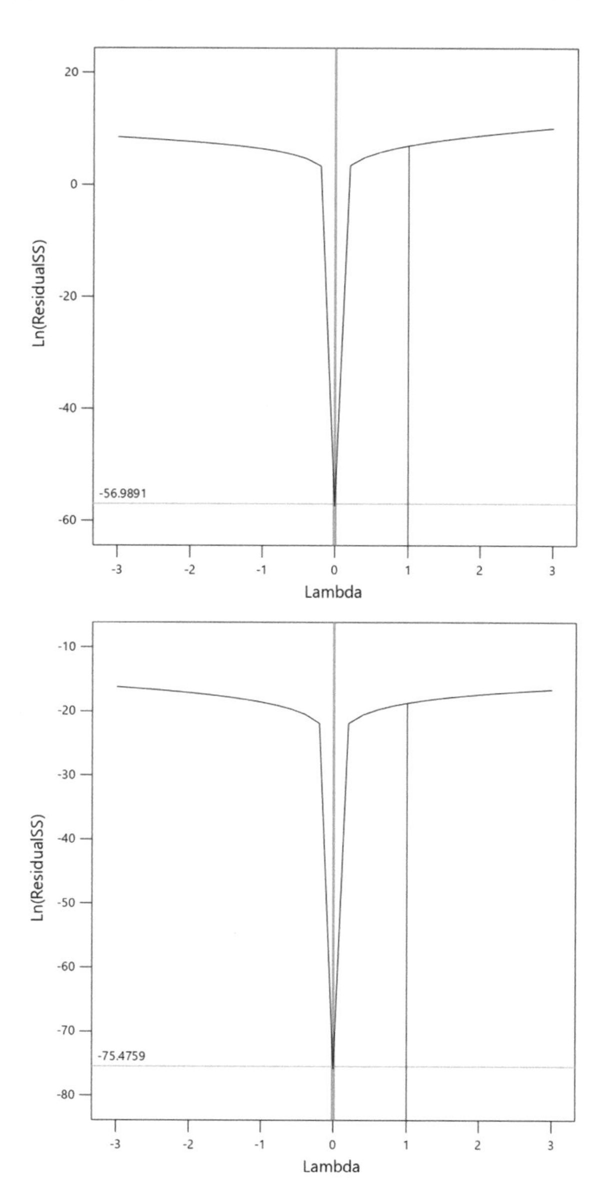

Fig. 4 Box–Cox plot for power transforms

Fig. 4 (continued)

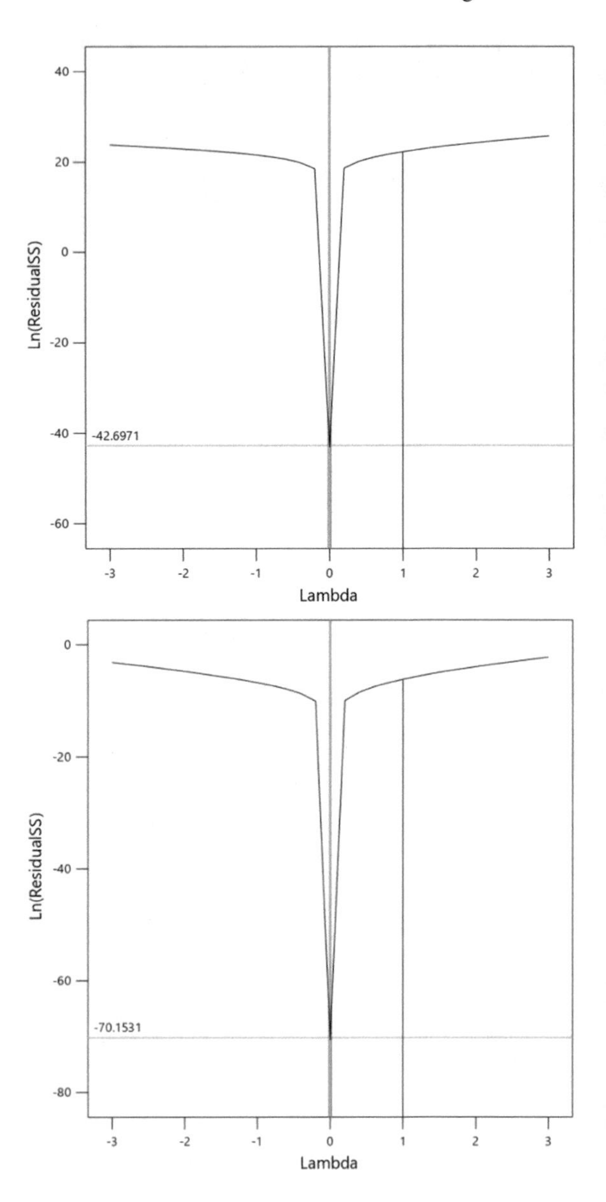

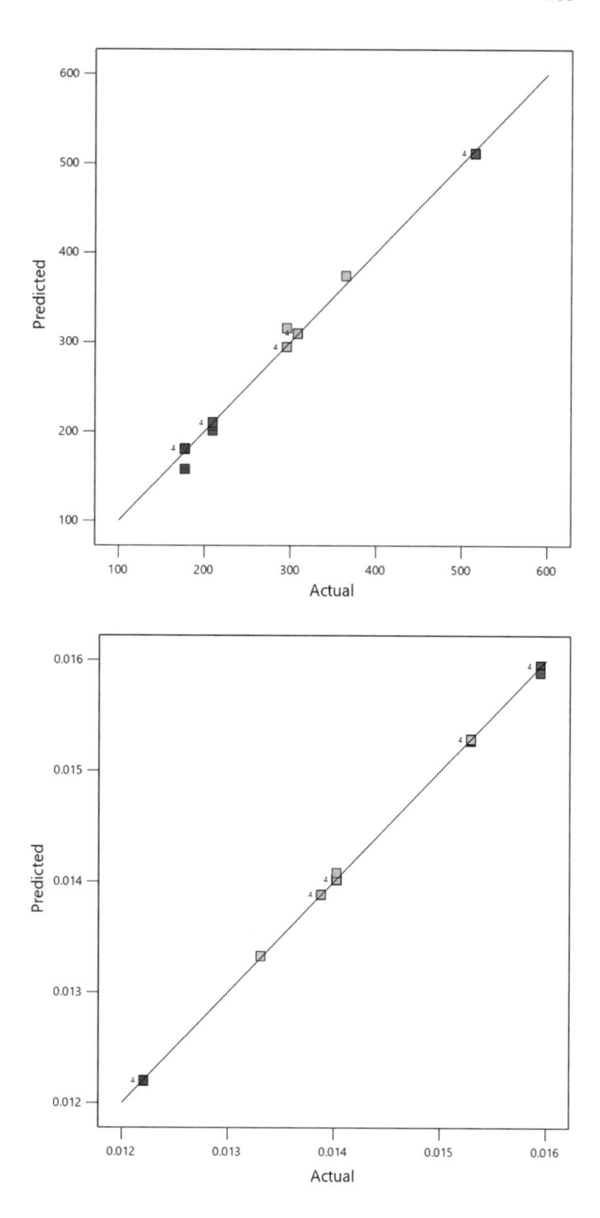

Fig. 5 Predicted versus actual

Fig. 5 (continued)

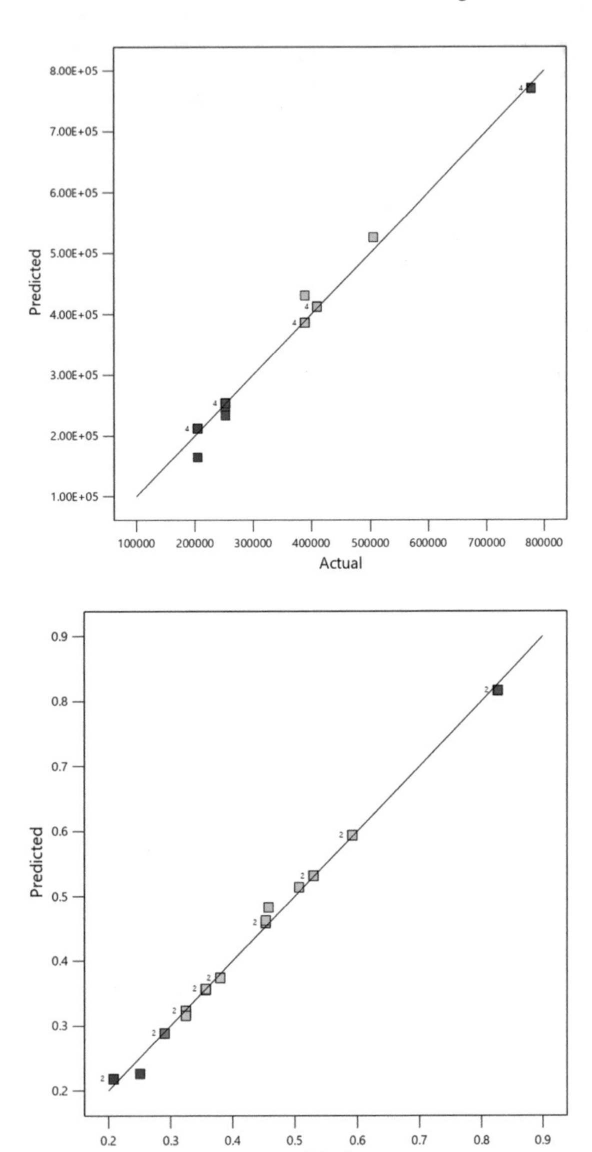

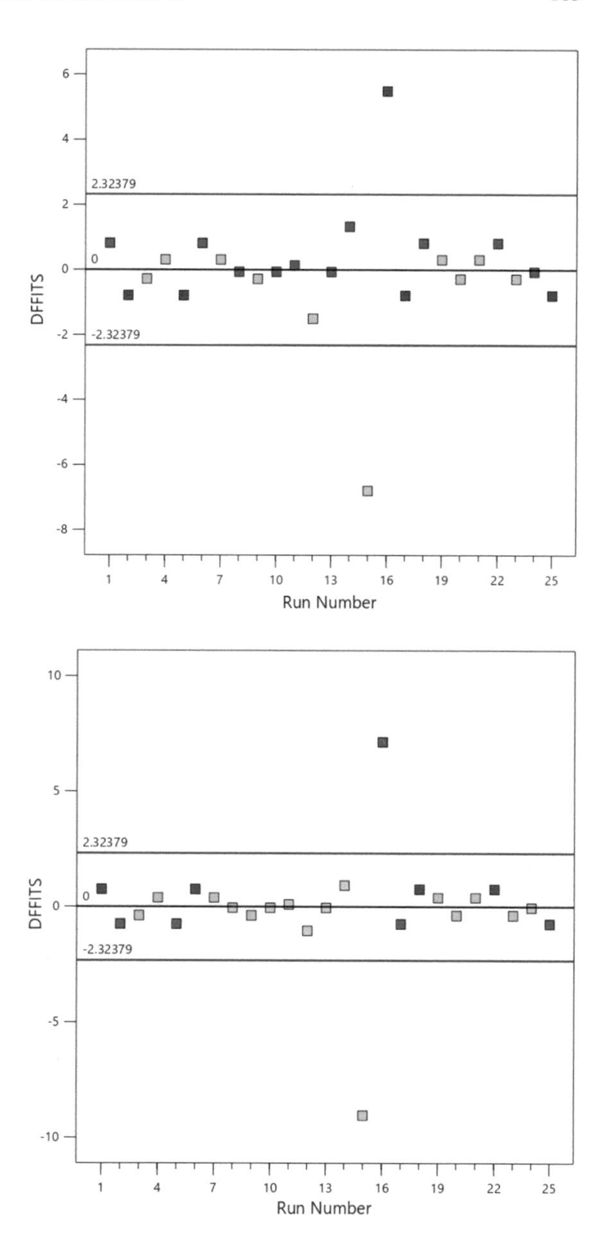

Fig. 6 DFFITS versus run

Fig. 6 (continued)

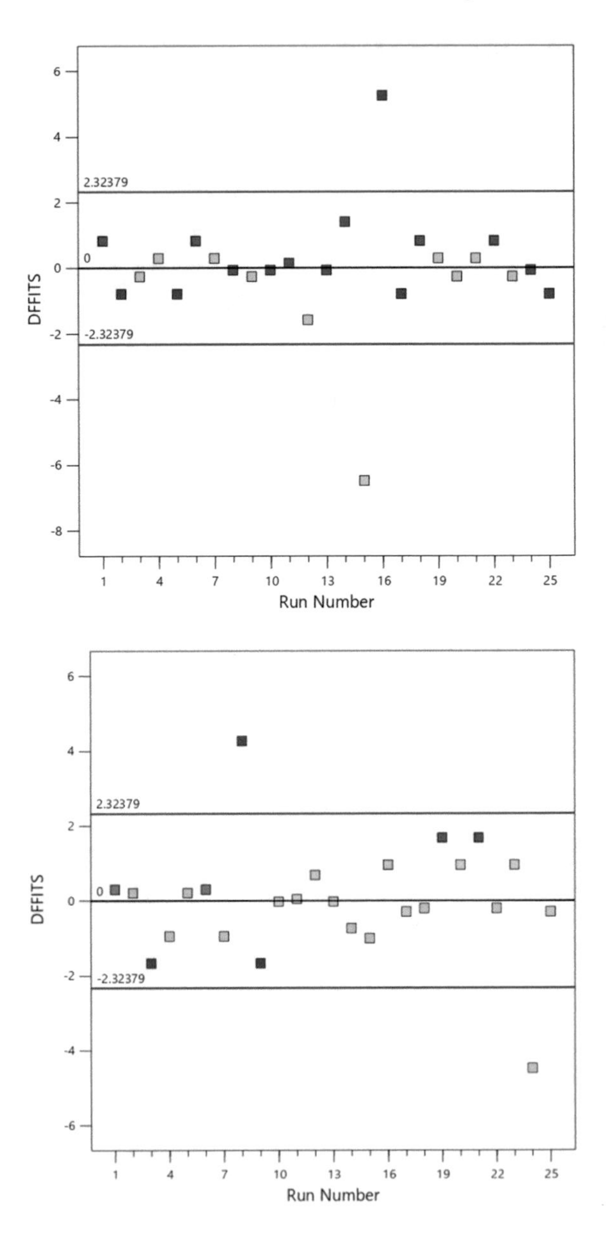

References

1. Gugulothu R, Sanke N, Gupta AVSSKS (2019) Numerical study of heat transfer characteristics in shell and tube heat exchanger. In: Lecture notes in mechanical engineering. Springer Nature Singapore Pte Ltd., pp 375–383
2. Gugulothu R, Somanchi NS, Reddy VK, Tirkey JV (2016) A review on design of baffles for shell and tube heat exchangers. In: Proceeding of ACGT-2016, Indian Institute of Technology Bombay, 14th–16th November 2016
3. Gugulothu R, Sanke N, Gupta AVSSKS, Jilugu RK (2018) A review on helical baffles for shell and tube heat exchangers. Int J Modern Eng Res Technol 5:129–139
4. Wang W, Cheng D, Liu T, Liu Y Performance comparison for oil-water heat transfer of circumferential overlap trisection helical baffle heat exchanger. J Central South Univ 2720–2727
5. Halle H, Chenoweth JM, Wambsganss MW (1988) Shell side water flow pressure drop distribution measurements in an industrial sized test heat exchanger. Trans ASME 110:60–67
6. Gugulothu R, Sanke N, Ahmed F, Jilugu RK (2020) Numerical study on shell and tube heat exchanger with segmental baffle. In: International joint conference on advances in computational intelligence (IJCACI-2020), Daffodil International University Dhaka, Bangladesh, during 20–21st November 2020
7. Stehlik P, Nemcansky J, Kral D, Swanson LW Comparison of correction factors for shell and tube heat exchangers with segmental or helical baffles. Heat Transfer Eng 15(1):55–65
8. Reddy KVK, Kumar BSP, Gugulothu R, Anuja K, Rao PV (2017) CFD Analysis of a helically coiled tube in tube heat exchanger. In: 5th international conference of materials processing and characterization (ICMPC-2016). Mater Today Proc 4:2341–2349
9. Gugulothu R, Reddy KVK, Somanchi NS, Adithya EL (2016) A review on enhancement of heat transfer techniques. In: 5th international conference of materials processing and characterization (ICMPC-2016). Mater Today Proc 4:1051–1056
10. Gugulothu R, Sanke N, Jilugu RK, Rangisetty SRD (2018) Numerical investigation on heat transfer of helical baffles shell and tube heat exchanger. Int J Modern Eng Res Technol 5:155–160
11. Ahmed F, Sumon MdM, Fuad M, Gugulothu R, Mollah AS (2021) Numerical simulation of heat exchanger for analyzing the performance of parallel and counter flow. WSEAS Trans Heat Mass Transfer 16:145–152
12. Gugulothu R, Sanke N (2022) Use of segmental baffle in shell and tube heat exchanger for nano emulsions. Heat Transfer 51(3):2645–2666. https://doi.org/10.1002/htj.22418
13. Gugulothu R, Sanke N (2020) Heat Transfer. https://doi.org/10.1002/htj.22474

Soliton Solutions of Dual-mode Kawahara Equation via Lie Symmetry Analysis

Sandeep Malik and Sachin Kumar

Abstract In this article, we investigate a newly proposed dual-mode Kawahara equation. Our main aim in this paper is to find out the soliton and periodic solutions of the Kawahara equation. Initially, we reduce the governing equation into an ordinary differential equation by applying the Lie symmetry analysis. Further, we derive the soliton and periodic solutions via three integration methods, namely sech-csch scheme, exp-expansion method, and modified F-expansion method.

Keywords Dual-mode Kawahara equation · Lie symmetry analysis · Exp-expansion method · Modified F-expansion method

1 Introduction

A new family of nonlinear differential equations has been seen as dual-mode partial differential equations (PDEs). The concept of dual-mode PDEs was started with Korsunsky [1], where he suggested an operator to derive the dual-mode KdV equation. Recently, Wazwaz [2–4] re-scaled Korsunsky's operator to convert the nonlinear equations from single mode to dual mode. His argument began with single-mode PDEs as:

$$w_t + N(w, w_x, \dots) + L(w_{xx}, w_{xxx}, \dots) = 0, \tag{1}$$

where N and L are nonlinear and linear operators, respectively. Then, Wazwaz [2–4] proposed the scale for the dual-mode PDEs as:

$$w_{tt} - s^2 w_{xx} + \left(\frac{\partial}{\partial t} - \lambda_1 s \frac{\partial}{\partial x}\right) N(w, w_x, \dots) + \left(\frac{\partial}{\partial t} - \lambda_2 s \frac{\partial}{\partial x}\right) L(w_{xx}, w_{xxx}, \dots) = 0, \tag{2}$$

where λ_1, λ_2, and s are parameters of nonlinearity, dispersion, and phase velocity, respectively.

S. Malik (✉) · S. Kumar
Department of Mathematics and Statistics, Central University of Punjab, Bathinda, Punjab 151401, India
e-mail: sndp796@gmail.com

S. S. Ray et al. (eds.), *Applied Analysis, Computation and Mathematical Modelling in Engineering*, Lecture Notes in Electrical Engineering 897,
https://doi.org/10.1007/978-981-19-1824-7_12

More recently, Alquran et al. [5] proposed a dual-mode Kawahara (DMK) equation of the form

$$w_{tt} - s^2 w_{xx} + \Big(\frac{\partial}{\partial t} - \lambda_1 s \frac{\partial}{\partial x}\Big)(a w^2 w_x) + \Big(\frac{\partial}{\partial t} - \lambda_2 s \frac{\partial}{\partial x}\Big)(b w_{xxx} - c w_{xxxxx}) = 0, \tag{3}$$

where a, b, c are free parameters and λ_1, λ_2, s are the parameters of nonlinearity, dispersion, and phase velocity, respectively, along with $|\lambda_1| \leq 1,\ |\lambda_2| \leq 1$, and $s \geq 0$. By taking $s = 0$ and integrating with respect to t, Eq. (3) converts into standard Kawahara equation

$$w_t + (a w^2 w_x) + (b w_{xxx} - c w_{xxxxx}) = 0, \tag{4}$$

which is used to describe the water waves with surface tension. Also, Kawahara equation is a model for plasma waves and magneto-acoustic waves [6–8]. Equation (3) indicates moving directional dual waves with linearity, dispersion, and phase velocity parameters.

The framework of rest of the article is as follows: In Sect. 2, two vector fields are obtained by implementing the Lie symmetry analysis. Then, by using transformation, the DMK equation is reduced into an ordinary differential equation. Sections 3 to 5 consist the soliton and periodic solutions of the governing equation. These solutions are constructed with the help of the sech-csch scheme, the exp-expansion method, and the modified F-expansion method. In the last, the conclusion is presented.

2 Symmetry Reduction

Applying the Lie symmetry approach [9–13], we get the following vector fields for Eq. (3)

$$V_1 = \frac{\partial}{\partial x}, \quad V_2 = \frac{\partial}{\partial t}. \tag{5}$$

The vector field

$$V_2 + k V_1 = \frac{\partial}{\partial t} + k \frac{\partial}{\partial x},$$

has the following form of characteristic equation

$$\frac{\mathrm{d}x}{k} = \frac{\mathrm{d}t}{1} = \frac{\mathrm{d}w}{0}, \tag{6}$$

where k is an arbitrary real number. On solving Eq. (6), we get the similarity variables as

$$\begin{aligned} \xi &= x - kt, \\ w(x,t) &= F(\xi), \end{aligned} \tag{7}$$

where F is a new dependent variable. Substituting Eq. (7) into Eq. (3) gives

$$[-a(k+\lambda_1 s)F^2 + (k^2 - s^2)]F'' - 2a(k+\lambda_1 s)F(F')^2 - (k+\lambda_2 s)[bF^{(4)} - cF^{(6)}] = 0, \tag{8}$$

where $(')$ indicates derivative w.r.t. ξ. Double integral of Eq. (8), with integral constants taking equal to zero, gives

$$-\frac{1}{3}a(k+\lambda_1 s)F^3 + (k^2 - s^2)F - (k+\lambda_2 s)[bF'' - cF^{(4)}] = 0. \tag{9}$$

3 Sech-csch Scheme

To obtain the solutions by sech-csch scheme [14–16], let us assume that

$$F(\xi) = \sum_{j=0}^{N} A_j \operatorname{sech}^j(m\xi), \tag{10}$$

is a solution of Eq. (9). By balancing the terms F^3 and $F^{(4)}$, we have $N = 2$. Substituting (10) in (9) and comparing the coefficients of the same power of the function sech equal to zero, we have the following system

$$\begin{aligned}
0 &= -\frac{1}{3}a\,(k+s\lambda_1)\,{a_2}^3 + 120\,cm^4\,(k+s\lambda_2)\,a_2, \\
0 &= 24\,cm^4\,(k+s\lambda_2)\,a_1 - a\,(k+s\lambda_1)\,a_1{a_2}^2, \\
0 &= 6\,bm^2\,(k+s\lambda_2)\,a_2 - 120\,cm^4\,(k+s\lambda_2)\,a_2 \\
&\quad - \frac{1}{3}\,a\,(k+s\lambda_1)\left(a_0{a_2}^2 + 2\,{a_1}^2a_2 + a_2\left(2\,a_0a_2 + {a_1}^2\right)\right), \\
0 &= 2\,bm^2\,(k+s\lambda_2)\,a_1 - 20\,cm^4\,(k+s\lambda_2)\,a_1 \\
&\quad - \frac{1}{3}\,a\,(k+s\lambda_1)\left(4\,a_0a_1a_2 + a_1\left(2\,a_0a_2 + {a_1}^2\right)\right), \\
0 &= \left(k^2 - s^2\right)a_2 - 4\,bm^2\,(k+s\lambda_2)\,a_2 \\
&\quad - \frac{1}{3}\,a\,(k+s\lambda_1)\left(a_0\left(2\,a_0a_2 + {a_1}^2\right) + 2\,{a_1}^2a_0 + a_2{a_0}^2\right) + 16\,cm^4\,(k+s\lambda_2)\,a_2, \\
0 &= -a\,(k+s\lambda_1)\,{a_0}^2a_1 - bm^2\,(k+s\lambda_2)\,a_1 + \left(k^2 - s^2\right)a_1 + cm^4\,(k+s\lambda_2)\,a_1, \\
0 &= -\frac{1}{3}\,a\,(k+s\lambda_1)\,{a_0}^3 + \left(k^2 - s^2\right)a_0.
\end{aligned} \tag{11}$$

By letting $\lambda_1 = \lambda_2 = \delta$, the solution of system (11) gives

$$\begin{aligned} &A_0 = A_1 = 0, \quad A_2 = \pm\frac{3b}{\sqrt{10ac}}, \quad m = \pm\sqrt{\frac{b}{20c}}, \\ &k = \frac{2b^2 \pm \sqrt{4b^4 + 625c^2s^2 + 100cs\delta b^2}}{25c}. \end{aligned} \tag{12}$$

Hence, the bright soliton solution of Eq. (3) becomes

$$w_1(x,t) = \pm\frac{3b}{\sqrt{10ac}}\ \mathrm{sech}^2\left[\sqrt{\frac{b}{20c}}\left(x - \frac{2b^2 \pm \sqrt{4b^4 + 625c^2s^2 + 100cs\delta b^2}}{25c}\,t\right)\right]. \tag{13}$$

In a similar manner, we derive the singular soliton solution of Eq. (3) as

$$w_2(x,t) = \pm\frac{3b}{\sqrt{10ac}}\ \mathrm{csch}^2\left[\sqrt{\frac{b}{20c}}\left(x - \frac{2b^2 \pm \sqrt{4b^4 + 625c^2s^2 + 100cs\delta b^2}}{25c}\,t\right)\right]. \tag{14}$$

4 Exp-expansion Method

A quick review of this method [17, 18] is described as:
Step-1: Let we have a governing model as

$$S(w, w_t, w_x, w_{tt}, w_{xt}, w_{xx}, \ldots) = 0. \tag{15}$$

It can be reduced to

$$G(U, U', U'', U''', \ldots) = 0, \tag{16}$$

with the help of transformation

$$w(x,t) = U(\xi), \quad \xi = x - kt. \tag{17}$$

Step-2: Let Eq. (16) satisfies by the solution

$$U(\xi) = \sum_{i=0}^{N} A_i\{\exp(-V(\xi))\}^i. \tag{18}$$

To obtain the number N, we have to apply the balancing principle in Eq. (16). The function $V(\xi)$ satisfies the following equation

$$V'(\xi) = \exp(-V(\xi)) + S\exp(V(\xi)) + R, \tag{19}$$

where S and R are constants. The corresponding solutions of Eq. (16) can be seen as

$$
\begin{aligned}
V(\xi) &= \ln\left[-\frac{R}{2S}-\frac{\sqrt{\mu}}{2S}\tanh\left(\frac{\sqrt{\mu}}{2}(\xi-\xi_0)\right)\right], \quad S\neq 0, \quad \mu>0,\\
V(\xi) &= \ln\left[-\frac{R}{2S}-\frac{\sqrt{\mu}}{2S}\coth\left(\frac{\sqrt{\mu}}{2}(\xi-\xi_0)\right)\right], \quad S\neq 0, \quad \mu>0,\\
V(\xi) &= \ln\left[-\frac{R}{2S}+\frac{\sqrt{-\mu}}{2S}\tan\left(\frac{\sqrt{-\mu}}{2}(\xi-\xi_0)\right)\right], \quad S\neq 0, \quad \mu<0,\\
V(\xi) &= \ln\left[-\frac{R}{2S}-\frac{\sqrt{-\mu}}{2S}\cot\left(\frac{\sqrt{-\mu}}{2}(\xi-\xi_0)\right)\right], \quad S\neq 0, \quad \mu<0,
\end{aligned}
\tag{20}
$$

where $\mu = R^2 - 4S$.
Step-3: By inserting Eqs. (18)–(19) into Eq. (16), we have the strategic equations. By solving these strategic equations, we have important results. Moreover, with the help of Eq. (20), we can obtain the solutions of Eq. (16).

4.1 Soliton and Other Solutions

In this section, solitons for Eq. (3) are derived via exp-expansion method [17, 18].

By implementing the balancing condition in Eqs. (9), (18) takes the form

$$U(\xi) = A_0 + A_1\exp(-V(\xi)) + A_2\exp(-2V(\xi)). \tag{21}$$

Inserting (21) into (9) and using (19), we have following system of equations:

$$
\begin{aligned}
0 &= -\frac{1}{3}aka_2{}^3 - \frac{1}{3}as\lambda_1a_2{}^3 + 120\,ca_2k + 120\,ca_2s\lambda_2,\\
0 &= 24\,ca_1k + 336\,ca_2Rk + 336\,ca_2Rs\lambda_2 + 24\,ca_1s\lambda_2 - as\lambda_1a_1a_2{}^2 - aka_1a_2{}^2,\\
0 &= 240\,ca_2Sk + 240\,ca_2Ss\lambda_2 - 6\,bka_2 - aka_1{}^2a_2 + 60\,ca_1Rs\lambda_2\\
&\quad + 60\,ca_1Rk - aka_0a_2{}^2 + 330\,ca_2R^2s\lambda_2 - as\lambda_1a_0a_2{}^2\\
&\quad - as\lambda_1a_1{}^2a_2 + 330\,ca_2R^2k - 6\,bs\lambda_2a_2,\\
0 &= 130\,ca_2R^3s\lambda_2 + 440\,ca_2SRk + 40\,ca_1Sk + 40\,ca_1Ss\lambda_2\\
&\quad - \frac{1}{3}as\lambda_1a_1{}^3 - 10\,bs\lambda_2a_2R + 50\,ca_1R^2k - \frac{1}{3}aka_1{}^3 + 50\,ca_1R^2s\lambda_2\\
&\quad + 130\,ca_2R^3k - 2\,bs\lambda_2a_1 - 2\,aka_0a_1a_2 - 2\,bka_1\\
&\quad + 440\,ca_2SRs\lambda_2 - 10\,bka_2R - 2\,as\lambda_1a_0a_1a_2,\\
0 &= 136\,ca_2S^2k - 8\,bs\lambda_2a_2S + 136\,ca_2S^2s\lambda_2 - as\lambda_1a_0{}^2a_2\\
&\quad - as\lambda_1a_0a_1{}^2 - 8\,bka_2S + 60\,ca_1SRs\lambda_2 - aka_0a_1{}^2 + 60\,ca_1SRk\\
&\quad - 3\,bs\lambda_2a_1R + 16\,ca_2R^4k + 15\,ca_1R^3s\lambda_2 + k^2a_2 - s^2a_2 + 15\,ca_1R^3k
\end{aligned}
$$

$$+\,16\,ca_2R^4s\lambda_2 + 232\,ca_2SR^2k + 232\,ca_2SR^2s\lambda_2$$
$$-\,3\,bka_1R - 4\,bka_2R^2 - aka_0{}^2a_2 - 4\,bs\lambda_2a_2R^2,$$
$$0 = -bka_1R^2 + 16\,ca_1S^2k + 30\,ca_2SR^3k - 6\,bka_2RS + ca_1R^4s\lambda_2$$
$$-\,2\,bs\lambda_2a_1S + 120\,ca_2S^2Rk + ca_1R^4k + 16\,ca_1S^2s\lambda_2 - s^2a_1$$
$$-\,bs\lambda_2a_1R^2 - as\lambda_1a_0{}^2a_1 + k^2a_1 + 22\,ca_1SR^2k + 120\,ca_2S^2Rs\lambda_2$$
$$-\,2\,bka_1S + 22\,ca_1SR^2s\lambda_2 + 30\,ca_2SR^3s\lambda_2 - 6\,bs\lambda_2a_2RS - aka_0{}^2a_1,$$
$$0 = -\frac{1}{3}\,as\lambda_1a_0{}^3 - \frac{1}{3}\,aka_0{}^3 + k^2a_0 - s^2a_0 - 2\,bka_2S^2 + 16\,ca_2S^3k$$
$$-\,2\,bs\lambda_2a_2S^2 + 16\,ca_2S^3s\lambda_2 + 8\,ca_1S^2Rk + 14\,ca_2S^2R^2k + 14\,ca_2S^2R^2s\lambda_2$$
$$+\,8\,ca_1S^2Rs\lambda_2 - bka_1RS + ca_1SR^3k - bs\lambda_2a_1RS + ca_1SR^3s\lambda_2.$$

Thus, by the assumption $\lambda_1 = \lambda_2 = \pm 1$, solutions of this system yields

$$R = \frac{A_1}{A_2},\quad S = \frac{A_0}{A_2},\quad k = \frac{4c(A_1{}^2 - 4A_0A_2)^2 \pm sA_2{}^4}{A_2{}^4},$$
$$a = \frac{360c}{A_2{}^2},\quad b = \frac{5c(A_1{}^2 - 4A_0A_2)}{A_2{}^2}. \tag{22}$$

Thus, $\mu = \dfrac{(A_1{}^2 - 4A_0A_2)}{A_2{}^2}$. Moreover, substituting Eq. (22) with Eq. (20) into Eq. (9) gives the dark soliton solution of Eq. (3) as

$$w_3(x,t) = -\frac{A_0(A_1{}^2 - 4A_0A_2)\left[1 - \tanh^2\left(\frac{1}{2}\frac{\sqrt{A_1{}^2-4A_0A_2}}{A_2}\left(x - \left(\frac{4c(A_1{}^2-4A_0A_2)^2 \pm sA_2{}^4}{A_2{}^4}\right)t\right)\right)\right]}{\left[A_1 + \sqrt{A_1{}^2 - 4A_0A_2}\tanh\left(\frac{1}{2}\frac{\sqrt{A_1{}^2-4A_0A_2}}{A_2}\left(x - \left(\frac{4c(A_1{}^2-4A_0A_2)^2 \pm sA_2{}^4}{A_2{}^4}\right)t\right)\right)\right]^2}, \tag{23}$$

with $(A_1{}^2 - 4A_0A_2) > 0$.

Singular soliton solution is (Figs. 1 and 2)

$$w_4(x,t) = -\frac{A_0(A_1{}^2 - 4A_0A_2)\left[1 - \coth^2\left(\frac{1}{2}\frac{\sqrt{A_1{}^2-4A_0A_2}}{A_2}\left(x - \left(\frac{4c(A_1{}^2-4A_0A_2)^2 \pm sA_2{}^4}{A_2{}^4}\right)t\right)\right)\right]}{\left[A_1 + \sqrt{A_1{}^2 - 4A_0A_2}\coth\left(\frac{1}{2}\frac{\sqrt{A_1{}^2-4A_0A_2}}{A_2}\left(x - \left(\frac{4c(A_1{}^2-4A_0A_2)^2 \pm sA_2{}^4}{A_2{}^4}\right)t\right)\right)\right]^2}, \tag{24}$$

with $(A_1{}^2 - 4A_0A_2) > 0$.

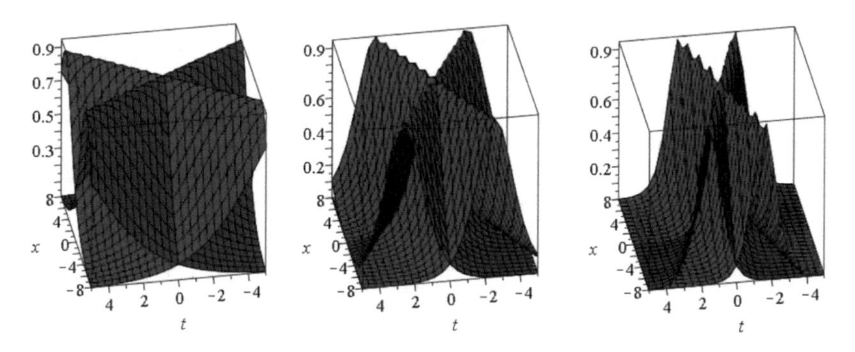

Fig. 1 Dual-wave interaction of (13) with phase velocity $s = 1, 3, 5$, respectively. The other values are $a = b = \delta = 1$ and $c = 0.1$

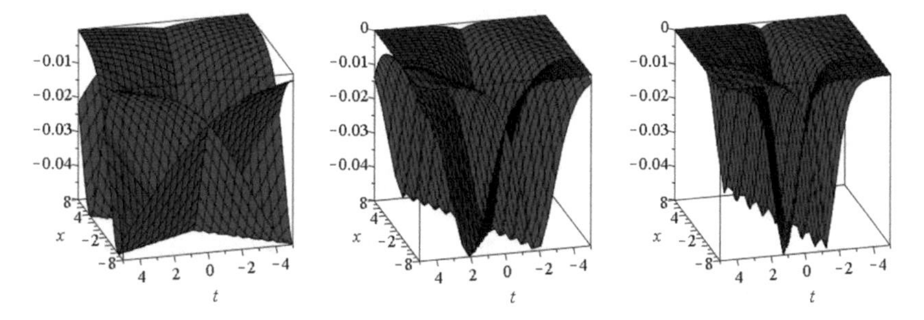

Fig. 2 Dual-wave interaction of (23) with phase velocity $s = 1, 3, 5$, respectively. The other values are $A_0 = 0.2$, $A_1 = A_2 = 1$, and $c = 0.1$

Singular periodic solutions are

$$w_5(x,t) = -\frac{A_0({A_1}^2 - 4A_0A_2)\left[1 + \tan^2\left(\frac{1}{2}\frac{\sqrt{-({A_1}^2-4A_0A_2)}}{A_2}\left(x - \left(\frac{4c({A_1}^2-4A_0A_2)^2 \pm s{A_2}^4}{{A_2}^4}\right)t\right)\right)\right]}{\left[A_1 + \sqrt{-({A_1}^2 - 4A_0A_2)}\tan\left(\frac{1}{2}\frac{\sqrt{-({A_1}^2-4A_0A_2)}}{A_2}\left(x - \left(\frac{4c({A_1}^2-4A_0A_2)^2 \pm s{A_2}^4}{{A_2}^4}\right)t\right)\right)\right]^2}, \tag{25}$$

$$w_6(x,t) = -\frac{A_0({A_1}^2 - 4A_0A_2)\left[1 + \cot^2\left(\frac{1}{2}\frac{\sqrt{-({A_1}^2-4A_0A_2)}}{A_2}\left(x - \left(\frac{4c({A_1}^2-4A_0A_2)^2 \pm s{A_2}^4}{{A_2}^4}\right)t\right)\right)\right]}{\left[A_1 + \sqrt{-({A_1}^2 - 4A_0A_2)}\cot\left(\frac{1}{2}\frac{\sqrt{-({A_1}^2-4A_0A_2)}}{A_2}\left(x - \left(\frac{4c({A_1}^2-4A_0A_2)^2 \pm s{A_2}^4}{{A_2}^4}\right)t\right)\right)\right]^2}, \tag{26}$$

with $({A_1}^2 - 4A_0A_2) < 0$.

5 Modified F-expansion Method

A quick review of this method [19, 20] is described as:
Step-1: Let we have a governing model as

$$S(w, w_t, w_x, w_{tt}, w_{xt}, w_{xx}, \dots) = 0. \tag{27}$$

It can be reduced to

$$G(F, F', F'', F''', \dots) = 0, \tag{28}$$

with the help of transformation

$$w(x, t) = F(\xi), \quad \xi = x - kt. \tag{29}$$

Step-2: Let Eq. (28) satisfies by the solution

$$F(\xi) = \sum_{i=0}^{N} A_i \, G^i(\xi) + \sum_{i=1}^{N} A_{-i} \, G^{-i}(\xi), \tag{30}$$

where for $i = 0, 1, \dots N$, A_i and A_{-i} are constants to be determined. To obtain the number N, we have to apply balancing principle in Eq. (28). The function $G(\xi)$ satisfies the following equation

$$G'(\xi) = A + BG(\xi) + CG(\xi)^2, \tag{31}$$

where A, B, and C are constants. The corresponding solutions of Eq. (28) are shown in Table 1.
Step-3: By inserting Eqs. (30)–(31) into Eq. (28), we have the strategic equations. By solving these strategic equations, we have important results. Moreover, with the help of Table 1, we can obtain the solutions of Eq. (28).

5.1 Solitons and Other Solutions

In this section, solitons for Eq. (3) are derived via modified F-expansion method [19, 20].

By implementing the balancing condition in Eqs. (9), (30) takes the form

$$F(\xi) = \frac{A_{-2}}{G^2(\xi)} + \frac{A_{-1}}{G(\xi)} + A_0 + A_1 G(\xi) + A_2 G^2(\xi). \tag{32}$$

Inserting (32) into (9) and using (31), we have a large system of algebraic equations. Thus, by the assumption $\lambda_1 = \lambda_2 = \pm 1$, derived system yields the following solutions:

Case 1:

$$\begin{aligned}
A_{-2} &= -\frac{6\sqrt{2}\,bA^2}{\sqrt{ab(B^2 - 4\,AC)}}, \quad A_{-1} = -\frac{6\sqrt{2}\,bAB}{\sqrt{ab(B^2 - 4\,AC)}}, \\
A_0 &= -\frac{6\sqrt{2}\,bAC}{\sqrt{ab(B^2 - 4\,AC)}}, \quad A_1 = A_2 = 0, \\
k &= \frac{4\,b(B^2 - 4\,AC)}{5} \pm s, \quad c = \frac{b}{5(B^2 - 4\,AC)}.
\end{aligned} \tag{33}$$

Case 2:

$$\begin{aligned}
A_{-2} &= A_{-1} = 0, \quad A_0 = -\frac{6\sqrt{2}\,bAC}{\sqrt{ab(B^2 - 4\,AC)}}, \\
A_1 &= -\frac{6\sqrt{2}\,bBC}{\sqrt{ab(B^2 - 4\,AC)}}, \quad A_2 = -\frac{6\sqrt{2}\,bC^2}{\sqrt{ab(B^2 - 4\,AC)}}, \\
k &= \frac{4\,b(B^2 - 4\,AC)}{5} \pm s, \quad c = \frac{b}{5(B^2 - 4\,AC)}.
\end{aligned} \tag{34}$$

So, corresponding to case (1), we can list the solutions of $w(x, t)$ as follows:

1. When $A = \frac{1}{2}$, $B = 0$, $C = -\frac{1}{2}$, then from Table 1, $G(\xi) = \coth(\xi) \pm \operatorname{csch}(\xi)$ or $\tanh(\xi) \pm \iota \operatorname{sech}(\xi)$, hence, solutions of Eq. (3) are

$$w_7(x, t) = \frac{3\sqrt{2}b}{\sqrt{ab}} \left\{ \frac{\operatorname{sech}\left[x - (\frac{4}{5}b \pm s)t\right]}{\operatorname{sech}\left[x - (\frac{4}{5}b \pm s)t\right] \pm 1} \right\}, \tag{35}$$

and

$$w_8(x, t) = -\frac{3\sqrt{2}b}{\sqrt{-ab}} \left\{ \frac{1}{\sinh\left[x - (\frac{4}{5}b \pm s)t\right] \pm \iota} \right\}. \tag{36}$$

2. When $A = 1$, $B = 0$, $C = -1$, then from Table 1, $G(\xi) = \tanh(\xi)$ or $\coth(\xi)$, hence, solutions of Eq. (3) are

$$w_9(x, t) = \frac{3\sqrt{2}b}{\sqrt{ab}} \left\{ 1 - \frac{1}{\tanh^2\left[x - (\frac{16}{5}b \pm s)t\right]} \right\}, \tag{37}$$

and

$$w_{10}(x, t) = \frac{3\sqrt{2}b}{\sqrt{ab}} \left\{ 1 - \frac{1}{\coth^2\left[x - (\frac{16}{5}b \pm s)t\right]} \right\}. \tag{38}$$

3. When $A = \frac{1}{2}$, $B = 0$, $C = \frac{1}{2}$, then from Table 1, $G(\xi) = \sec(\xi) + \tan(\xi)$ or $\csc(\xi) - \cot(\xi)$, hence, solutions of Eq. (3) are

$$w_{11}(x,t) = -\frac{3\sqrt{2}b}{\sqrt{-ab}}\left\{\frac{1}{1+\sin\left[x+(\frac{4}{5}b\mp s)t\right]}\right\}, \tag{39}$$

and

$$w_{12}(x,t) = -\frac{3\sqrt{2}b}{\sqrt{-ab}}\left\{\frac{1}{1-\cos\left[x+(\frac{4}{5}b\mp s)t\right]}\right\}. \tag{40}$$

4. When $A = -\frac{1}{2}$, $B = 0$, $C = -\frac{1}{2}$, then from Table 1, $G(\xi) = \sec(\xi) - \tan(\xi)$ or $\csc(\xi) + \cot(\xi)$, hence, solutions of Eq. (3) are

$$w_{13}(x,t) = \frac{3\sqrt{2}b}{\sqrt{-ab}}\left\{\frac{1}{1-\sin\left[x+(\frac{4}{5}b\mp s)t\right]}\right\}, \tag{41}$$

and

$$w_{14}(x,t) = -\frac{3\sqrt{2}b}{\sqrt{-ab}}\left\{\frac{1}{1+\cos\left[x+(\frac{4}{5}b\mp s)t\right]}\right\}. \tag{42}$$

5. When $A = 1\ (-1)$, $B = 0$, $C = 1\ (-1)$, then from Table 1, $G(\xi) = \tan(\xi)$, $(\cot(\xi))$, hence, solutions of Eq. (3) are

$$w_{15}(x,t) = -\frac{3\sqrt{2}b}{\sqrt{-ab}}\left\{1+\frac{1}{\tan^2\left[x+(\frac{16}{5}b\mp s)t\right]}\right\}, \tag{43}$$

and

$$w_{16}(x,t) = -\frac{3\sqrt{2}b}{\sqrt{-ab}}\left\{1+\frac{1}{\cot^2\left[x+(\frac{16}{5}b\mp s)t\right]}\right\}. \tag{44}$$

Now, we have enlisted the solutions $w(x,t)$ of Eq. (3), corresponding to case (2), as follows:

1. When $A = 0$, $B = 1$, $C = -1$, then from Table 1, $G(\xi) = \frac{1}{2} + \frac{1}{2}\tanh(\frac{\xi}{2})$, hence, solution of Eq.(3) becomes

$$w_{17}(x,t) = \frac{3b}{\sqrt{2ab}}\left\{\operatorname{sech}^2\left[\frac{1}{2}\left(x-\left(\frac{4}{5}b\pm s\right)t\right)\right]\right\}. \tag{45}$$

2. When $A = 0$, $B = -1$, $C = 1$, then from Table 1, $G(\xi) = \frac{1}{2} - \frac{1}{2}\coth(\frac{\xi}{2})$, hence, solution of Eq. (3) becomes

$$w_{18}(x,t) = -\frac{3b}{\sqrt{2ab}}\left\{\operatorname{csch}^2\left[\frac{1}{2}\left(x - \left(\frac{4}{5}b \pm s\right)t\right)\right]\right\}. \tag{46}$$

3. When $A = \frac{1}{2}$, $B = 0$, $C = -\frac{1}{2}$, then from Table 1, $G(\xi) = \coth(\xi) \pm \operatorname{csch}(\xi)$ or $\tanh(\xi) \pm \iota \operatorname{sech}(\xi)$, hence, solutions of Eq. (3) are

$$\begin{aligned} w_{19}(x,t) = \frac{3\sqrt{2}b}{\sqrt{ab}}&\left\{\operatorname{csch}^2\left[x - \left(\frac{4}{5}b \pm s\right)t\right]\right. \\ &\left.\mp \coth\left[x - \left(\frac{4}{5}b \pm s\right)t\right]\operatorname{csch}\left[x - \left(\frac{4}{5}b \pm s\right)t\right]\right\}, \end{aligned} \tag{47}$$

and

$$\begin{aligned} w_{20}(x,t) = \frac{3\sqrt{2}b}{\sqrt{ab}}&\left\{\operatorname{sech}^2\left[x - \left(\frac{4}{5}b \pm s\right)t\right]\right. \\ &\left.\mp \iota\tanh\left[x - \left(\frac{4}{5}b \pm s\right)t\right]\operatorname{sech}\left[x - \left(\frac{4}{5}b \pm s\right)t\right]\right\}. \end{aligned} \tag{48}$$

4. When $A = 1$, $B = 0$, $C = -1$, then from Table 1, $G(\xi) = \tanh(\xi)$ or $\coth(\xi)$, hence, solutions of Eq. (3) are

$$w_{21}(x,t) = \frac{3\sqrt{2}b}{\sqrt{ab}}\left\{\operatorname{sech}^2\left[x - \left(\frac{16}{5}b \pm s\right)t\right]\right\}, \tag{49}$$

and

$$w_{22}(x,t) = -\frac{3\sqrt{2}b}{\sqrt{ab}}\left\{\operatorname{csch}^2\left[x - \left(\frac{16}{5}b \pm s\right)t\right]\right\}. \tag{50}$$

5. When $A = \frac{1}{2}$, $B = 0$, $C = \frac{1}{2}$, then from Table 1, $G(\xi) = \sec(\xi) + \tan(\xi)$ or $\csc(\xi) - \cot(\xi)$, hence, solutions of Eq. (3) are

$$\begin{aligned} w_{23}(x,t) = -\frac{3\sqrt{2}b}{\sqrt{-ab}}&\left\{\sec^2\left[x + \left(\frac{4}{5}b \mp s\right)t\right]\right. \\ &\left.+ \tan\left[x + \left(\frac{4}{5}b \mp s\right)t\right]\sec\left[x + \left(\frac{4}{5}b \mp s\right)t\right]\right\}, \end{aligned} \tag{51}$$

and

$$\begin{aligned} w_{24}(x,t) = -\frac{3\sqrt{2}b}{\sqrt{-ab}}&\left\{\csc^2\left[x + \left(\frac{4}{5}b \mp s\right)t\right]\right. \\ &\left.- \csc\left[x + \left(\frac{4}{5}b \mp s\right)t\right]\cot\left[x + \left(\frac{4}{5}b \mp s\right)t\right]\right\}. \end{aligned} \tag{52}$$

6. When $A=-\frac{1}{2}$, $B=0$, $C=-\frac{1}{2}$, then from Table 1, $G(\xi)=\sec(\xi)-\tan(\xi)$ or $\csc(\xi)+\cot(\xi)$, hence, solutions of Eq.(3) are

$$\begin{aligned} w_{25}(x,t)=&-\frac{3\sqrt{2}b}{\sqrt{-ab}}\left\{\sec^2\left[x+\left(\frac{4}{5}b\mp s\right)t\right]\right.\\ &\left.-\tan\left[x+\left(\frac{4}{5}b\mp s\right)t\right]\sec\left[x+\left(\frac{4}{5}b\mp s\right)t\right]\right\}, \end{aligned} \tag{53}$$

and

$$\begin{aligned} w_{26}(x,t)=&-\frac{3\sqrt{2}b}{\sqrt{-ab}}\left\{\csc^2\left[x+\left(\frac{4}{5}b\mp s\right)t\right]\right.\\ &\left.+\csc\left[x+\left(\frac{4}{5}b\mp s\right)t\right]\cot\left[x+\left(\frac{4}{5}b\mp s\right)t\right]\right\}. \end{aligned} \tag{54}$$

7. When $A=1\ (-1)$, $B=0$, $C=1\ (-1)$, then from Table 1, $G(\xi)=\tan(\xi)$, $(\cot(\xi))$, hence, solutions of Eq.(3) are

$$w_{27}(x,t)=-\frac{3\sqrt{2}b}{\sqrt{-ab}}\left\{\sec^2\left[x+\left(\frac{16}{5}b\mp s\right)t\right]\right\}, \tag{55}$$

and

$$w_{28}(x,t)=-\frac{3\sqrt{2}b}{\sqrt{-ab}}\left\{\csc^2\left[x+\left(\frac{16}{5}b\mp s\right)t\right]\right\}. \tag{56}$$

Table 1 Solutions to Eq. (31) given in [20]

A	B	C	$G(\xi)$
0	1	-1	$\frac{1}{2}+\frac{1}{2}\tanh\left(\frac{1}{2}\xi\right)$
0	-1	1	$\frac{1}{2}-\frac{1}{2}\coth\left(\frac{1}{2}\xi\right)$
$\frac{1}{2}$	0	$-\frac{1}{2}$	$\coth(\xi)\pm\mathrm{csch}(\xi),\ \tanh(\xi)\pm\iota\,\mathrm{sech}(\xi)$
1	0	-1	$\tanh(\xi),\ \coth(\xi)$
$\frac{1}{2}$	0	$\frac{1}{2}$	$\sec(\xi)+\tan(\xi),\ \csc(\xi)-\cot(\xi)$
$-\frac{1}{2}$	0	$-\frac{1}{2}$	$\sec(\xi)-\tan(\xi),\ \csc(\xi)+\cot(\xi)$
$1(-1)$	0	$1(-1)$	$\tan(\xi)(\cot(\xi))$
0	0	$\neq 0$	$-\frac{1}{C(\xi)+m}$
Arbitrary	0	0	$A\xi$
Arbitrary	$\neq 0$	0	$\frac{\exp(B)-A}{B}$

6 Conclusion

The article presents various solitons and periodic solutions of a newly proposed dual-mode Kawahara equation that indicates moving directional dual waves with linearity, dispersion, and phase velocity parameters. The authors have utilized the Lie classical method that reduces the governing equation into an ODE. Subsequently, three analytical approaches, namely sech-csch scheme, exp-expansion method, and modified F-expansion method, have been employed to get the results. Moreover, the graphical representation of some solutions is presented, which may be helpful to understand the physical structure of the model.

Acknowledgements Sandeep Malik thankfully acknowledges CSIR SRF Grant: 09/1051(0028)/2018-EMR-I.

References

1. Korsunsky SV (1994) Soliton solutions for a second-order KdV equation. Phys Lett A 185(2):174–176
2. Wazwaz AM (2017) A two-mode Burgers equation of weak shock waves in a fluid: multiple kink solutions and other exact solutions. Int J Appl Comput Math 3(4):3977–3985
3. Wazwaz AM (2017) Multiple soliton solutions and other exact solutions for a two-mode KdV equation. Math Methods Appl Sci 40(6):2277–2283
4. Wazwaz AM (2018) Two-mode Sharma-Tasso-Olver equation and two-mode fourth-order Burgers equation: multiple kink solutions. Alexandria Eng J 57(3):1971–1976
5. Alquran M, Jaradat I, Ali M, Baleanu D (2020) The dynamics of new dual-mode Kawahara equation: interaction of dual-waves solutions and graphical analysis. Phys Scr 95(4), Article ID 045216
6. Kawahara T (1972) Oscillatory solitary waves in dispersive media. J Phys Soc Jpn 33(1):260–264
7. Sirendaoreji (2004) New exact travelling wave solutions for the Kawahara and modified Kawahara equations. Chaos Solitons Fractals 19(1):147–150
8. Assas LMB (2009) New exact solutions for the Kawahara equation using exp-function method. J Comput Appl Math 233(2):97–102
9. Bira B, Sekhar TR (2013) Exact solutions to magnetogasdynamics using Lie point symmetries. Meccanica 48(5):1023–1029
10. Bira B, Sekhar TR (2013) Symmetry group analysis and exact solutions of isentropic magnetogasdynamics. Indian J Pure Appl Math 44(2):153–165
11. Bira B, Sekhar TR (2015) Exact solutions to drift-flux multiphase flow models through Lie group symmetry analysis. Appl Math Mech 36(8):1105–1112
12. Kumar S, Malik S, Biswas A, Zhou Q, Moraru L, Alzahrani AK, Belic MR (2020) Optical solitons with Kudryashov's equation by Lie symmetry analysis. Phys Wave Phenom 28(3):299–304
13. Bluman G, Anco S (2008) Symmetry and integration methods for differential equations, vol 154. Springer Science & Business Media
14. Kumar S, Malik S, Biswas A, Yildirim Y, Alshomrani AS, Belic MR (2020) Optical solitons with generalized anti-cubic nonlinearity by Lie symmetry. Optik 206, Article ID 163638
15. Jaradat HM (2018) Two-mode coupled Burgers equation: multiple-kink solutions and other exact solutions. Alexandria Eng J 57(3):2151–2155

16. Malik S, Almusawa H, Kumar S, Wazwaz AM, Osman MS (2021) A (2+1)-dimensional Kadomtsev-Petviashvili equation with competing dispersion effect: Painlevé analysis, dynamical behavior and invariant solutions. Results Phys 23, Article ID 104043
17. Yildirim Y, Biswas A, Ekici M, Triki H, Gonzalez-Gaxiola O, Alzahrani AK, Belic MR (2020) Optical solitons in birefringent fibers for Radhakrishnan-Kundu-Lakshmanan equation with five prolific integration norms. Optik 208, Article ID 164550
18. Biswas A, Ekici M, Sonmezoglu A, Belic MR (2019) Highly dispersive optical solitons with cubic-quintic-septic law by exp-expansion. Optik 186:321–325
19. Cai G, Wang Q, Huang J (2006) A modified F-expansion method for solving breaking soliton equation. Int J Nonlinear Sci 2(2):122–128
20. Ebadi G, Fard NY, Bhrawy AH, Kumar S, Triki H, Yildirim A, Biswas A (2013) Solitons and other solutions to the (3+1)-dimensional extended Kadomtsev-Petviashvili equation with power law nonlinearity. Rom Rep Phys 65(1):27–62

Soliton Solutions of (2+1)-Dimensional Modified Calogero-Bogoyavlenskii-Schiff (mCBS) Equation by Using Lie Symmetry Method

Shivam Kumar Mishra

Abstract In the present article, the Lie transformation method has been used to find out the group-invariant solutions of (2+1)-dimensional modified Calogero-Bogoyavlenskii-Schiff(mCBS) equation. The equation has been reduced to ordinary differential equations (ODEs) using the method. The method plays a significant role in the reduction of the number of independent variables in the system by one in each proceeding stage and, eventually, forms an ODE whose exact solutions can be established. Moreover, solutions derived here contain some arbitrary constants and functions. These solutions are mainly multisoliton, single soliton, periodic or quasi-periodic and evolutionary wave types. Finally, with the adjustments in these arbitrary parameters and functions, some graphs have been plotted.

Keywords mCBS equation · Lie group theory · Similarity transformation method · Invariant solutions

1 Introduction

The nonlinear phenomenon has increasingly wide use in the world around us. These nonlinear processes are described by means of their governing equations which are nothing but nonlinear partial differential equations (NPDEs). The NPDEs are being used in various fields like physics, chemical sciences, biological sciences to cure diseases, finance and business management for optimality, astrophysics, engineering and technology and many more. To examine the integrability and exact solutions of these equations are the most important to study their natural and dynamical behaviour. To find the attractive features of these nonlinear evolution equations (NLEEs) [1–35] gives rise to the soliton solutions. Solitons are generally such kind of wavepackets that retain and are capable to maintain their shape and size throughout the processes. It was first observed by Russell in 1834, while he was on his boat which was drawn

S. K. Mishra (✉)
School of Basic Sciences, Indian Institute of Technology Mandi, Mandi, Himachal Pradesh 175005, India
e-mail: shivammishra1807@gmail.com

S. S. Ray et al. (eds.), *Applied Analysis, Computation and Mathematical Modelling in Engineering*, Lecture Notes in Electrical Engineering 897,
https://doi.org/10.1007/978-981-19-1824-7_13

by the horses along narrow Union Canal in Scotland. He then experimented with all his laboratory observations in the water tank and called it 'waves of translation'. This theory of solitary wave attained researchers more concern and attention in the field of science and technology. Its application can be widely seen in the fields of plasma physics, optical lattice, biophysics, quantum dynamics. The most interesting application of it can be viewed in the field of biology as it carries the molecules of adenosine triphosphate (ATP) along with x-helical protein molecules without serious losses.

The present work mainly focuses on the following (2 + 1)-dimensional nonlinear modified Calogero-Bogoyavlenskii-Schiff (mCBS) equation [25].

$$v_t + v_{xxy} + 4v^2 v_y + 4v_x w = 0, \tag{1.1}$$

$$v v_y = w_x. \tag{1.2}$$

where $v(x, y, t)$ and $w(x, y, t)$ are the amplitude function of scaled temporal coordinate t, spatial coordinates x and y. Also, subscripts here are used to denote the respective partial derivatives. Equations (1.1) and (1.2) are also mentioned in [25], where another form of new integrable system was constructed by using the recursion operator and inverse operator of modified Calogero-Bogoyavlenskii-Schiff(mCBS) equation. This mCBS equation is the model for evolutionary shallow water waves and other complex physical phenomena. Therefore, it has significant role in study of hydrostatics, hydrodynamics and climatology. The fluid flow associated with this equations can be studied by some experimental tools and techniques, thus had great frequency in all parts of mathematical physics or any applied disciplines.

There are a number of methods that have been used to find out the exact solutions of NLEEs, such as the Painlevé expansion method [1], the multiple exp-function methods [2], Bäcklund and Darboux transform [3], Jacobi elliptic function method [4], the simple equation method [5], the inverse scattering transform [6], the tanh-function method [7], the soliton ansatz method [8]. To study the elastic collision and their nonlinear interaction, [11] had investigated the various forms of the CBS equation. This equation was firstly introduced by Bogoyavlenskii and Schiff [11].

$$v_t + v v_y + \frac{1}{2} v_x \partial_x^{-1} v_y + \frac{1}{4} v_{xxy} = 0. \tag{1.3}$$

Bogoyavlenskii applied the modified Lax formalism; on the other hand, by reducing the self-dual Yang-Mills equation, Schiff obtained the Eq. (1.3). Another form of mCBS equation was investigated by Kumar and Kumar [15], where Lie symmetry and optimal subalgebras were used to find the exact solutions. The equation was expressed as

$$a v_{tx} + b v_{ty} + c(4 v_{xx} v_{xy} + v_{xxxy}) - v_{yy} = 0, \tag{1.4}$$

where a, b and c are arbitrary constants. Equation (1.4) is the modified version of CBS equation. Now, with $a = 4,\ b = 8$ and $c = 4$, Eq. (1.4) reduces to modified

CBS equation [27, 29]. Also, with $a = 4,\ b = 4$ and $c = 2$, Eq. (1.4) reduces to mCBS equation [28]. Further, with $a = 1,\ b = -4$ and $c = -2$, Eq. (1.4) gives breaking soliton [31]. In [9], two distinct types of lump solutions were established by using Hirota bilinear method.

In the early nineteenth century, Lie discovered that the exact solutions for linear or first order ODEs are based on the groups of symmetries. Therefore, by the systematic use of Lie transformations, one can derive the exact solutions of these NPDEs. This method provides a systematic technique to figure out the group-invariant solutions of NLEEs. Later on, the contributions of Cartan, Herman, Klein, Ovsiannikov, Engel, Bluman, Noether to this field made it more inter-related to the other areas as well. The basic idea behind the method was simple; i.e. if there exists a reduction transformation, then the system of differential equations becomes invariant under Lie group of transformation. Therefore, the number of independent variables can be reduced by one in each progressing step. This makes our original system of differential equations easier to solve. Consequently, abundant number of group-invariant solutions can be derived. The applications of Lie groups can be seen in quantum physics, computer vision for pattern recognition, pulse propagation in nerves, fluid dynamics, topology, algebraic geometry, etc. Motivated from the reductions and previous findings [9, 27–29, 31], the exact solutions of Eqs. (1.1) and (1.2) were derived by using Lie symmetry approach.

The whole article has five sections. The first section gives a brief of mCBS equation. The second section is about the Lie symmetry method. The third section explains the invariant solutions of the mCBS equation by Lie symmetry reductions. The fourth section explains the analysis part of the solutions by MATLAB simulation. Finally, the fifth section details the concluding remarks.

2 Lie Symmetry Analysis

The present section elaborates the infinitesimal generated by using Lie symmetry. The one-parameter ($\bar{\epsilon}$) Lie group of transformation is given by Olver [12]

$$\begin{aligned}
x^* &= x + \bar{\epsilon}\ \xi(x, y, t, v, w) + O(\bar{\epsilon}^2),\\
y^* &= y + \bar{\epsilon}\ \eta(x, y, t, v, w) + O(\bar{\epsilon}^2),\\
t^* &= t + \bar{\epsilon}\ \tau(x, y, t, v, w) + O(\bar{\epsilon}^2),\\
v^* &= v + \bar{\epsilon}\ \phi(x, y, t, v, w) + O(\bar{\epsilon}^2)\\
w^* &= w + \bar{\epsilon}\ \psi(x, y, t, v, w) + O(\bar{\epsilon}^2),
\end{aligned} \tag{2.1}$$

where $\xi,\ \eta,\ \tau,\ \phi$ and ψ depend on x, y, t, v and w resulted by the infinitesimal symmetries.

The vector field associated with Eqs. (1.1) and (1.2) is defined as

$$\begin{aligned}\mathbf{w} &= \xi(x, y, t, v, w)\frac{\partial}{\partial x} + \eta(x, y, t, v, w)\frac{\partial}{\partial y} \\ &+ \tau(x, y, t, v, w)\frac{\partial}{\partial t} + \phi(x, y, t, v, w)\frac{\partial}{\partial v} + \psi(x, y, t, v, w)\frac{\partial}{\partial w} .\end{aligned} \quad (2.2)$$

Equation (2.2) is the base for the generation of Lie symmetry of Eqs. (1.1) and (1.2). Moreover, using the invariance surface condition [10, 12], $\mathrm{Pr}^{(3)}\mathbf{w}(\Delta) = 0$ and $\mathrm{Pr}^{(1)}\mathbf{w}(\Delta) = 0$ along with $\Delta = 0$, produces symmetry condition for Eqs. (1.1) and (1.2) mentioned below

$$\mathrm{Pr}^{(3)}\mathbf{w}[v_t + v_{xxy} + 4v^2 v_y + 4v_x w] = 0, \quad \mathrm{Pr}^{(1)}\mathbf{w}[v v_y - w_x] = 0,$$

which signifies

$$\phi_t + \phi_{xxy} + 4v^2\phi_y + 8v v_y \phi + 4w\phi_x + 4v_x\psi = 0, \quad v\phi_y + v_y\phi = \psi_x, \quad (2.3)$$

where ϕ_t , ϕ_x , ϕ_y , ψ_x and ϕ_{xxy} are the coefficient of $Pr^{(3)}\mathbf{w}$ and $Pr^{(1)}\mathbf{w}$. These are expressed as

$$\begin{aligned}\phi_t &= D_t\phi - v_x\, D_t\xi - v_y\, D_t\eta - v_t\, D_t\tau \\ \phi_x &= D_x\,\phi - v_x\, D_x\xi - v_y\, D_x\eta - v_t\, D_x\tau \\ \phi_y &= D_y\phi - v_x\, D_y\xi - v_y\, D_y\eta - v_t\, D_y\tau \\ \psi_x &= D_x\,\psi - w_x\, D_x\xi - w_y\, D_x\eta - w_t\, D_x\tau \\ \phi_{xxy} &= D_x^2 D_y\phi - v_x D_x^2 D_y\xi - v_{xy} D_x^2\xi - 2v_{xx} D_x D_y\xi \\ &\quad - 2v_{xxy} D_x\xi - v_{xxx} D_y\xi - v_y D_x^2 D_y\eta - \\ &\quad v_{yy} D_x^2\eta - 2v_{xy} D_x D_y\eta - 2v_{xxy} D_x\eta - v_{xxy} D_y\eta \\ &\quad - v_t D_x^2 D_y\tau - v_{yt} D_x^2\tau - 2v_{xt} D_x D_y\tau - \\ &\quad 2v_{xyt} D_x\tau - v_{xxt} D_y\tau,\end{aligned}$$

where operators D_x, D_y and D_t denote the total derivative of variables x, y and t, respectively. For example one out of them can be denoted as

$$D_x = \frac{\partial}{\partial x} + v_x\frac{\partial}{\partial v} + v_{xx}\frac{\partial}{\partial v_x} + v_{xy}\frac{\partial}{\partial v_y} + v_{xt}\frac{\partial}{\partial v_t} + \cdots.$$

After all these substitutions together with the total derivative operator, the following system of determining equations is obtained simply by equating the coefficient terms of the occurring different partial derivatives

$$\begin{aligned}
\xi_y &= \xi_v = \eta_x = \eta_t = \eta_v = \eta_{y,y} = \tau_x = \tau_y = \tau_v \\
&= \tau_{t,t} = \tau_w = 2\xi_x - \tau_t + \eta_y = v\phi_w + \eta_x = \psi_v + v\eta_y \\
&= \psi_x - v\phi_y = v\xi_w - \eta_v = \eta_t - 2w(\tau_t + \eta_y) = 2\phi + v(\tau_t - \eta_y) \\
&= v^2\phi_{v,v} + v^2\eta_{v,x} - v\phi_v - \phi = v\phi_{v,x} \\
&+ v\eta_{x,x} + \phi_x = v\phi_{v,y} + \xi_{x,y} + \phi_y = v\psi_w - v\xi_x - v\phi_v + v\eta_y - \phi = 0. \quad (2.4)
\end{aligned}$$

Thus, the solutions of the system (2.4) provide the following infinitesimal generators

$$\begin{aligned}
\xi &= m_1h_1 + m_2h_2 + m_3h_3 + m_4h_4 + m_4x \\
\eta &= m_2 + (m_3 - 2m_4)y \\
\tau &= m_1 + m_3t \\
\phi &= -m_4v \\
\psi &= \frac{1}{4}(m_1\bar{h}_1 + m_2\bar{h}_2 + m_3\bar{h}_3 + m_4\bar{h}_4) + (m_4 - m_3)w, \quad (2.5)
\end{aligned}$$

where $h_i (1 \le i \le 4)$ is arbitrary smooth functions of t and $m_i (1 \le i \le 4)$ is arbitrary constants. Also, bar denotes their respective derivatives.

The spanned vector field of Eqs. (1.1) and (1.2) is represented as

$$\mathbf{w} = \mathbf{w_1}(f) + \mathbf{w_2} + \mathbf{w_3} + \mathbf{w_4} + \mathbf{w_5}.$$

The vectors $\mathbf{w_1}(f)$, $\mathbf{w_2}$, $\mathbf{w_3}$, $\mathbf{w_4}$ and $\mathbf{w_5}$ are respective infinitesimal generators. Here, $f(t) = m_1h_1 + m_2h_2 + m_3h_3 + m_4h_4$. Thus, they all can be represented as

$$\begin{aligned}
&\mathbf{w_1}(f) = f(t)\partial_x + \frac{1}{4}\bar{f}(t)\partial_w, \quad \mathbf{w_2} = \partial_t, \quad \mathbf{w_3} = \partial_y, \quad \mathbf{w_4} = y\partial_y + t\partial_t - w\partial_w \\
&\mathbf{w_5} = x\partial_x - 2y\partial_y - v\partial_v + w\partial_w. \quad (2.6)
\end{aligned}$$

Further, the set of all formed infinitesimal symmetries of the BKP system form a Lie algebra under Lie bracket $[\mathbf{w_i}, \mathbf{w_j}] = \mathbf{w_i}\mathbf{w_j} - \mathbf{w_j}\mathbf{w_i}$ that preserves skew-symmetry property with all diagonal entries zero Table 1.

Table 1 Commutator table of Lie algebra for mCBS equation

$[\mathbf{w_i}, \mathbf{w_j}]$	$\mathbf{w_1}$	$\mathbf{w_2}$	$\mathbf{w_3}$	$\mathbf{w_4}$	$\mathbf{w_5}$
$\mathbf{w_1}$	0	$-\mathbf{w_1}(\bar{f})$	0	$-\mathbf{w_1}(t\bar{f})$	$\mathbf{w_1}(f)$
$\mathbf{w_2}$	$\mathbf{w_1}(\bar{f})$	0	0	$\mathbf{w_2}$	0
$\mathbf{w_3}$	0	0	0	$\mathbf{w_3}$	$-2\mathbf{w_3}$
$\mathbf{w_4}$	$\mathbf{w_1}(t\bar{f})$	$-\mathbf{w_2}$	$-\mathbf{w_3}$	0	0
$\mathbf{w_5}$	$-\mathbf{w_1}(f)$	0	$2\mathbf{w_3}$	0	0

3 Exact Solutions by Similarity Reductions

To examine the exact solutions of mCBS Eqs. (1.1) and (1.2), some particular choices of the arbitrary occurring functions for further integration have been considered . Thus, the corresponding characteristic equation is formed as

$$\begin{aligned}\frac{\mathrm{d}v}{-m_4 v} &= \frac{\mathrm{d}w}{\frac{1}{4}(m_1\bar{h}_1 + m_2\bar{h}_2 + m_3\bar{h}_3 + m_4\bar{h}_4) + (m_4 - m_3)w} \\ &= \frac{\mathrm{d}x}{m_1 h_1 + m_2 h_2 + m_3 h_3 + m_4 h_4 + m_4 x} \\ &= \frac{\mathrm{d}y}{m_2 + (m_3 - 2m_4)y} = \frac{\mathrm{d}t}{m_1 + m_3 t}. \end{aligned} \tag{3.1}$$

Ultimately, this further follows the following cases for Eq. (3.1)

Case 1: Suppose that $m_3 \neq 0$ and $\frac{1}{m_3}(m_1 h_1 + m_2 h_2 + m_3 h_3 + m_4 h_4) = f_1(t)$. Also, set $M_1 = \frac{m_1}{m_3}$, $M_2 = \frac{m_2}{m_3}$ and $M_3 = \frac{m_4}{m_3}$. Thus, Eq. (3.1) takes the form as

$$\frac{\mathrm{d}v}{-M_3 v} = \frac{\mathrm{d}w}{\frac{1}{4}\bar{f}_1 + (M_3 - 1)w} = \frac{\mathrm{d}x}{M_3 x + f_1} = \frac{\mathrm{d}y}{M_2 + (1 - 2M_3)y} = \frac{\mathrm{d}t}{M_1 + t}. \tag{3.2}$$

Therefore, the similarity form of mCBS equations becomes

$$v = \frac{V(\xi_1, \eta_1)}{(M_1 + t)^{M_3}} \tag{3.3}$$

$$w = \left(W(\xi_1, \eta_1) + \frac{1}{4} \int \frac{\bar{f}_1(t)}{(M_1 + t)^{M_3}} \mathrm{d}t \right) (M_1 + t)^{(M_3 - 1)}, \tag{3.4}$$

where $V(\xi_1, \eta_1)$ and $W(\xi_1, \eta_1)$ are the similarity function in terms of similarity variables ξ_1 and η_1 provided by

$$\xi_1 = \frac{x}{(M_1 + t)^{M_3}} - \int \frac{f_1}{(M_1 + t)^{(M_3 + 1)}} \mathrm{d}t,$$

$$\eta_1 = \left(y - \frac{M_2}{2M_3 - 1} \right) (M_1 + t)^{(2M_3 - 1)}, \text{ with } M_3 \neq \frac{1}{2}.$$

Ultimately, similarity reduction of this case provides the following PDEs

$$V_{\xi_1 \xi_1 \eta_1} + (2M_3 - 1)\eta_1 V_{\eta_1} + 4V^2 V_{\eta_1} + 4V_{\xi_1} W - M_3 \xi_1 V_{\xi_1} - M_3 V = 0, \tag{3.5}$$

$$V V_{\eta_1} = W_{\xi_1}. \tag{3.6}$$

Moving further, the generated infinitesimals $\hat{\xi}$, $\hat{\eta}$, $\hat{\phi}$ and $\hat{\psi}$ are as follows

$$\hat{\xi} = m_5\xi_1 + m_6, \quad \hat{\eta} = -2m_5\eta_1 \quad \hat{\phi} = -m_5 V \quad \text{and} \quad \hat{\psi} = m_5 W + \frac{m_6 M_3}{4},$$

where m_5 and m_6 are the arbitrary constants. Thus, the characteristic equation is formed as

$$\frac{\mathrm{d}V}{-m_5 V} = \frac{\mathrm{d}W}{m_5 W + \frac{m_6 M_3}{4}} = \frac{\mathrm{d}\xi_1}{m_5\xi_1 + m_6} = \frac{\mathrm{d}\eta_1}{-2m_5\eta_1}. \tag{3.7}$$

Furthermore, solutions can be explored for the case mentioned below

Case 1a: If $m_5 \neq 0$, then Eq. (3.7) is furnished as

$$\frac{\mathrm{d}V}{-V} = \frac{\mathrm{d}W}{W + \frac{M_3 M_4}{4}} = \frac{\mathrm{d}\xi_1}{\xi_1 + M_4} = \frac{\mathrm{d}\eta_1}{-2\eta_1},$$

where $M_4 = \frac{m_6}{m_5}$. Thus,the similarity forms are given by

$$V = \sqrt{\eta_1} V_1(\xi_2), \quad W = \frac{M_3 M_4}{4} + \frac{W_1(\xi_2)}{\sqrt{\eta_1}}, \quad \text{with } \xi_2 = \eta_1(\xi_1 + M_4)^2. \tag{3.8}$$

Substituting the values of V and W in Eqs. (3.5) and (3.6), the attained values are

$$4\xi_2^2 \bar{\bar{\bar{V}}}_1 + 12\xi_2 \bar{\bar{V}}_1 + 4\xi_2 V_1^2 \bar{V}_1 + 8\sqrt{\xi_2} W_1 \bar{V}_1 - \xi_2 \bar{V}_1 + 3\bar{V}_1 + 2V_1^3 - \frac{V_1}{2} = 0, \tag{3.9}$$

$$2\xi_2 V_1 \bar{V}_1 + V_1^2 - 4\sqrt{\xi_2}\bar{W}_1 = 0. \tag{3.10}$$

The bar here stands for the differentiation w.r.t variable ξ_2. Therefore, the derived results for Eqs. (3.9) and (3.10) are

$$V_1 = \pm\frac{1}{2}, \quad W_1 = \frac{\sqrt{\xi_2}}{8} + c_1, \quad \text{where } c_1 \text{ is any arbitrary constant.} \tag{3.11}$$

Hence, solutions of Eq. (1.1) and (1.2) become

$$v = \pm\frac{\sqrt{\eta_1}}{2(M_1 + t)^{M_3}} \tag{3.12}$$

$$w = \left(\frac{M_3 M_4}{4} + \frac{(\xi_1 + M_4)}{8} + \frac{c_1}{\sqrt{\eta_1}} + \frac{1}{4}\int \frac{\bar{f}_1}{(M_1 + t)^{M_3}}\mathrm{d}t\right)(M_1 + t)^{(M_3 - 1)}, \tag{3.13}$$

where $\xi_1 = \frac{x}{(M_1+t)^{M_3}} - \int \frac{f_1}{(M_1+t)^{(M_3+1)}}\mathrm{d}t$ and $\eta_1 = \left(y - \frac{M_2}{2M_3 - 1}\right)(M_1 + t)^{(2M_3 - 1)}$.

Now, if $M_3 = \frac{1}{2}$ in Eq. (3.2), the following case arises:

Case 1b: If $2m_4 = m_3$, Eq. (3.2) becomes

$$\frac{2\mathrm{d}v}{-v} = \frac{4vw}{\bar{f_1} - 2w} = \frac{\mathrm{d}x}{f_1 + \frac{x}{2}} = \frac{\mathrm{d}y}{M_2} = \frac{\mathrm{d}t}{M_1 + t}.$$

Finally, the similarity form becomes

$$v = \frac{V(\xi_1, \eta_1)}{(M_1 + t)^{\frac{1}{2}}} \tag{3.14}$$

$$w = \frac{1}{4(M_1 + t)^{\frac{1}{2}}} \int \frac{\bar{f_1}(t)}{(M_1 + t)^{\frac{1}{2}}} \mathrm{d}t + \frac{W(\xi_1, \eta_1)}{(M_1 + t)^{\frac{1}{2}}}, \tag{3.15}$$

where ξ_1 and η_1 are provided as

$$\xi_1 = \frac{x}{(M_1 + t)^{\frac{1}{2}}} - \int \frac{f_1}{(M_1 + t)^{\frac{3}{2}}} \mathrm{d}t$$
$$\eta_1 = y - M_2 log(M_1 + t).$$

Thus, the first reductions are attained as

$$V_{\xi_1\xi_1\eta_1} + 4V^2 V_{\eta_1} + 4V_{\xi_1} W - \xi_1 \frac{V_{\xi_1}}{2} - M_2 V_{\eta_1} - \frac{V}{2} = 0, \quad V V_{\eta_1} = W_{\xi_1}. \tag{3.16}$$

Moving further, the generated infinitesimals $\hat{\xi}$, $\hat{\eta}$, $\hat{\phi}$ and $\hat{\psi}$ are as follows

$$\hat{\xi} = m_7, \quad \hat{\eta} = m_8 \quad \hat{\phi} = 0 \quad \text{and} \quad \hat{\psi} = \frac{m_7}{8},$$

where m_7 and m_8 are the arbitrary constants. Therefore, the Lagrange equation is formed as

$$\frac{\mathrm{d}V}{0} = \frac{8\mathrm{d}W}{m_7} = \frac{\mathrm{d}\xi_1}{m_7} = \frac{\mathrm{d}\eta_1}{m_8},$$

when $m_7 \neq 0$, the similarity form takes the form as

$$V = V_1(\xi_2), \quad W = \frac{\xi_1}{8} + W_1(\xi_2), \tag{3.17}$$

with similarity variable as $\xi_2 = \eta_1 - M_5\xi_1$, where $M_5 = \frac{m_8}{m_7}$. Substituting the values of V and W from Eq. (3.17) into Eq. (3.16), the obtained system of ODEs is

$$M_5^2 \bar{\bar{\bar{V}}}_1 + 4V_1^2 \bar{V_1} - 4M_5 W_1 \bar{V_1} - M_2 \bar{V_1} - \frac{V_1}{2} = 0, \qquad 8V_1 \bar{V_1} + 8M_5 \bar{V_1} - 1 = 0.$$

Thus, the required primitives are

$$V_1 = 0, \quad W_1 = \frac{\xi_2}{8M_5} + c_2, \text{ where } c_2 \text{ is an arbitrary constant.} \tag{3.18}$$

Thus, the required solutions for mCBS equation are

$$v = 0, \quad w = \frac{1}{4(M_1+t)^{\frac{1}{2}}} \int \frac{\bar{f}_1}{(M_1+t)^{\frac{1}{2}}} dt + \left(\frac{\xi_1}{8} + \frac{\eta_1 - M_5\xi_1}{8M_5} + c_2\right)(M_1+t)^{\frac{1}{2}}, \tag{3.19}$$

where $\xi_1 = \frac{x}{(M_1+t)^{\frac{1}{2}}} - \int \frac{f_1}{(M_1+t)^{\frac{3}{2}}} dt$ and $\eta_1 = y - M_2 log(M_1+t)$.

Case 2: For $m_3 = 0$ and $m_1 \neq 0$, suppose that $\frac{1}{m_1}(m_1h_1 + m_2h_2 + m_4h_4) = f_2(t)$, thus, the Lagrange equation is formed as

$$\frac{dv}{-M_7 v} = \frac{dw}{\frac{\bar{f}_2}{4} + M_7 w} = \frac{dx}{f_2 + M_7 x} = \frac{dy}{M_6 - 2M_7 y} = \frac{dt}{1},$$

where $M_6 = \frac{m_2}{m_1}$ and $M_7 = \frac{m_4}{m_1}$. This leads to the following similarity form

$$v = V(\xi_1, \eta_1)e^{-M_7 t}, \quad w = W(\xi_1, \eta_1)e^{M_7 t} + \frac{e^{M_7 t}}{4} \int \bar{f}_2(t)e^{-M_7 t} dt, \tag{3.20}$$

with the similarity variables defined by

$$\xi_1 = xe^{-M_7 t} - \int f_2 e^{-M_7 t} dt, \quad \eta_1 = \left(y - \frac{M_6}{2M_7}\right)e^{2M_7 t}. \tag{3.21}$$

Now, using the values of Eqs. (3.20) and (3.21) to Eqs. (1.1) and (1.2), the second stage reduction becomes

$$V_{\xi_1\xi_1\eta_1} + 4V^2 V_{\eta_1} + 4WV_{\xi_1} + 2M_7\eta_1 V_{\eta_1} - M_7\xi_1 V_{\xi_1} - M_7 V = 0, \quad VV_{\eta_1} = W_{\xi_1}. \tag{3.22}$$

Further, the following infinitesimals are established

$$\hat{\xi} = m_9\xi_1 + m_{10}, \quad \hat{\eta} = -2m_9\eta_1, \quad \hat{\phi} = -m_9 V \quad \text{and} \quad \hat{\psi} = m_9 W + \frac{a_{10}M_7}{4},$$

where m_9 and m_{10} are the arbitrary constants. Therefore, the Lagrange equation is formed as

$$\frac{dV}{-V} = \frac{dW}{W + \frac{M_7 M_8}{4}} = \frac{d\xi_1}{\xi_1 + M_8} = \frac{d\eta_1}{-2\eta_1},$$

where $M_8 = \frac{m_{10}}{m_9}$, thus, the similarity form achieved is

$$V = \sqrt{\eta_1} V_1(\xi_2), \qquad W = -\frac{M_7 M_8}{4} + \frac{W_1(\xi_2)}{\sqrt{\eta_1}}, \tag{3.23}$$

with $\xi_2 = \eta_1(\xi_1 + M_8)^2$, therefore, the second stage reduction becomes

$$4\xi_2^2 \bar{\bar{\bar{V}}}_1 + 12\xi_2 \bar{\bar{V}}_1 + 4\xi_2 V_1^2 \bar{V}_1 + 8\sqrt{\xi_2} W_1 \bar{V}_1 + 3\bar{V}_1 + 2V_1^3 = 0, \tag{3.24}$$

$$2\xi_2 V_1 \bar{V}_1 + V_1^2 - 4\sqrt{\xi_2} \bar{W}_1 = 0. \tag{3.25}$$

Thus, the solutions derived for Eqs. (3.24) and (3.25) are

$$V_1 = \frac{c_3}{\sqrt{\xi_2}}, \quad W_1 = 0,$$

where c_3 is any arbitrary constant. Subsequently, the solutions becomes

$$v = \frac{c_3}{(\xi_1 + M_8)} e^{-M_7 t}, \qquad w = -\frac{M_7 M_8}{4} e^{M_7 t} + \frac{e^{M_7 t}}{4} \int \bar{f}_2 e^{-M_7 t} \mathrm{d}t. \tag{3.26}$$

Case 3: For $m_3 = m_4 = 0$ and $m_1 \neq 0$, suppose that $\frac{1}{m_1}(m_1 h_1 + m_2 h_2) = f_3(t)$, Eq. (3.1) gives

$$\frac{\mathrm{d}v}{0} = \frac{4\mathrm{d}w}{\bar{f}_3} = \frac{\mathrm{d}x}{f_3} = \frac{\mathrm{d}y}{M_6}.$$

Thus, the following similarity reductions are formed

$$v = V(\xi_1, \eta_1), \qquad w = \frac{f_3(t)}{4} + W(\xi_1, \eta_1),$$

with similarity variables as

$$\xi_1 = x - \int f_3 \mathrm{d}t, \qquad \eta_1 = y - M_6 t.$$

Ultimately, these transformations gives

$$V_{\xi_1 \xi_1 \eta_1} + 4V^2 V_{\eta_1} + 4W V_{\xi_1} - M_6 V_{\eta_1} = 0 \tag{3.27}$$

$$V V_{\eta_1} = W_{\xi_1}. \tag{3.28}$$

Further, the following infinitesimals are established

$$\hat{\xi} = 1, \quad \hat{\eta} = H_1(\eta_1) \quad \hat{\phi} = 0 \quad \text{and} \quad \hat{\psi} = -\bar{H}_1(\eta_1) V.$$

Therefore, similarity forms are made as

$$V = V_1(\xi_2), \quad W = \frac{W_1(\xi_2)}{H_1(\eta_1)},$$

with $\xi_2 = \xi_1 - \int \frac{1}{H_1} \mathrm{d}\eta_1$. Finally, the second stage reduction becomes

$$\bar{\bar{\bar{V}}}_1 + 4V_1^2\bar{V}_1 - 4W_1\bar{V}_1 - M_6\bar{V}_1 = 0, \tag{3.29}$$

$$V_1\bar{V}_1 + \bar{W}_1 = 0. \tag{3.30}$$

On integrating, Eq. (3.30), w.r.t ξ_2 the value of W_1 is

$$W_1 = c_4 - \frac{1}{2}V_1^2. \tag{3.31}$$

Now, integrating Eq. (3.29), obtained ODE becomes

$$\bar{V}_1^2 + V_1^4 - M_9V_1^2 - c_5V_1 - c_6 = 0, \tag{3.32}$$

where $M_9 = 4c_4 + M_6$, c_4, c_5 and c_6 are integration constant. By making some adjustments in the constants, different solutions can be derived as

Case 3a: For $M_9 = c_5 = c_6 = 0$, Eq. (3.32) gives

$$\bar{V}_1^2 + V_1^4 = 0.$$

Therefore, the solutions are

$$V_1 = \pm\frac{i}{c_7 - \xi_2}, \qquad W_1 = c_4 + \frac{1}{2(c_7 - \xi_2)^2}, \quad c_7 \text{ is an arbitrary constant.} \tag{3.33}$$

So, the solutions of mCBS equation are

$$v = \pm\frac{i}{(c_7 - \xi_1 + \int \frac{1}{H_1}\mathrm{d}\eta_1)}, \qquad w = \frac{2c_4(c_7 - \xi_1 + \int \frac{1}{H_1}\mathrm{d}\eta_1)^2 - 1}{2(c_7 - \xi_1 + \int \frac{1}{H_1}\mathrm{d}\eta_1)^2 H_1} + \frac{f_3}{4}. \tag{3.34}$$

Case 3b: For $c_5 = 0$ and $c_6 = -\frac{M_9^2}{4}$, Eq. (3.32) is reduced to

$$\bar{V}_1^2 + \left(V_1^2 - \frac{M_9}{2}\right)^2 = 0. \tag{3.35}$$

Now, if $M_9 = -2k_1^2$, then the solutions for Eq. (3.35) are attained as

$$V_1 = \pm i k_1 \tan(k_1 \xi_2 + c_8), \qquad W_1 = c_4 \pm \frac{1}{2} k_1^2 \tan^2(k_1 \xi_2 + c_8).$$

Therefore, the required solutions fot Eq. (1.1) and (1.2) are

$$v = \pm i k_1 \tan\left[k_1(\xi_1 - \int \frac{1}{H_1} \mathrm{d}\eta_1) + c_8\right],$$

$$w = \frac{2c_4 \pm k_1^2 \tan^2\left[k_1(\xi_1 - \int \frac{1}{H_1} \mathrm{d}\eta_1) + c_8\right]}{2H_1} + \frac{f_3}{4}. \tag{3.36}$$

Similarly, for $M_9 = 2k_2^2$, solutions are attained as

$$V_1 = \pm k_2 \tanh(i k_2 \xi_2 + c_9), \qquad W_1 = c_4 - \frac{1}{2} k_2^2 \tanh^2(i k_2 \xi_2 + c_9).$$

Therefore, solutions of Eq. (1.1) and (1.2) are

$$v = \pm k_2 \tanh\left[i k_2(\xi_1 - \int \frac{1}{H_1} \mathrm{d}\eta_1) + c_9\right],$$

$$w = \frac{2c_4 - k_2^2 \tanh^2\left[i k_2(\xi_1 - \int \frac{1}{H_1} \mathrm{d}\eta_1) + c_9\right]}{2H_1} + \frac{f_3}{4}. \tag{3.37}$$

Another form of the solutions takes the form as

$$V_1 = \pm k_2 \coth(i k_2 \xi_2 + c_{10}), \qquad W_1 = c_4 - \frac{1}{2} k_2^2 \coth^2(i k_2 \xi_2 + c_{10}).$$

Ultimately, the solutions of mCBS equation are

$$v = \pm k_2 \coth\left[i k_2(\xi_1 - \int \frac{1}{H_1} \mathrm{d}\eta_1) + c_{10}\right],$$

$$w = \frac{2c_4 - k_2^2 \coth^2\left[i k_2(\xi_1 - \int \frac{1}{H_1} \mathrm{d}\eta_1) + c_{10}\right]}{2H_1} + \frac{f_3}{4}. \tag{3.38}$$

where $\xi_1 = x - \int f_3 \mathrm{d}t$, $\eta_1 = y - M_6 t$, c_8, c_9 and c_{10} are any arbitrary constants.

The derived solutions provided by Eqs. (3.34), (3.37) and (3.38) are complex wave or current solutions. It can be easily verified by direct substituting these values in their respective equations. Likewise, we can express the exact solutions both as real wave solutions and complex wave solutions for the wave equation. Mathematically, the complex solutions are more succinct than the real solutions, while on the other hand, the complex representation also makes wave superposition's easier to handle. In [17–20], some exact complex wave solutions of nonlinear differential equations have been established. To determine the behaviour of these solutions, in [18], the absolute values of the solutions have been plotted. The complex functions have essential concrete applications in applied sciences such as control theory, signal processing, vibration analysis, cartography, electromagnetism, quantum mechanics, and many others. The complex solutions here are periodic or quasi-periodic types.

Comparing the derived results with the previous findings in [15], they had considered another form of mCBS equation. The derived solutions in [15] are corresponding to each subalgebra made. Thus, they established the results for some classes, while the present article deals with the more general case for tracing out the solutions. In [14], they had investigated the integrable equation by combining the modified Calogero-Bogoyavlenskii-Schiff equation with its negative order (mCBS-nmCBS). The mCBS-nmCBS equation was successfully derived by combining modified CBS recursion operator with its inverse form. Thus, the results established in the present article are different from that of [14, 15].

The derived invariant solutions can be set as the standard benchmarks for finding the numerical solutions by the computational algorithms. These solutions are then compared with different numerical techniques to reduce the error's occurred in the solutions. Therefore, the exact solutions provided here can be used as a comparison tool.

4 Discussions

In order to understand the physical richness of the established results, it is important to know the nature of the results. The behaviour of the derived results is studied by the MATLAB simulation. In this section, the space domains are set as $-20 \leq x, y \leq 20$. The complex wave solutions given by Eqs. (3.34), (3.37) and (3.38) are plotted with their absolute values. Now, due to the presence of various arbitrary constants and functions, these exact solutions provide rich physical structure. By assuming the appropriate choices for the arbitrary constants and functions, execution of codes is made. These assumptions are made on the basis of some factors like looking the defined domain for the existence of the solutions and its singularities. For example the solutions mentioned in Eqs. (3.36) and (3.36) are not defined for $\tan\frac{\pi}{2}$. The traced solutions are mainly of soliton types like evolutionary wave, multisoliton and single soliton. The details have been illustrated as follows:

Figure 1: Expressions v and w at $t = 0.6892$ given by Eqs. (3.12) and (3.13) have been traced by making $c_1 = 0.5499,\ m_1 = 0.1656,\ m_2 = 0.6020,\ m_3 = 0.2630,\ m_4 = 0.6541,\ b_1 = 0.7482,\ b_2 = 0.4505,\ b_3 = 0.0838$, and the value for arbitrary function is made as $f_1(t) = (M_1 + t)^{(M_3+1)}(2b_1t + b_2)sec^2(b_1t^2 + b_2t + b_3)$. Thus, both the profiles show dynamical structure.

Figure 2: Expression v represented by Eq. (3.26) has been plotted with the assumptions of arbitrary constants as $m_1 = 0.8034,\ m_4 = 0.0605,\ m_{10} = 0.4509,\ m_9 = 0.5470,\ c_3 = 0.7447$. The graph shows the soliton behaviour. Meanwhile, 2D plot of w has been traced for $f_2 = e^{M_7t}$, keeping the rest values same as taken for v.

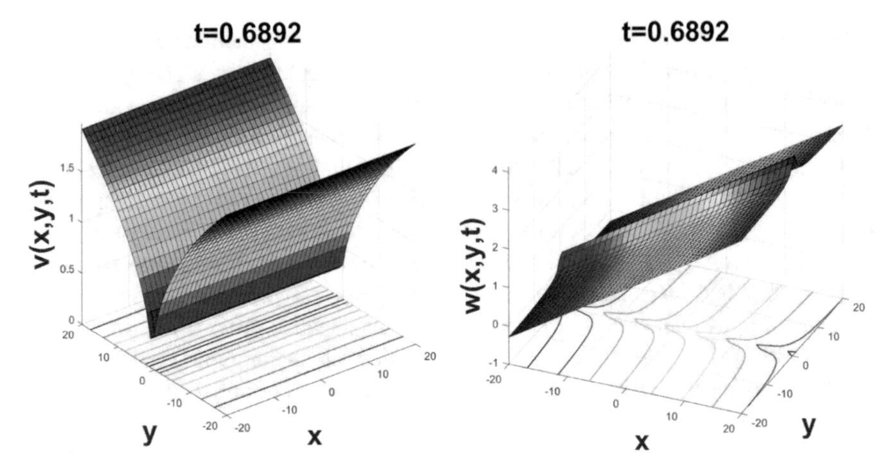

Fig. 1 Single soliton wave profile given by Eq. (3.12) and (3.13) with their respective contour plots at $t = 0.6892$

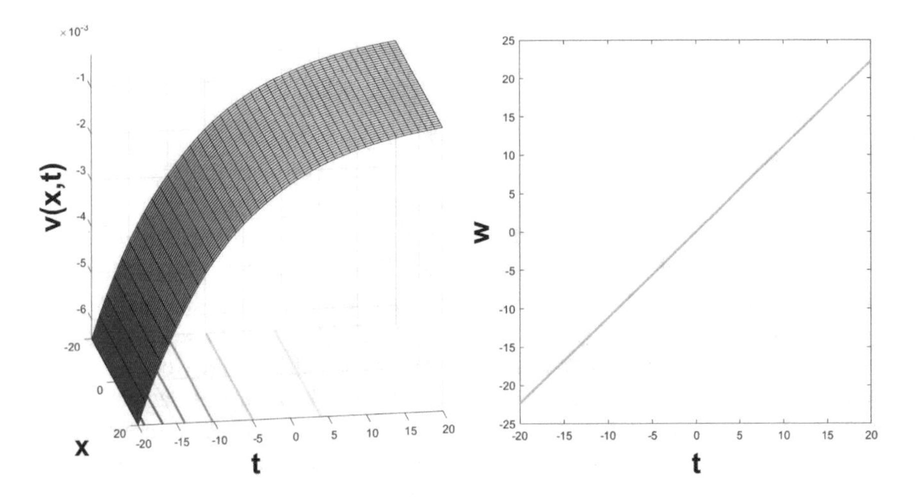

Fig. 2 Evolutionary wave profiles of v and w given by Eq. (3.26) along with its contour plot

Figure 3: At $t = 0.4587$, the dynamical Eqs. (3.34) and (3.36) with the absolute values of v and w have been pictured by setting arbitrary constants as $c_7 = 0.4714$, $c_8 = 0.5225$, $k_1 = 0.9937$, $c_4 = 0.0358$, $m_2 = 0.6073$, $m_1 = 0.4501$, $b_1 = 0.4735$, $b_2 = 0.1527$, $b_3 = 0.3411$ as well as arbitrary functions as $f_3(t) = (2b_1t + b_2)e^{(b_1t^2+b_2t+b_3)}$ and $H_1(Y) = \frac{\cos^2(b_1Y^2+b_2Y+b_3)}{(2b_1Y+b_2)}$. The graph shows the multi-soliton and single soliton behaviour of the expressions.

Figure 4: Expressions of v and w given by Eqs. (3.37) and (3.38) with their absolute values have been plotted by assuming the arbitrary constants as $c_9 = 0.1097$, $k_2 = 0.4484$ and $c_{10} = 0.4046$, while other constants and arbitrary functions remain same as Fig. (3). The quasi-periodic and multisoliton behaviours can be easily seen.

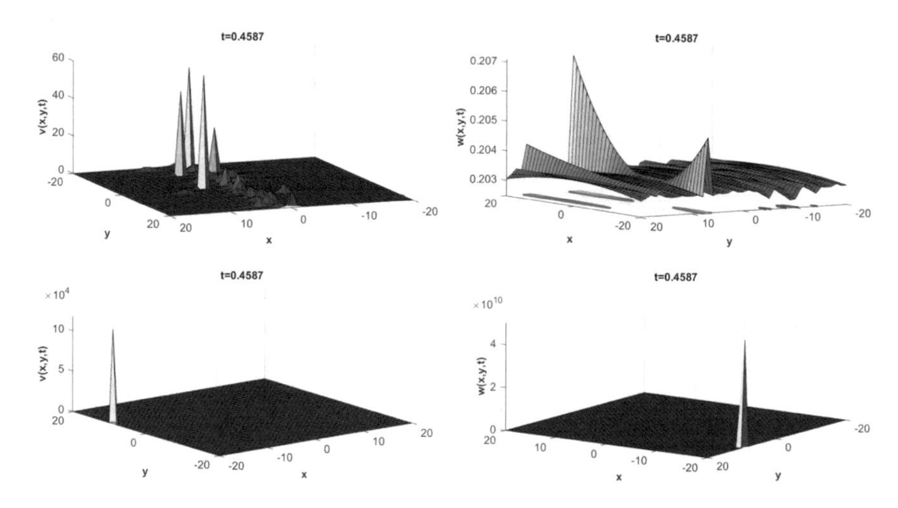

Fig. 3 Multisoliton and single soliton wave expressions $|v|$ and $|w|$ given by Eqs. (3.34) and (3.36) at $t = 0.4587$

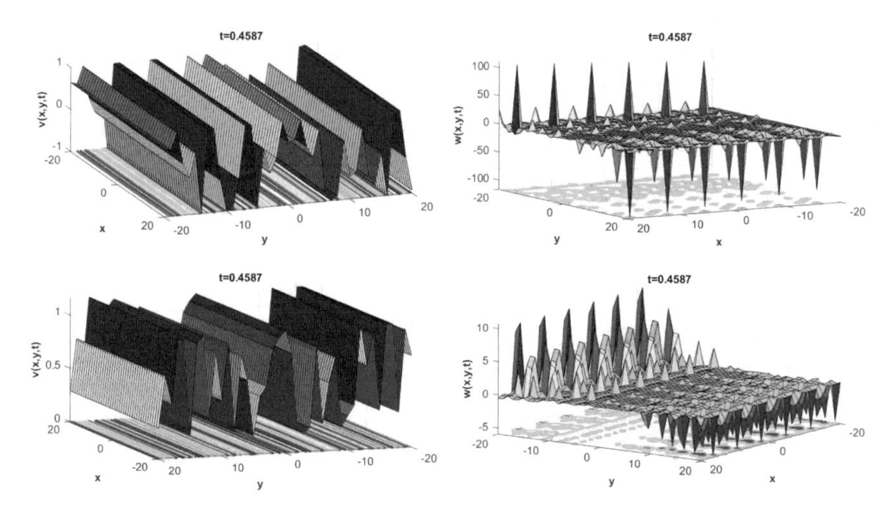

Fig. 4 Quasi-periodic and multisoliton expressions $|v|$ and $|w|$, respectively, given by Eqs. (3.37) and (3.38) along with its contour plots at $t = 0.4587$

5 Conclusions

The present work provides some invariant solutions of (2+1)-dimensional nonlinear modified Calogero-Bogoy-avlenskii-Schiff(mCBS) equation. The exact solutions are derived by using Lie symmetry reduction. Lie algebra, vector fields of the mCBS equation are examined. The analytical solutions obtained here contain different arbitrary constants and functions; thus, their evolutionary profiles are traced through numerical simulations. The derived results are mainly soliton type like evolutionary wave, single soliton and multisoliton, which explain the mathematical physical processes. All the solutions are different from that of [15] and are absolutely new. The established invariant solutions play a key role in the study of fibre optics, plasma physics, chemical sciences and other applied fields. The Lie symmetry method used here can be used in future for finding other soliton solutions like rogue wave, breather wave and others. The numerical solutions can be considered as a subset of these invariant solutions.

Conflict of interest The author declares that there is no conflict of interest.

References

1. Russo M, Choudhury SR (2017) Analytic solutions of a microstructure PDE and the KdV and Kadomtsev–Petviashvili equations by invariant Painlevé analysis and generalized Hirota techniques. Appl Math Comput 311:228
2. Zayed EME, Elshater MEM (2017) Jacobi elliptic solutions, soliton solutions and other solutions to four higher-order nonlinear Schrodinger equations using two mathematical methods. Optik 131:1044
3. Matveev VB, Salle MA (1991) Darboux transformations and solitons. Springer-Verlag, Berlin
4. Yomba E (2010) Jacobi elliptic function solutions of the generalized Zakharov–Kuznetsov equation with nonlinear dispersion and t-dependent coefficients. Phys Lett A 374:1611
5. Lu D, Seadawy AR, Arshad M (2017) Applications of extended simple equation method on unstable nonlinear Schrödinger equations. Optik 140:136
6. Ablowitz MJ, Ladik JK (1977) On the solution of a class of nonlinear partial difference equations. Appl Math 57:1
7. Wang ML, Li XZ, Zhang JL (2007) Sub-ODE method and solitary wave solutions for higher order nonlinear Schrödinger equation. Phys Lett A 363:96
8. Tariq KUH, Seadawy AR (2017) Bistable Bright-Dark solitary wave solutions of the (3 + 1)-dimensional Breaking soliton, Boussinesq equation with dual dispersion and modified Korteweg–de Vries–Kadomtsev–Petviashvili equations and their applications. Results Phys 7:1143
9. Manukure S, Zhou Y (2019) A (2 + 1)-dimensional shallow water equation and its explicit lump solutions. Int J Mod Phys B 33:1950038
10. Bluman GW, Cole JD (1974) Similarity methods for differential equations, 1st edn. Springer-Verlag, New York
11. Schiff J (1992) Painlevé transcendents: their asymptotics and physical applications, 1st edn. Springer, New York
12. Olver PJ (1993) Applications of lie groups to differential equations, 1st edn. Springer-Verlag, New York
13. Bluman GW, Kumei S (1989) Symmetries and differential equations, 1st edn. Springer-Verlag, New York
14. Kumar S, Kumar D (2019) Lie symmetry analysis, complex and singular solutions of (2 + 1)-dimensional combined MCBS–nMCBS equation. Int J Dynam Control 7:496
15. Kumar S, Kumar D (2020) Lie symmetry analysis and dynamical structures of soliton solutions for the (2 + 1)-dimensional modified CBS equation. Int J Modern Phys B 34:2050221
16. Kumar M, Manju K (2021) Solitary wave solutions of mKdV–Calogero–Bogoyavlenskii–Schiff equation by using Lie symmetry analysis. Int J Geomet Methods Mod Phys 18:2150028
17. Zayed EME, Gepreel KA, El Horbaty MM (2008) Exact solutions for some non-linear differential equations using complex hyperbolic function methods. Appl Anal Int J 87:509

18. Sulaiman TA et al (2020) Nonautonomous complex wave solutions to the $(2 + 1)$-dimensional variable-coefficients nonlinear Chiral Schrödinger equation. Results Phys 19:103604
19. Yadav S, Chauhan A, Arora R (2021) Invariance analysis, optimal system and conservation laws of $(2 + 1)$-dimensional non-linear Vakhnenko equation. Pramana J Phys 95:8
20. Chauhan A, Sharma K, Arora R (2020) Lie symmetry analysis, optimal system, and generalized group invariant solutions of the $(2 + 1)$-dimensional Date–Jimbo–Kashiwara–Miwa equation. Math Meth Appl Sci 43:8823
21. Satapathy P, Sekhar TR, Zeidan D (2020) Codimension two Lie invariant solutions of the modified Khokhlov–Zabolotskaya–Kuznetsov equation. Math Meth Appl Sci 44:1
22. Satapathy P, Sekhar TR (2018) Optimal system, invariant solutions and evolution of weak discontinuity for isentropic drift flux model. Appl Math Comput 334:107
23. Sil S, Sekhar TR, Zeidan D (2020) Nonlocal conservation laws, nonlocal symmetries and exact solutions of an integrable soliton equation. Chaos Sol Fract 139:110010
24. Sil S, Sekhar TR (2021) Nonclassical symmetry analysis, conservation laws of one-dimensional macroscopic production model and evolution of nonlinear waves. J Math Anal Appl 97:124847
25. Wazwaz AM (2018) Painlevé analysis for a new integrable equation combining the modified Calogero–Bogoyavlenskii–Schiff (MCBS) equation with its negative-order form. Nonlinear Dyn 91:877
26. Kumar M, Manju K (2020) Closed form invariant solutions of $(2 + 1)$-dimensional extended shallow water wave equation via Lie approach. Eur Phys J Plus 135:803
27. Calogero F (1975) A method to generate solvable nonlinear evolution equations. Lett Nuovo Cimento 14:443
28. Wazwaz AM (2008) Erratum to: Multiple-soliton solutions for the ninth-order KdV equation and sixth-order Boussinesq equation. Appl Math Comput 203:592
29. Bogoyavlenskii OI (1990) Breaking solitons in $(2 + 1)$-dimensional integrable equations. Russ Math Surv 45:1
30. Ray SS (2018) Lie symmetry analysis and reduction for exact solution of (2+1)-dimensional Bogoyavlensky–Konopelchenko equation by geometric approach. Modern Phys Lett B 32(11):1850127
31. Bruzon MS et al (2003) The Calogero–Bogoyavlenskii–Schiff equation in $(2 + 1)$ dimensions. Theor Math Phys 137:1367
32. Kumar R (2016) Application of Lie-group theory for solving Calogero–Bogoyavlenskii–Schiff equation. IOSR J Math (IOSR-JM) 12:144
33. Kumar R, Kumar M, Tiwari AK (2014) Dynamics of some more invariant solutions of $(3 + 1)$-Burgers system. Int J Comput Methods Eng Sci Mech. https://doi.org/10.1080/15502287.2021.1916693.
34. Kumar M, Kumar R, Kumar A (2021). Some more invariant solutions of $(2 + 1)$-water waves. Int J Appl Comput Math. https://doi.org/10.1007/s40819-020-00945-9
35. Vinita S, Ray SS (2020) On the invariant analysis, symmetry reduction with group-invariant solution and the conservation laws for $(2 + 1)$-dimensional modified Heisenberg ferromagnetic system. Int J Modern Phys B 34(31):2050305

Estimation and Classification Using Samples from Two Logistic Populations with a Common Scale Parameter

Pushkal Kumar and Manas Ranjan Tripathy

Abstract This paper deals with the estimation and classification of two logistic populations with a common scale and different location parameters. Utilizing the Metropolis–Hastings method, we compute the Bayes estimators of the associated unknown parameters. For this purpose, we consider gamma priors for the common scale parameter and normal priors for two location parameters. These Bayes estimators are compared with some of the existing estimators in terms of their bias and the mean squared error numerically. Moreover, utilizing these estimators for the associated parameters, we construct some classification rules in order to classify a single observation into one of the two logistic populations under the same model. The performances of each of the classification rules are evaluated through expected probability of misclassification, numerically. Finally, two real-life data sets have been considered in order to show the potential application of the model problem.

Keywords Bayes estimator · Classification rules · Expected probability of misclassification · Metropolis–Hastings procedure · Numerical comparison

1 Introduction

Suppose we have random samples from two logistic populations with a common scale parameter σ and possibly distinct and unknown location parameters μ_1 and μ_2. Specifically, let $\underset{\sim}{X} = (X_1, X_2, \ldots, X_m)$ and $\underset{\sim}{Y} = (Y_1, Y_2, \ldots, Y_n)$ be m and n independent random samples taken from two logistic populations Logistic(μ_1, σ) and Logistic(μ_2, σ) respectively. Here we denote Logistic(μ_i, σ); $i = 1, 2$ as the logistic population having the probability density function

P. Kumar (✉) · M. R. Tripathy
Department of Mathematics, National Institute of Technology, Rourkela, Odisha 769008, India
e-mail: pushkalkumar0812@gmail.com

M. R. Tripathy
e-mail: manas@nitrkl.ac.in

S. S. Ray et al. (eds.), *Applied Analysis, Computation and Mathematical Modelling in Engineering*, Lecture Notes in Electrical Engineering 897,
https://doi.org/10.1007/978-981-19-1824-7_14

$$f(x; \mu_i, \sigma) = \frac{\frac{1}{\sigma}\exp\{-(\frac{x-\mu_i}{\sigma})\}}{[1+\exp\{-(\frac{x-\mu_i}{\sigma})\}]^2}, \quad -\infty < x < \infty, \quad -\infty < \mu_i < \infty, \ \sigma > 0 \tag{1.1}$$

The problem is to classify a new observation into one of the two logistic populations using some of the estimators of the common scale parameter and two different location parameters. First, we will derive certain estimators of the associated parameters such as σ, and the other two nuisance parameters μ_1 and μ_2 from a Bayesian perspective. Using these estimators, we will construct some classification rules to classify a new observation into one of the two logistic populations using training samples. The performances of all the estimators will be evaluated through their bias

$$E(d-\theta) \tag{1.2}$$

and the mean squared error (MSE)

$$E(d-\theta)^2, \tag{1.3}$$

where 'd' is an estimator for estimating the parameter θ.

In a general framework, the problem of classification is to classify a new observation, say z into one of the several populations/groups, say $\pi_1, \pi_2, \ldots, \pi_k$, using certain methodologies and training data. The classification problem can be seen in machine learning for image classification, text classification, electrical science for signal classification, military surveillance for recognizing speech, in biomedical sciences for taxonomic study. It also has an application in computer science for classifying the object using receiver operating characteristic (ROC) curve.

Researchers in the past have considered the problem of classification under the equality assumption of parameters. For example, [12] considered the classification problem under equality assumption on the location parameters using training samples from several shifted exponential populations. Utilizing some well-known estimators of the common location parameter, the authors proposed several classification rules. The performances of all the rules have been evaluated using the expected probability of misclassification. Further, [13] considered the classification rules for the two Gaussian populations with a common mean. It is noted that the classification problem was first considered by Fisher in the year 1936 in a general framework. However, under multivariate normal model, the problem of classification was considered by [1, 2, 16]. Basu [5] proposed the classification rules for the two shifted exponential distribution. Note that [2] considered the classification problem under multivariate normal set up with a common co-variance matrix. This model reduces the problem under common standard deviation (scale) in the case of univariate normal.

Contrary to the above, significantly less attention has been paid to constructing classification rules and studying their performances when the distribution is not normal and exponential. However, there are many practical scenarios where the emerging data sets satisfactorily modeled by logistic distribution. In view of this, we take two logistic populations having a common scale parameter (which is also the standard error) and two different location parameters and construct several classification rules. In this regard, we have considered two real-life data sets from civil engineering science, and the other is related to food science, which has been satisfactorily modeled by using logistic distribution. The details of which have been discussed in Sect. 5. Using the estimators under the same model problem, we construct specific classification rules that may classify a new observation. The applications of logistic distribution are seen in several fields of study. Logistic distribution is use to describe the modeling agriculture production data, growth model in a demographic study (see [4]), analysis of biomedical data, life testing data and bio-assay (see [14, 21]).

Several authors have considered the problem of estimating parameters under equality restrictions in the past. For recent detailed and updates literature on estimating parameters under equality restriction under the various probabilistic model, we refer to [18–20, 23–26]. Recently, [20] studied the estimation of common scale parameter (σ) of two logistic populations with the unknown and unequal location parameters (μ_i) from Bayesian point of view. They derived the MLEs and individual approximate Bayes estimators of the parameter using non-informative and informative priors, and using the Monte Carlo simulation study; they compare the performances of all the proposed Bayes estimators numerically.

The researchers studied the point and interval estimation of logistic population parameters, namely the location and scale parameters, from classical and Bayesian points of view. The first attempt to estimate the logistic distribution parameters was made by [9], where they derived the linear unbiased estimators based on sample quantiles and some approximate unbiased linear estimators and compared their efficiencies numerically. One may refer to [3, 7, 11, 17] for results on estimating scale and location parameters of a logistic distribution.

The remaining contribution of the current research work can be described as follows. In Sect. 2, we discuss the Bayes classification rule in a general framework and then obtained the classification rule for our proposed model set up. Consequently, using the maximum likelihood estimators (MLEs) of the associated parameters, we construct a classification rule. In Sect. 3, we consider Bayes estimators of the associated parameters using Markov chain Monte Carlo methods. Using these estimators, classification rule also has been constructed. In Sect. 4, we carry out a detailed simulation study and compute the bias and the mean squared error of the proposed estimators. Moreover, these estimators are compared with some of the existing estimators previously proposed by Nagamani [20] for the associated parameters. Further classification rules are constructed using the estimators proposed by Nagamani [20] and compared with the rules using MLEs and the MCMC method. In Sect. 5, we present two real-life data sets, which has been satisfactorily modeled by logistic distributions with common scale parameter. A new observation has been classified into one of the data sets, using the proposed classification rules. Concluding remarks are given in Sect. 6.

2 Classification Rule Using the MLE for Two Logistic Populations

In this section, we construct a classification rule using the MLEs of the associated parameters, σ, μ_1 and μ_2 to the model (1.1).

Suppose we have training samples $\underset{\sim}{X} = (X_1, X_2, \ldots, X_m)$ and $\underset{\sim}{Y} = (Y_1, Y_2, \ldots, Y_n)$ from two Logistic populations Π_1 and Π_2 respectively, where Π_i is the population having density $f_i(x; \mu_i, \sigma)$, $i = 1, 2$. Now a new observation, say z is classified into the population Π_1 if

$$\log\left(\frac{f_1(z)}{f_2(z)}\right) \geq \log\left(\frac{C(2|1) * q_2}{C(1|2) * q_1}\right) \tag{2.1}$$

and into the population Π_2 if

$$\log\left(\frac{f_1(z)}{f_2(z)}\right) < \log\left(\frac{C(2|1) * q_2}{C(1|2) * q_1}\right). \tag{2.2}$$

Here $C(i|j)$ is the cost of misclassification when an observation is from Π_j and is misclassified into the population Π_i, $i \neq j$, $i, j = 1, 2$. If we assume that $C(2|1) = C(1|2)$ and $q_2 = q_1$, where q_i is the prior probability associated to the population Π_i then, the classification function for classifying an observation z into Logistic(μ_1, σ) or Logistic(μ_2, σ) obtained as

$$\begin{aligned} W(z) = \frac{1}{\sigma}\Big((z-\mu_2) - (z-\mu_1)\Big) + 2\log\left(1 + \exp\left(-\frac{(z-\mu_2)}{\sigma}\right)\right) \\ -2\log\left(1 + \exp\left(-\frac{(z-\mu_1)}{\sigma}\right)\right). \end{aligned} \tag{2.3}$$

When the parameters are unknown, a natural approach is to replace the parameters by their respective estimators and obtain the classification rule. Thus utilizing the classification function $W(z)$, we propose classification Rule R as: classify z into the population Π_1 if $W(z) \geq 0$, else classify z into the population Π_2.

Let us denote $P(i|j, R)$, $(i \neq j,\ i, j = 1, 2)$ as the probability of misclassification for an observation from Π_j when it is misclassified into the population Π_i. The new observation z is classified into one of the two populations, such that the probability of misclassification becomes the minimum. We refer to Anderson [1], for the details of the classification rules using prior information.

In order to obtain the MLEs of the associated parameters, let us consider the log-likelihood function under the current model setup which is given by

$$L(\underset{\sim}{y}; \underset{\sim}{x}, \underset{\sim}{y}) = -\log \sigma(m+n) - \sum_{i=1}^{m}\Big(\frac{x_i - \mu_1}{\sigma}\Big) - \sum_{j=1}^{n}\Big(\frac{y_j - \mu_2}{\sigma}\Big)$$
$$-2\sum_{i=1}^{m}\log\Big(1 + \exp\Big\{-\Big(\frac{x_i - \mu_1}{\sigma}\Big)\Big\}\Big)$$
$$-2\sum_{j=1}^{n}\log\Big(1 + \exp\Big\{-\Big(\frac{y_j - \mu_2}{\sigma}\Big)\Big\}\Big), \quad (2.4)$$

where $-\infty < \mu_i < \infty (i = 1, 2); \sigma > 0$.

Note that, under the same model setup the MLEs of the parameters σ, μ_1 and μ_2 have been numerically obtained by Nagamani [20]. Let us denote the MLEs of the parameters σ, μ_1 and μ_2 respectively by $\hat{\sigma}_{ml}$, $\hat{\mu}_{1ml}$ and $\hat{\mu}_{2ml}$. For computing the MLEs, using Newton–Raphson method, we have taken the initial guess as suggested by Nagamani [20].

Now using the MLEs of the parameters μ_1, μ_2 and σ, we define a classification function, say $W_{ML}(z)$ as,

$$W_{ML}(z) = \frac{1}{\hat{\sigma}_{ml}}\Big((z - \hat{\mu}_{2ml}) - (z - \hat{\mu}_{1ml})\Big) + 2\log\Big(1 + \exp\Big(-\frac{(z - \hat{\mu}_{2ml})}{\hat{\sigma}_{ml}}\Big)\Big)$$
$$-2\log\Big(1 + \exp\Big(-\frac{(z - \hat{\mu}_{1ml})}{\hat{\sigma}_{ml}}\Big)\Big). \quad (2.5)$$

Using this classification function $W_{ML}(z)$, we propose a classification rule, say R_{ML} as: classify z into Π_1 if $W_{ML}(z) \geq 0$, else classify z into the Π_2.

3 Bayesian Estimation of Parameters Using MCMC Approach and Classification Rule

It is noted that certain Bayes estimators of parameters using Lindley's approximation method and different types of priors with respect to the squared error loss function and the LINUX loss function have been obtained by Nagamani [20]. Here, we consider Bayes estimators of the associated parameters using MCMC method along with the Metropolis–Hastings algorithm. Moreover, utilizing these estimators we propose a classification rule. Metropolis–Hastings algorithm is an important class of MCMC method, which is applied to get the samples from the posterior distribution effectively in a systematic manner.

A special case of Metropolis–Hastings method is the symmetric random walk Metropolis (RWM) algorithm. In this method, we take the Markov chain described by c alone to be a simple symmetric random walk, so that $c(\delta, \delta^*) = c(\delta^* - \delta)$ that is the probability density c is a symmetric function. Suppose that a target distribution has probability density function π. Thus for given δ_n, (at n^{th} state) we generate

δ^*_{n+1} from a specified probability density, say $q(\delta_n, \delta^*)$ and is then accepted with probability $\alpha(\delta_n, \delta^*_{n+1})$ which is given by

$$\alpha(\delta, \delta^*) = \min\left(\frac{\pi(\delta^*)}{\pi(\delta)}, 1\right). \tag{3.1}$$

If the proposed value is accepted, we set $\delta_{n+1} = \delta^*_{n+1}$, otherwise we reject it and assign $\delta_{n+1} = \delta_n$. The function $\alpha(\delta, \delta^*)$ is chosen precisely to ensure that the Markov chain $\delta_0, \delta_1, \ldots, \delta_n$ is reversible with respect to the target probability density $\pi(\delta^*)$, in the sense that the target density is stationary for the chain. We refer to [6, 8, 10] for details of the procedure of the Metropolis–Hastings algorithm.

In order to estimate the parameters using Metropolis–Hastings algorithm, in the current model, we need to specify the prior distributions for the unknown parameters σ, μ_1 and μ_2. We assume that the three parameters are independent having the prior probability densities respectively as

$$\mu_i \sim N(a_i, b_i^2),\ \sigma \sim \Gamma(c, d);\ i = 1, 2, \tag{3.2}$$

where $N(a_i, b_i^2)$ denotes the normal distribution having mean a_i and variance b_i^2 and $\Gamma(c, d)$ is the gamma distribution having c as shape and d as scale parameters, respectively. These hyperparameters are known. The posterior probability density of the parameters is seen to be of the following forms.

$$\mu_1|(\mu_2, \sigma, \underset{\sim}{x}, \underset{\sim}{y}) \propto \exp\left\{\mu_1\left(\frac{m}{\sigma} + \frac{a_1}{b_1^2}\right) - \frac{\mu_1^2}{2b_1^2}\right\}\prod_{i=1}^{m}\left(1 + g(\mu_1, \sigma, x_i)\right)^{-2}$$

$$\mu_2|(\mu_1, \sigma, \underset{\sim}{x}, \underset{\sim}{y}) \propto \exp\left\{\mu_2\left(\frac{m}{\sigma} + \frac{a_2}{b_2^2}\right) - \frac{\mu_2^2}{2b_2^2}\right\}\prod_{i=1}^{m}\left(1 + h(\mu_2, \sigma, y_i)\right)^{-2}$$

$$\sigma|(\mu_1, \mu_2, \underset{\sim}{x}, \underset{\sim}{y}) \propto \frac{1}{\sigma_2^{n+m+1-c}} \exp\left\{-\frac{K_1 + K_2}{\sigma} - \frac{\sigma}{d}\right\}$$
$$\prod_{j=1}^{n}\left(1 + g(\mu_1, \sigma, x_j)\right)^{-2}\prod_{j=1}^{n}\left(1 + h(\mu_2, \sigma, y_j)\right)^{-2},$$

where we denote $g(\mu_1, \sigma_1, x_i) = \exp\{-(\frac{x_i - \mu_1}{\sigma})\}$, $h(\mu_2, \sigma, y_j) = \exp\{-(\frac{y_j - \mu_2}{\sigma})\}$, $K_1 = \sum_{i=1}^{m}(x_i - \mu)$, and $K_2 = \sum_{j=1}^{n}(y_j - \mu)$. Note that, the posterior densities of μ_1, μ_2 and σ do not have any known distributional form. Hence, to generate samples from the posterior probability distributions of μ_1, μ_2 and σ, we use the well-known random walk Metropolis–Hastings algorithm (RWM) which is a MCMC procedure. The details of the computational steps to generate μ_1, μ_2 and σ simultaneously using Metropolis–Hastings algorithm can be described as follows.

Step-1: Let the kth state of Markov chain consists of $(\mu_1^{(k-1)}, \mu_2^{(k-1)}, \sigma^{(k-1)})$.

Step-2: In order to update the parameters using random walk Metropolis–Hastings algorithm (RWM), we generate $\varepsilon_1 \sim N(0, \sigma_{\mu_1}^2)$, $\varepsilon_2 \sim N(0, \sigma_{\mu_2}^2)$ and $\varepsilon_3 \sim N(0, \sigma_\sigma^2)$. Set $\mu_1^{(*)} = \mu_1^{(k-1)} + \varepsilon_1$, $\mu_2^{(*)} = \mu_2^{(k-1)} + \varepsilon_2$, $\sigma^{(*)} = \sigma^{(k-1)} + \varepsilon_3$.

Step-3: Calculate the term $\alpha(\mu_1^{(*)}, \mu_2^{(*)}, \sigma^{(*)})$ as given (3.1).

Step-4: Define a term $S = \min(1, \alpha(\mu_1^{(*)}, \mu_2^{(*)}, \sigma^{(*)}))$. Generate a random number $v \sim U(0, 1)$. If $v \leq S$, accept $(\mu_1^{(*)}, \mu_2^{(*)}, \sigma^*)$ and update the parameters as $\mu_1^{(k)} = \mu_1^{(*)}$, $\mu_2^{(k)} = \mu_2^{(*)}$ and $\sigma^{(k)} = \sigma^{(*)}$, otherwise reject $\mu_1^{(*)}$, $\mu_2^{(*)}$ and $\sigma^{(*)}$ and set $\mu_1^{(k)} = \mu_1^{(k-1)}$, $\mu_2^{(k)} = \mu_2^{(k-1)}$ and $\sigma^{(k)} = \sigma^{(k-1)}$. Repeat this step for $k = 1, 2, \ldots, K$, where K is a large number suitably fixed.

In this computational method, the most important step is to choose the values of $\sigma_{\mu_1}^2$, $\sigma_{\mu_2}^2$ and σ_σ^2. One may refer to [6] regarding choice of these variances. The values of $\sigma_{\mu_1}^2$, $\sigma_{\mu_2}^2$ and σ_σ^2 are chosen in such a way that a acceptance ratio should lie within the range from 20% to 30%. Now using these MCMC samples, we estimate the parameters μ, σ_1 and σ_2 respectively as

$$\hat{\mu}_{1mc} = \frac{1}{M - M_0} \sum_{i=M_0+1}^{M} \mu_1^{(i)}, \ \hat{\mu}_{2mc} = \frac{1}{M - M_0} \sum_{i=M_0+1}^{M} \mu_2^{(i)} \text{ and } \hat{\sigma} = \frac{1}{M - M_0} \sum_{i=M_0+1}^{M} \sigma^{(i)}, \tag{3.3}$$

where M_0 is the burn-in period for the RWM method.

Utilizing these Bayes estimators of the parameters μ_1, μ_2 and σ, we construct a classification function, say $W_{MC}(z)$ as

$$W_{MC}(z) = \frac{1}{\hat{\sigma}}\Big((z - \hat{\mu}_{2mc}) - (z - \hat{\mu}_{1mc})\Big) + 2\log\left(1 + \exp\left(-\frac{(z - \hat{\mu}_{2mc})}{\hat{\sigma}_{mc}}\right)\right)$$
$$-2\log\left(1 + \exp\left(-\frac{(z - \hat{\mu}_{1mc})}{\hat{\sigma}_{mc}}\right)\right). \tag{3.4}$$

Using this classification function $W_{MC}(z)$, we propose a classification rule, say R_{MC} as: classify z into Π_1 if $W_{MC}(z) \geq 0$, else classify z into the Π_2.

Remark 1 We note that, all the existing estimators such as the MLE, Bayes estimators using different priors for the current model are compared with the Bayes estimator using MCMC approach in terms of MSE and bias numerically in Sect. 4. It has been noticed that the MLEs and the Bayes estimators using MCMC approach outperform the earlier estimators for all sample sizes. Hence, in the numerical comparison section, we only consider classification rules for comparison purpose based on the MLEs and the Bayes estimators using the MCMC approach.

4 Numerical Comparison of Classification Rules

In this section, we will evaluate and compare the performances of all the classification rules in terms of probability of misclassification, numerically. Note that none of the estimators, and hence the classification rules, could be obtained in closed form expressions. It is not possible to compare their performances analytically. However, utilizing the advanced computational facilities available nowadays, we have compared the performances of all the proposed rules numerically, which will be handy in certain practical applications.

In Sect. 2, we have proposed the classification rule R_{ML} using the MLEs of the associated parameters for the two logistic populations. In Sect. 3, we have proposed the classification rule R_{MC} utilizing the Bayes estimators through MCMC approach. In our simulation study, we have also included estimators proposed by Nagamani [20] and compared their MSE and bias with the Bayes estimator that uses MCMC approach (our proposed estimator). It has been noticed that the MLEs as well as the proposed Bayes estimator through MCMC approach always have minimum MSE and bias. Hence, we have not considered the classification rules based on the Bayes estimators proposed by Nagamani [20], for numerical comparison in terms of probability of misclassification.

In order to compute the probability of misclassification of the classification rules R_{ML} and R_{MC}, we proceed in the following manner.

Step-1: Generate training samples of size m from the logistic population Logistic (μ_1, σ). Then, similarly, we generate training samples of size n from another logistic population Logistic(μ_2, σ). Utilizing these training samples, we estimate the parameters μ_1, μ_2 and σ. In particular, we compute the MLEs and the Bayes estimators using MCMC approach that is we compute $\hat{\mu}_{1ml}$, $\hat{\mu}_{2ml}$, $\hat{\sigma}_{ml}$, $\hat{\mu}_{1mc}$, $\hat{\mu}_{2mc}$, $\hat{\sigma}_{mc}$, which are involved in the classification functions $W_{ML}(z)$ and $W_{MC}(z)$.

Step-2: Generate a new observation, say z from the logistic population Logistic (μ_1, σ) and substitute it in the classification rules R_{ML} and R_{MC}, then we check whether it belongs to the population Π_1 or not. Similarly, we generate a new sample z from the logistic population Π_2, and substitute it in the classification rules, then check whether it belongs to the population Π_2 or not.

The above procedure is carried out using the well-known Monte Carlo simulation method with 20, 000 replications. A high level of accuracy has been achieved in the sense that the standard error in simulation is of the order of 10^{-4}. . The probability of misclassification $P(1|2)$ and $P(2|1)$ using the probability frequency approach has been computed. In the computation of Bayes estimator using MCMC approach, the hyperparameters involved in the conjugate prior have been suitably taken as $c = 2.5$, $d = 1$ (for the common scale parameter), $a_1 = 2.5$, $b_1 = 3$, $a_2 = 2$, $b_2 = 1$ (in the case of location parameters). The computation also has been done using some other choices of hyperparameters; however, the probability of misclassification of

R_{MC} changes insignificantly. In the MCMC method that uses Metropolis–Hastings algorithm, we have generated 50, 000 MCMC samples from the posterior density of σ, μ_1 and μ_2 with burn-in period $M_0 = 5000$. In the simulation study, we have seen that for some rules, the $P(1|2)$ may be small but $P(2|1)$ may be high. In view of this and to have a better picture of the performance, we also define the expected probability of misclassification (EPM), given by

$$E(R) = \frac{P(1|2, R) + P(2|1, R)}{2}$$

along with the probability of misclassifications, for comparing the rules. We consider the cost and the prior probabilities are equal, that is $C(1|2) = C(2|1)$ and $q_1 = q_2$, for convenience.

The simulation study has been carried out by considering various combinations of sample sizes and different choices of parameters. However, for illustration purpose we present the expected probability of misclassification of the rules R_{ML} and R_{MC} for some specific choices of sample sizes and parameters. In Table 1, we have presented the EPM of the rules R_{ML} and R_{MC} for equal and unequal sample sizes, for fixed values of μ_1 and μ_2 with variation in scale parameter σ. The following observations were made from our simulation study as well as the Table 1, regarding the performances of the classification rules.

Table 1 Expected probability of misclassification (EPM) for the proposed rules with $\mu_1 = 1$ and $\mu_2 = 2$ and Various Values of σ

(m, n)	(σ)	$E(R_{ML})$	$E(R_{MC})$	(m, n)	$E(R_{ML})$	$E(R_{MC})$
	0.5	0.274	0.270		0.268	0.267
(10,10)	1.5	0.442	0.428	(10,20)	0.460	0.450
	2.5	0.472	0.465		0.482	0.452
	3.5	0.490	0.488		0.495	0.475
	0.5	0.274	0.272		0.296	0.296
(15,15)	1.5	0.432	0.426	(20,10)	0.444	0.438
	2.5	0.452	0.449		0.492	0.482
	3.5	0.477	0.475		0.475	0.496
	0.5	0.274	0.272		0.279	0.280
(20,20)	1.5	0.432	0.426	(20,30)	0.434	0.427
	2.5	0.452	0.449		0.497	0.495
	3.5	0.475	0.477		0.478	0.476
	0.5	0.272	0.270		0.268	0.269
(30,30)	1.5	0.430	0.424	(30,20)	0.426	0.424
	2.5	0.450	0.446		0.478	0.483
	3.5	0.474	0.470		0.470	0.474

- The expected probability of misclassification for the proposed rules decrease as the sample sizes increase. It is also noticed that for different location points (μ_i) and fixing the scale parameter σ, the EPM remains almost same. It seems that the rules are location invariant.
- The rule R_{MC} based on the MCMC procedure always performs better than the rule R_{ML} based on MLE in terms of expected probability of misclassification. It is further noticed that when σ is close to 4, the maximum values for EPM is 0.5 for both the rules.
- Considering the EPM as measure of performance, the classification rule R_{MC} has the lowest EPM value among all the proposed rules. Hence, we recommend to use the classification rule R_{MC} for classifying a new observation into one of the two logistic populations.

5 Application with Real-Life Data Sets

In this section, we consider two real-life data sets which are related to compressive strength of bricks and dietary fiber content in food. It has been shown that the logistic distribution is a reasonable fit to these data sets. Further, Levene's test with significance level 0.05 indicates that the equality of scale parameters cannot be rejected (see [20]).

Example 1 Nagamani [20] considered the data sets related to compressive strength (MPa) of clay bricks and fly ash bricks. The experiment was conducted by Teja [22] to determine the mechanical properties, such as initial rate of absorption (IRA), water absorption (WA), dry density (DD), and compressive strength (CS) of 50 brick units from each type. The compressive strength (MPa) of clay bricks and fly ash bricks is given as follows.

Clay Brick: 8.02, 7.31, 7.31, 7.87, 7.09, 9.75, 5.90, 6.76, 6.58, 7.70, 5.70, 6.56, 8.14, 7.28, 5.70, 7.38, 7.02, 5.95, 5.90, 5.67, 7.22, 6.76, 7.73, 8.65, 8.08, 7.98, 5.60, 8.66, 9.67, 8.18, 8.70, 4.86, 9.33, 7.77, 6.21, 7.74, 11.31, 9.13, 8.28, 7.09, 5.62, 11.88, 5.73, 9.21, 7.03, 9.07, 7.81, 6.70, 9.97, 8.85

Fly Ash Brick: 3.62, 4.74, 9.88, 5.93, 6.09, 6.94, 6.32, 5.30, 5.14, 4.55, 4.03, 7.36, 3.57, 3.98, 4.03, 4.74, 7.32, 3.23, 5.38, 7.18, 6.07, 3.62, 6.64, 5.58, 5.23, 3.95, 5.86, 5.58, 6.97, 5.05, 4.35, 4.55, 4.79, 4.03, 4.74, 7.58, 3.62, 6.01, 3.99, 6.04, 4.74, 7.21, 3.61, 5.69, 7.21, 6.40, 3.55, 8.70, 4.35, 7.51.

It is our interest to classify a new observation, say z into Clay Brick or Fly Ash Brick. So, we compute the estimators for the parameters (μ_1, μ_2, σ) as $(\hat{\mu}_{1ml}, \hat{\mu}_{2ml}, \hat{\sigma}_{ml}) = (7.53694, 5.35400, 0.84857)$ and $(\hat{\mu}_{1mc}, \hat{\mu}_{2mc}, \hat{\sigma}_{mc}) = (7.02195, 4.86868, 0.78183)$. Utilizing these estimators, we compute the classification functions $W_{ML}(z)$ and $W_{MC}(z)$.

Suppose we have a new observation, say $z = 6.5$ and we want to classify this observation into one of the two types of bricks using our proposed classification rules. The values of classification functions are computed as $W_{ML} = 0.07285$ and

$W_{MC} = -0.91663$. The rule R_{ML} classifies $z = 6.5$ into the clay brick, whereas the rule R_{MC} classifies it into the fly ash brick.

Example 2 These data sets are related to the percentage of total dietary fiber (TDF) content in foods, such as fruits, vegetables, and purified polysaccharides. Li and Cardozo [15] conducted an inter laboratory (nine laboratories involved) study using a procedure as described by Association of Official Agricultural Chemists (AOAC) international to determine the percentage of total dietary fiber (TDF) content in food samples. The experiment was conducted for six different types of food samples, such as (a) apples, (b) apricots, (c) cabbage, (d) carrots, (e) onions, and (f) FIBRIM 1450 (soy fiber). The percentage of TDF using non-enzymatic-gravimetric method from nine laboratories for apples and carrots is given as, Apple: 12.44, 12.87, 12.21, 12.82, 13.18, 12.31, 13.11, 14.29, 12.08; Carrots: 29.71, 29.38, 31.26, 29.41, 30.11, 27.02, 30.06, 31.30, 28.37.

Nagamani [20] shown that logistic distribution fits these two data sets reasonably well. Their Levene's test also confirms the equality of the scale parameters with level of significance 0.05. In order to classify an observation into these two types of foods, we compute the estimators of the parameters (μ_1, μ_2, σ) as $(\hat{\mu}_{1ml}, \hat{\mu}_{2ml}, \hat{\sigma}_{ml}) = (12.765, 29.6497, 0.5604)$ and $(\hat{\mu}_{1mc}, \hat{\mu}_{2mc}, \hat{\sigma}_{mc}) = (12.65066, 29.47925, 0.63596)$. Utilizing these estimators, we compute the classification functions $W_{ML}(z)$ and $W_{MC}(z)$.

Suppose we have a new observation, say $z = 21.2$ and we want to classify this observation into one of the two types of foods using our proposed classification rules. The values of classification functions are computed as $W_{ML} = 0.026588$ and $W_{MC} = -0.314322$. The rule R_{ML} classifies $z = 21.2$ into Apples, whereas the rule R_{MC} classifies it into the Carrots.

6 Concluding Remarks

In this note, we have considered the problem of classification into one of the two logistic populations with a common scale parameter and possibly different location parameters. It is worth mentioning that the same model was previously considered by Nagamani [20] and estimated the associated parameters. Specifically the authors had derived certain Bayes estimators using different types of priors and Lindley's approximations. Utilizing their proposed estimators, and the one we have proposed (Bayes estimators using MCMC approach), we have constructed several classification rules. It has been seen that the proposed rule R_{MC} outperforms all other rules in terms of expected probability of misclassification. The application of our model problem has been explained using two real-life data sets.

Acknowledgements The authors would like to express gratitude to the two anonymous reviewers for valuable comments that led to improvements in the presentation of this article.

References

1. Anderson TW (1951) Classification by multivariate analysis. Psychometrika 16(1):31–50
2. Classification into two multivariate normal distributions with different covariance matrices. Ann Math Stat 33(2):420–431
3. Asgharzadeh A, Valiollahi R, Abdi M (2016) Point and interval estimation for the logistic distribution based on record data. SORT: Stat Oper Res Trans 40(1):0089–112
4. Balakrishnan N (1991) Handbook of the logistic distribution. CRC Press, Dekker, New York
5. Basu AP, Gupta AK (1976) Classification rules for exponential populations: two parameter case. Theor Appl Reliab Emph Bayesian Non parametric Methods 1:507–525
6. Chib S, Greenberg E (1995) Understanding the Metropolis-Hastings algorithm. Am Stat 49(4):327–335
7. Eubank RL (1981) Estimation of the parameters and quantiles of the logistic distribution by linear functions of sample quantiles. Scandinavian Actuarial J 1981(4):229–236
8. Gilks WR (1996) Introducing Markov chain Monte Carlo. Markov chain Monte Carlo in practice, Chapman and Hall, London
9. Gupta SS, Gnanadesikan M (1966) Estimation of the parameters of the logistic distribution. Biometrika 53(3–4):565–570
10. Hastings WK (1970) Monte Carlo sampling methods using Markov chains and their applications. Biometrika 57(1):97–109
11. Howlader H, Weiss G (1989) Bayes estimators of the reliability of the logistic distribution. Commun Stat-Theor Methods 18(4):1339–1355
12. Jana N, Kumar S (2016) Classification into two-parameter exponential populations with a common guarantee time. Am J Math Manage Sci 35(1):36–54
13. Jana N, Kumar S (2017) Classification into two normal populations with a common mean and unequal variances. Commun Stat-Simul Comput 46(1):546–558
14. Kotz S, Balakrishnan N, Johnson NL (2004) Continuous multivariate distributions, Volume 1: models and applications. Wiley, New York
15. Li BW, Cardozo MS (1994) Determination of total dietary fiber in foods and products with little or no starch, nonenzymatic-gravimetric method: collaborative study. J AOAC Int 77(3):687–689
16. Long T, Gupta RD (1998) Alternative linear classification rules under order restrictions. Commun Stat-Theor Methods 27(3):559–575
17. Muttlak HA, Abu-Dayyeh W, Al-Sawi E, Al-Momani M (2011) Confidence interval estimation of the location and scale parameters of the logistic distribution using pivotal method. J Stat Comput Simul 81(4):391–409
18. Nagamani N, Tripathy MR (2017) Estimating common scale parameter of two gamma populations: a simulation study. Am J Math Manage Sci 36(4):346–362
19. Nagamani N, Tripathy MR (2018) Estimating common dispersion parameter of several inverse Gaussian populations: a simulation study. J Stat Manage Syst 21(7):1357–1389
20. Nagamani N, Tripathy MR, Kumar S (2020) Estimating common scale parameter of two logistic populations: a Bayesian study. Am J Math Manage Sci. https://doi.org/10.1080/01966324.2020.1833794
21. Rashad A, Mahmoud M, Yusuf M (2016) Bayes estimation of the logistic distribution parameters based on progressive sampling. Appl Math Inf Sci 10(6):2293–2301
22. Teja PRR (2015) Studies on mechanical properties of brick masonry. M. Tech. thesis, Department of Civil Engineering, National Institute of Technology Rourkela

23. Tripathy MR, Kumar S (2015) Equivariant estimation of common mean of several normal populations. J Stat Comput Simul 85(18):3679–3699
24. Tripathy MR, Kumar S, Misra N (2014) Estimating the common location of two exponential populations under order restricted failure rates. Am J Math Manage Sci 33(2):125–146
25. Tripathy MR, Nagamani N (2017) Estimating common shape parameter of two gamma populations: a simulation study. J Stat Manage Syst 20(3):369–398
26. Yang Z, Lin DK (2007) Improved maximum-likelihood estimation for the common shape parameter of several Weibull populations. Appl Stochastic Models Bus Industry 23(5):373–383

Testing Quantiles of Two Normal Populations with a Common Mean and Order Restricted Variances

Habiba Khatun and Manas Ranjan Tripathy

Abstract In this study, we consider the problem of testing the hypothesis for the quantile $\theta = \mu + \eta\sigma_1$ (η is known) when independent random samples are available from two normal populations with a common mean μ and ordered restricted variances. Utilizing some of the popular estimators of the common mean under order restricted variances and the generalized p-value approach, we propose several test procedures for the quantiles. All the proposed test procedures are evaluated through their sizes and powers using the Monte Carlo simulation procedure. It has been observed that the proposed tests compete with each other. Finally, two datasets have been considered for illustrating the testing procedures.

Keywords Common mean · Generalized p-value · Generalized test variable · Numerical comparison · Order restricted variances · Power · Size · Testing quantile

1 Introduction

Suppose two normal populations with a common mean μ and unknown different variances σ_1^2 and σ_2^2 are available. Further, it is known in advance that the variances are ordered. Particularly, let $(X_{11}, X_{12}, \ldots, X_{1n_1})$ and $(X_{21}, X_{22}, \ldots, X_{2n_2})$ be independent observations of sizes n_1 and n_2 available from $N(\mu, \sigma_1^2)$ and $N(\mu, \sigma_2^2)$, respectively. It is known that $(\bar{X}_1, \bar{X}_2, S_1^2, S_2^2)$ is the minimal sufficient for $(\mu, \sigma_1^2, \sigma_2^2)$, where $\bar{X}_1 = \sum_{i=1}^{n_1} X_{1i}/n_1 \sim N(\mu, \sigma_1^2/n_1)$, $\bar{X}_2 = \sum_{j=1}^{n_2} X_{2j}/n_2 \sim N(\mu, \sigma_2^2/n_2)$, $S_1^2 = \sum_{i=1}^{n_1}(X_{1i} - \bar{X}_1)^2 \sim \sigma_1^2\chi^2_{n_1-1}$ and $S_2^2 = \sum_{j=1}^{n_2}(X_{2j} - \bar{X}_2)^2 \sim \sigma_2^2\chi^2_{n_2-1}$. All the random variables $\bar{X}_1$, $\bar{X}_2$, S_1^2 and S_2^2 are stochastically independent.

The problem discussed here is to test the hypothesis regarding the quantile $\theta = \mu + \eta\sigma_1$ under the belief that the variances are ordered, that is, $\sigma_1^2 \leq \sigma_2^2$. Here,

H. Khatun (✉) · M. R. Tripathy
Department of Mathematics, National Institute of Technology Rourkela, Rourkela, Odisha 769008, India
e-mail: habibakhatun7860@gmail.com

M. R. Tripathy
e-mail: manas@nitrkl.ac.in

S. S. Ray et al. (eds.), *Applied Analysis, Computation and Mathematical Modelling in Engineering*, Lecture Notes in Electrical Engineering 897,
https://doi.org/10.1007/978-981-19-1824-7_15

$\eta = \Phi^{-1}(p)$, $p \in (0, 1)$, and Φ is the cdf of a standard normal random variable. Specifically, we consider the hypothesis testing

$$H_0 : \theta = \theta_0 \text{ against } H_1 : \theta \neq \theta_0, \quad (1)$$

where $\theta_0 \in \mathbb{R}$ is a predefined constant.

Note that the quantities like—median, quartiles, deciles, percentiles are obtained from quantiles by considering the different values of η; hence, we consider the testing of the quantile. Several literature pieces are available in the problem of estimating the quantiles of normal and exponential distributions. The first work in this direction was probably considered by Zidek [25], who addressed the problem of estimating quantiles in the case of the normal population and obtained some decision-theoretic results. Further, Rukhin [20] discussed the estimation problem on the quantile of normal populations in decision-theoretic viewpoint and gives some application.

Researchers also considered estimating quantiles in the presence of more than one independent normal populations with an equal mean. For some results in this direction, one can refer to Kumar and Tripathy [12], Tripathy et al. [23] and the references cited therein. Recently, under the order restriction on variances, Nagamani and Tripathy [17] considered estimating quantiles of two Gaussian populations with equal mean. Comparison of quantiles of two or several distributions is also useful, and the same has been considered by Guo and Krishnamoorthy [6] and Li et al. [13]. The problem of interval estimation and testing for quantiles of two normal populations with equal mean and unrestricted variances has been recently investigated by Khatun et al. [10]. The application of quantiles can be seen in the study of life testing, reliability, survival analysis, statistical quality control and related areas. We refer to Saleh [21], Keating and Tripathi [8] and the references cited therein for some application of quantiles.

The current problem has its importance in the sense that the test procedures obtained for the quantiles are based on the estimators of the common mean under order restricted variances. In a particular case, by choosing $\eta = 0$, one can write all the test procedures easily. The problem of estimating common mean of two or more normal populations with unrestricted variances has been considered by Elfessi and Pal [4], Jena et al. [7] and Misra and van der Meulen [15]. The problem of hypothesis testing on the common mean of normal populations without order restriction on variances has drawn several researchers' attention in the past, and probably, Cohen and Sackrowitz [3] was the first to consider this problem. Krishnamoorthy and Lu [11] considered testing the common mean of two Gaussian populations using the generalized variable method. Further, Lin and Lee [14] extended their results to several normal populations. We refer to Chang and Pal [2] and Pal et al. [18] for some review on testing common mean of several Gaussian populations.

The rest of our contributions can be described as follows. In Sect. 2, we discuss some well-known results on estimating common mean μ of two normal populations with and without order restrictions on variances. In Sect. 3, we propose some generalized pivot variable and using these constructed generalized test variable for testing the quantile θ. The generalized test variables have been constructed using some of the well-known estimators of the common mean under order restricted variances. A comprehensive simulation study has been done in Sect. 4 in order to compare the performances of all the proposed tests in terms of size and power for several combinations of sample sizes and parameters. Finally, the article is concluded with some real-life applications.

2 Some Basic Results

In literature, several estimators are available for the common mean of two normal populations when there is no restriction on the variances and also when there is order restriction on variances, that is $\sigma_1^2 \leq \sigma_2^2$.

When there is no order restriction on the variances, some well-known estimators for the common mean μ are proposed by Graybill and Deal [5], Khatri and Shah [9], Moore and Krishnamoorthy [16], Tripathy and Kumar [22] and the grand sample mean, which are given as follows,

$$\mu_{\text{GD}} = \frac{(n_1-1)n_1 S_2^2 \bar{X}_1 + (n_2-1)n_2 S_1^2 \bar{X}_2}{(n_1-1)n_1 S_2^2 + (n_2-1)n_2 S_1^2}$$

$$\mu_{\text{KS}} = \frac{(n_1-3)n_1 S_2^2 \bar{X}_1 + (n_2-3)n_2 S_1^2 \bar{X}_2}{(n_1-3)n_1 S_2^2 + (n_2-3)n_2 S_1^2}$$

$$\mu_{\text{MK}} = \frac{\sqrt{(n_1-1)n_1 S_2^2}\bar{X}_1 + \sqrt{(n_2-1)n_2 S_1^2}\bar{X}_2}{\sqrt{(n_1-1)n_1 S_2^2} + \sqrt{(n_2-1)n_2 S_1^2}}$$

$$\mu_{\text{TK}} = \frac{\sqrt{n_1 S_2^2 b_{n_2}}\bar{X}_1 + \sqrt{n_2 S_1^2 b_{n_1}}\bar{X}_2}{\sqrt{n_1 S_2^2 b_{n_2}} + \sqrt{n_2 S_1^2 b_{n_1}}}$$

$$\mu_{\text{GM}} = \frac{n_1 \bar{X}_1 + n_2 \bar{X}_2}{n_1 + n_2}$$

where $b_{n_1} = \frac{\Gamma((n_1-1)/2)}{\sqrt{2}\Gamma(n_1/2)}$ and $b_{n_2} = \frac{\Gamma((n_2-1)/2)}{\sqrt{2}\Gamma(n_2/2)}$.

Further, Elfessi and Pal [4] proposed an estimator for the common mean when the variances are ordered as $\sigma_1^2 \leq \sigma_2^2$ or equivalently $\sigma_1 \leq \sigma_2$ which is

$$\hat{\mu}_{\text{GD}} = \begin{cases} (1-C)\bar{X}_1 + C\bar{X}_2, & \text{if } \frac{S_1^2}{n_1-1} \leq \frac{S_2^2}{n_2-1} \\ C^*\bar{X}_1 + (1-C^*)\bar{X}_2, & \text{if } \frac{S_1^2}{n_1-1} > \frac{S_2^2}{n_2-1}, \end{cases} \tag{2}$$

where

$$C = \frac{n_2(n_2-1)S_1^2}{n_1(n_1-1)S_2^2 + n_2(n_2-1)S_1^2}$$

and

$$C^* = \begin{cases} C, & \text{if } n_1 = n_2 \\ \frac{n_1}{n_1+n_2}, & \text{if } n_1 \neq n_2. \end{cases}$$

When $\sigma_1^2 \leq \sigma_2^2$, the estimator $\hat{\mu}_{\text{GD}}$ dominates μ_{GD} stochastically and universally for two normal populations. The results have been extended for several Gaussian populations by Misra and van der Meulen [15].

Jena et al. [7] considered the same problem and proposed some alternative estimators of the common mean μ which improve upon the estimators μ_{KS}, μ_{MK} and μ_{TK} when $\sigma_1^2 \leq \sigma_2^2$. These improved estimators are given by

$$\hat{\mu}_{\text{KS}} = \begin{cases} (1-C_1)\bar{X}_1 + C_1\bar{X}_2, & \text{if } \frac{S_1^2}{S_2^2} \leq \frac{n_1-3}{n_2-3} \\ C_1^*\bar{X}_1 + (1-C_1^*)\bar{X}_2, & \text{if } \frac{S_1^2}{S_2^2} > \frac{n_1-3}{n_2-3}, \end{cases} \tag{3}$$

$$\hat{\mu}_{\text{MK}} = \begin{cases} (1-C_2)\bar{X}_1 + C_2\bar{X}_2, & \text{if } \frac{\sqrt{(n_2-1)S_1^2}}{\sqrt{(n_1-1)S_2^2}} \leq \frac{\sqrt{n_2}}{\sqrt{n_1}} \\ C_2^*\bar{X}_1 + (1-C_2^*)\bar{X}_2, & \text{if } \frac{\sqrt{(n_2-1)S_1^2}}{\sqrt{(n_1-1)S_2^2}} > \frac{\sqrt{n_2}}{\sqrt{n_1}}, \end{cases} \tag{4}$$

$$\hat{\mu}_{\text{TK}} = \begin{cases} (1-C_3)\bar{X}_1 + C_3\bar{X}_2, & \text{if } \frac{S_1}{S_2} \leq \frac{\sqrt{n_2}b_{n_2}}{\sqrt{n_1}b_{n_1}} \\ C_3^*\bar{X}_1 + (1-C_3^*)\bar{X}_2, & \text{if } \frac{S_1}{S_2} > \frac{\sqrt{n_2}b_{n_2}}{\sqrt{n_1}b_{n_1}}, \end{cases} \tag{5}$$

where $C_1 = \frac{n_2(n_2-3)S_1^2}{n_1(n_1-3)S_2^2+n_2(n_2-3)S_1^2}$, $C_2 = \frac{\sqrt{n_2(n_2-1)}S_1}{\sqrt{n_1(n_1-1)}S_2+\sqrt{n_2(n_2-1)}S_1}$,
$C_3 = \frac{\sqrt{n_2}b_{n_1}S_1}{\sqrt{n_1}b_{n_2}S_2+\sqrt{n_2}b_{n_1}S_1}$ and

$$C_i^* = \begin{cases} C_i, & \text{if } n_1 = n_2 \\ \frac{n_1}{n_1+n_2}, & \text{if } n_1 \neq n_2. \end{cases}$$

for $i = 1, 2, 3$.

They proved that these estimators dominate the unrestricted estimators μ_{KS}, μ_{MK} and μ_{TK} in terms of stochastic domination and Pitman nearness criteria. Applying Brewster and Zidek [1] technique, they have also improved the estimators μ_{GD}, μ_{KS}, μ_{MK} and μ_{TK} under order restricted variances. These improved estimators are, respectively, given by

$$\mu_{\text{GD}}^a = \begin{cases} \mu_{\text{GD}}, & \text{if } \frac{S_1^2}{n_1-1} \leq \frac{S_2^2}{n_2-1} \\ \mu_{\text{GM}}, & \text{if } \frac{S_1^2}{n_1-1} > \frac{S_2^2}{n_2-1}, \end{cases} \tag{6}$$

$$\mu_{\text{KS}}^a = \begin{cases} \mu_{\text{KS}}, & \text{if } \frac{S_1^2}{n_1-3} \leq \frac{S_2^2}{n_2-3} \\ \mu_{\text{GM}}, & \text{if } \frac{S_1^2}{n_1-3} > \frac{S_2^2}{n_2-3}, \end{cases} \tag{7}$$

$$\mu_{\text{MK}}^a = \begin{cases} \mu_{\text{MK}}, & \text{if } \sqrt{\frac{S_1^2}{n_1-1}} \leq \sqrt{\frac{S_2^2}{n_2-1}} \\ \mu_{\text{GM}}, & \text{if } \sqrt{\frac{S_1^2}{n_1-1}} > \sqrt{\frac{S_2^2}{n_2-1}}, \end{cases} \tag{8}$$

$$\mu_{\text{TK}}^a = \begin{cases} \mu_{\text{TK}}, & \text{if } \frac{S_1}{S_2} \leq \frac{\sqrt{n_2}b_{n_2}}{\sqrt{n_1}b_{n_1}} \\ \mu_{\text{GM}}, & \text{if } \frac{S_1}{S_2} > \frac{\sqrt{n_2}b_{n_2}}{\sqrt{n_1}b_{n_1}}. \end{cases} \tag{9}$$

Jena et al. [7] noted that for unequal sample sizes $(n_1 \neq n_2)$, $\mu_{\text{GD}}^a = \hat{\mu}_{\text{GD}}$, $\mu_{\text{KS}}^a = \hat{\mu}_{\text{KS}}$, $\mu_{\text{MK}}^a = \hat{\mu}_{\text{MK}}$ and $\mu_{\text{TK}}^a = \hat{\mu}_{\text{TK}}$.

In the next section, we propose some generalized test variables using these improved estimators under order restriction on the variances, to test the quantile $\theta = \mu + \eta\sigma_1$.

3 Generalized Test Variable with P-Value

In this section, we will apply the generalized variable method proposed by Tsui and Weerahandi [24] to test the hypothesis (1) regarding the quantile θ. We note that Krishnamoorthy and Lu [11] and Lin and Lee [14] successfully applied this method to test the common mean of two and more than two normal populations. In order to obtain the test statistics using this approach, we first state the following definitions.

Let X be any random variable, and the distribution of X only depends on (δ, β), where we want to test the parameter δ, which is the parameter of interest and β is the nuisance parameter. Further, suppose one is interested in testing the hypothesis

$$H_0^* : \delta \leq \delta_0 \text{ against } H_1^* : \delta > \delta_0, \tag{10}$$

where δ_0 is a known constant.

Definition 1 A random variable $P = P(X; x, \delta, \beta)$ will be called a generalized pivot variable of δ if it satisfies the following conditions

(a) The distribution of $P(X; x, \delta, \beta)$ is free of all unknown parameters for a fixed $X = x$.
(b) The value of P at $X = x$, is δ, that is, $P(X; x, \delta, \beta) = \delta$, the parameter of interest.

Definition 2 A variable $T = T(X; x, \delta, \beta)$ will be called a generalized test variable for testing the hypothesis (10), if it satisfies the conditions (a)–(c).

(a) The distribution of $T = T(X; x, \delta, \beta)$ is free from the nuisance parameter β for a given x.
(b) The value of $T = T(X; x, \delta, \beta)$ is free of any unknown parameters when $X = x$ fixed.

(c) For fixed x and β, the distribution of $T = T(X; x, \delta, \beta)$ is either stochastically increasing or stochastically decreasing as a function of δ.

where x is the observed value of X.

Definition 3 Let $t = T(x; x, \delta, \beta)$, the value of T for fixed $X = x$. In the case of stochastically increasing T with respect to δ, the generalized p-value for testing the hypothesis (10) is given by

$$\sup_{H_0^*} \Pr\{T(X; x, \delta, \beta) \geq t\} = \Pr\{T(X; x, \delta_0, \beta) \geq t\}, \tag{11}$$

and if $T(X; x, \delta, \beta)$ is stochastically decreasing in δ, the generalized p-value for testing the hypothesis (10) is given by

$$\sup_{H_0^*} \Pr\{T(X; x, \delta, \beta) \leq t\} = \Pr\{T(X; x, \delta_0, \beta) \leq t\}. \tag{12}$$

3.1 Generalized Test Variable for the Quantile

In this subsection, we propose generalized test variables using the estimators of common mean μ to test the hypothesis (1). Let the observed value of $(\bar{X}_1, \bar{X}_2, S_1^2, S_2^2)$ is $(\bar{x}_1, \bar{x}_2, s_1^2, s_2^2)$. Further, suppose $Z_1 = (\bar{X}_1 - \mu)/(\sigma_1/\sqrt(n_1))$ and $Z_2 = (\bar{X}_2 - \mu)/(\sigma_2/\sqrt(n_2))$ where Z_1 and Z_2 follows $N(0, 1)$. Now, denote $U_1^2 = S_1^2/\sigma_1^2 \sim \chi^2_{n_1-1}$ and $U_2^2 = S_2^2/\sigma_2^2 \sim \chi^2_{n_2-1}$ which are independent of Z_1 and Z_2.

When sample sizes are equal ($n_1 = n_2$) using the alternative estimators of common mean μ given in equations (2) to (5), we propose the generalized pivot variable for the quantile $\theta = \mu + \eta\sigma_1$ as

$$P_{\rm GD} = \begin{cases} \bar{\mu}_{\rm GD} - \frac{(\sqrt{(n_1-1)n_1}s_1 s_2^2 t_1) + (\sqrt{(n_2-1)n_2}s_1^2 s_2 t_2)}{((n_2-1)n_2 s_1^2) + ((n_1-1)n_1 s_2^2)} + \eta\frac{s_1}{U_1}, & \text{if } \frac{s_1^2}{n_1-1} \leq \frac{s_2^2}{n_2-1} \\ \bar{\mu}_{\rm GD} - \frac{((n_1-1)n_1 s_2^3 t_2/\sqrt{(n_2-1)n_2}) + ((n_2-1)n_2 s_1^3 t_1/\sqrt{(n_1-1)n_1})}{((n_2-1)n_2 s_1^2) + ((n_1-1)n_1 s_2^2)} + \eta\frac{s_1}{U_1}, & \text{if } \frac{s_1^2}{n_1-1} > \frac{s_2^2}{n_2-1}, \end{cases}$$

$$P_{\rm KS} = \begin{cases} \bar{\mu}_{\rm KS} - \frac{(\sqrt{n_1/(n_1-1)}(n_1-3)s_1 s_2^2 t_1) + (\sqrt{n_2/(n_2-1)}(n_2-3)s_1^2 s_2 t_2)}{((n_2-3)n_2 s_1^2) + ((n_1-3)n_1 s_2^2)} + \eta\frac{s_1}{U_1}, & \text{if } \frac{s_1^2}{n_1-3} \leq \frac{s_2^2}{n_2-3} \\ \bar{\mu}_{\rm KS} - \frac{((n_1-3)n_1 s_2^3 t_2/\sqrt{(n_2-1)n_2}) + ((n_2-3)n_2 s_1^3 t_1/\sqrt{(n_1-1)n_1})}{((n_2-3)n_2 s_1^2) + ((n_1-3)n_1 s_2^2)} + \eta\frac{s_1}{U_1}, & \text{if } \frac{s_1^2}{n_1-3} > \frac{s_2^2}{n_2-3}, \end{cases}$$

$$P_{\rm MK} = \begin{cases} \bar{\mu}_{\rm MK} - \frac{(t_1+t_2)s_1 s_2}{(\sqrt{(n_2-1)n_2}s_1) + (\sqrt{(n_1-1)n_1}s_2)} + \eta\frac{s_1}{U_1}, & \text{if } \frac{\sqrt{(n_2-1)s_1^2}}{\sqrt{(n_1-1)s_2^2}} \leq \frac{\sqrt{n_2}}{\sqrt{n_1}} \\ \bar{\mu}_{\rm MK} - \frac{(\sqrt{(n_2/n_1)((n_2-1)/(n_1-1))}s_1^2 t_1) + (\sqrt{(n_1/n_2)((n_1-1)/(n_2-1))}s_2^2 t_2)}{(\sqrt{(n_2-1)n_2}s_1) + (\sqrt{(n_1-1)n_1}s_2)} + \eta\frac{s_1}{U_1}, & \text{if } \frac{\sqrt{(n_2-1)s_1^2}}{\sqrt{(n_1-1)s_2^2}} > \frac{\sqrt{n_2}}{\sqrt{n_1}}, \end{cases}$$

$$P_{\rm TK} = \begin{cases} \bar{\mu}_{\rm TK} - \frac{(b_{n_2} s_1 s_2 t_1/\sqrt{n_1-1}) + (b_{n_1} s_1 s_2 t_2/\sqrt{n_2-1})}{(\sqrt{n_2}s_1 b_1) + (\sqrt{n_1}s_2 b_2)} + \eta\frac{s_1}{U_1}, & \text{if } \frac{s_1}{s_2} \leq \frac{\sqrt{n_2}b_{n_2}}{\sqrt{n_1}b_{n_1}} \\ \bar{\mu}_{\rm TK} - \frac{(\sqrt{n_2/(n_1-1)}b_{n_1} s_1 t_1) + (\sqrt{n_1/(n_2-1)}b_{n_2} s_2 t_2)}{(\sqrt{n_2}s_1 b_1) + (\sqrt{n_1}s_2 b_2)} + \eta\frac{s_1}{U_1}, & \text{if } \frac{s_1}{s_2} > \frac{\sqrt{n_2}b_{n_2}}{\sqrt{n_1}b_{n_1}}, \end{cases}$$

where $\bar{\mu}_{\rm GD}$, $\bar{\mu}_{\rm KS}$, $\bar{\mu}_{\rm MK}$ and $\bar{\mu}_{\rm TK}$ are the observed values of $\mu_{\rm GD}$, $\mu_{\rm KS}$, $\mu_{\rm MK}$ and $\mu_{\rm TK}$, respectively. Observe that all the four statistics $P_{\rm GD}$, $P_{\rm KS}$, $P_{\rm MK}$ and $P_{\rm TK}$ satisfy the conditions given in Definition 1. Now, we construct the generalized test variables for the quantile θ using these four alternative estimators of the common mean μ under the condition $\sigma_1^2 \leq \sigma_2^2$ as $T_{\text{GD}} = P_{\text{GD}} - \theta$, $T_{\text{KS}} = P_{\text{KS}} - \theta$, $T_{\text{MK}} = P_{\text{MK}} - \theta$ and $T_{\text{TK}} = P_{\text{TK}} - \theta$. These test variables satisfy the conditions (a) and (b) in Definition 2 and also stochastically decreasing in θ. Utilizing the Definition 3, we compute the p-values of all the generalized test variables as

$$2\min(\Pr\{P_{\text{GD}} \geq \theta_0\}, \Pr\{P_{\text{GD}} \leq \theta_0\}) \tag{13}$$

$$2\min(\Pr\{P_{\text{KS}} \geq \theta_0\}, \Pr\{P_{\text{KS}} \leq \theta_0\}) \tag{14}$$

$$2\min(\Pr\{P_{\text{MK}} \geq \theta_0\}, \Pr\{P_{\text{MK}} \leq \theta_0\}) \tag{15}$$

$$2\min(\Pr\{P_{\text{TK}} \geq \theta_0\}, \Pr\{P_{\text{TK}} \leq \theta_0\}). \tag{16}$$

The null hypothesis H_0 will be rejected if the p-values are less than α, the level of significance.

Next, we construct the generalized pivot variables and generalized test variables for the quantile θ utilizing the improved estimators of the common mean under order restriction $\sigma_1^2 \leq \sigma_2^2$ given in (6)–(9). The generalized pivot variables are given by

$$P_{\text{GD}}^a = \begin{cases} \bar{\mu}_{\text{GD}}^a - \frac{(\sqrt{(n_1-1)n_1}s_1s_2^2t_1)+(\sqrt{(n_2-1)n_2}s_1^2s_2t_2)}{((n_2-1)n_2s_1^2)+((n_1-1)n_1s_2^2)} + \eta\frac{s_1}{U_1}, & \text{if } \frac{s_1^2}{n_1-1} \leq \frac{s_2^2}{n_2-1} \\ \bar{\mu}_{\text{GD}}^a - \frac{(\sqrt{(n_1-1)n_1}s_1t_1)+(\sqrt{(n_2-1)n_2}s_2t_2)}{n_1+n_2} + \eta\frac{s_1}{U_1}, & \text{if } \frac{s_1^2}{n_1-1} > \frac{s_2^2}{n_2-1}, \end{cases}$$

$$P_{\text{KS}}^a = \begin{cases} \bar{\mu}_{\text{KS}}^a - \frac{(\sqrt{n_1/(n_1-1)}(n_1-3)s_1s_2^2t_1)+(\sqrt{n_2/(n_2-1)}(n_2-3)s_1^2s_2t_2)}{((n_2-3)n_2s_1^2)+((n_1-3)n_1s_2^2)} + \eta\frac{s_1}{U_1}, & \text{if } \frac{s_1^2}{n_1-3} \leq \frac{s_2^2}{n_2-3} \\ \bar{\mu}_{\text{KS}}^a - \frac{(\sqrt{(n_1-1)n_1}s_1t_1)+(\sqrt{(n_2-1)n_2}s_2t_2)}{n_1+n_2} + \eta\frac{s_1}{U_1}, & \text{if } \frac{s_1^2}{n_1-3} > \frac{s_2^2}{n_2-3}, \end{cases}$$

$$P_{\text{MK}}^a = \begin{cases} \bar{\mu}_{\text{MK}}^a - \frac{(t_1+t_2)s_1s_2}{(\sqrt{(n_2-1)n_2}s_1)+(\sqrt{(n_1-1)n_1}s_2)} + \eta\frac{s_1}{U_1}, & \text{if } \sqrt{\frac{s_1^2}{n_1-1}} \leq \sqrt{\frac{s_2^2}{n_2-1}} \\ \bar{\mu}_{\text{MK}}^a - \frac{(\sqrt{(n_1-1)n_1}s_1t_1)+(\sqrt{(n_2-1)n_2}s_2t_2)}{n_1+n_2} + \eta\frac{s_1}{U_1}, & \text{if } \sqrt{\frac{s_1^2}{n_1-1}} > \sqrt{\frac{s_2^2}{n_2-1}}, \end{cases}$$

$$P_{\text{TK}}^a = \begin{cases} \bar{\mu}_{\text{TK}}^a - \frac{(b_{n_2}s_1s_2t_1/\sqrt{n_1-1})+(b_{n_1}s_1s_2t_2/\sqrt{n_2-1})}{(\sqrt{n_2}s_1b_1)+(\sqrt{n_1}s_2b_2)} + \eta\frac{s_1}{U_1}, & \text{if } \frac{s_1}{s_2} \leq \frac{\sqrt{n_2}b_{n_2}}{\sqrt{n_1}b_{n_1}} \\ \bar{\mu}_{\text{TK}}^a - \frac{(\sqrt{(n_1-1)n_1}s_1t_1)+(\sqrt{(n_2-1)n_2}s_2t_2)}{n_1+n_2} + \eta\frac{s_1}{U_1}, & \text{if } \frac{s_1}{s_2} > \frac{\sqrt{n_2}b_{n_2}}{\sqrt{n_1}b_{n_1}}. \end{cases}$$

where $\bar{\mu}_{\text{GD}}^a$, $\bar{\mu}_{\text{KS}}^a$, $\bar{\mu}_{\text{MK}}^a$ and $\bar{\mu}_{\text{TK}}^a$ are the observed values of μ_{GD}^a, μ_{KS}^a, μ_{MK}^a and μ_{TK}^a, respectively. In a similar manner, as discussed before, we construct the generalized test variables for the quantile θ to test hypothesis (1), which are given by $T_{\text{GD}}^a = P_{\text{GD}}^a - \theta$, $T_{\text{KS}}^a = P_{\text{KS}}^a - \theta$, $T_{\text{MK}}^a = P_{\text{MK}}^a - \theta$ and $T_{\text{TK}}^a = P_{\text{TK}}^a - \theta$. All these test variables are stochastically decreasing in θ and satisfy the first two conditions in Definition 2. The p-values for these tests are computed as

$$2\min(\Pr\{P_{\text{GD}}^a \geq \theta_0\}, \Pr\{P_{\text{GD}}^a \leq \theta_0\}) \tag{17}$$

$$2\min(\Pr\{P_{\text{KS}}^a \geq \theta_0\}, \Pr\{P_{\text{KS}}^a \leq \theta_0\}) \tag{18}$$

$$2\min(\Pr\{P_{\text{MK}}^a \geq \theta_0\}, \Pr\{P_{\text{MK}}^a \leq \theta_0\}) \tag{19}$$

$$2\min(\Pr\{P_{\text{TK}}^a \geq \theta_0\}, \Pr\{P_{\text{TK}}^a \leq \theta_0\}). \tag{20}$$

If the p-values are less than significance level α, one should reject the null hypothesis H_0 given in (1), otherwise accept it.

For unequal sample sizes $(n_1 \neq n_2)$, $T_{\text{GD}} = T^a_{\text{GD}}$, $T_{\text{KS}} = T^a_{\text{KS}}$, $T_{\text{MK}} = T^a_{\text{MK}}$ and $T_{\text{TK}} = T^a_{\text{TK}}$ as their corresponding estimators are equal.

Remark 1 Using the above pivot variables, one can also construct the generalized confidence intervals. The $(1 - \psi)100\%$ generalized confidence interval is $(P(\psi/2), P(1 - \psi/2))$.

4 Simulation Study

In Sect. 3, we have proposed several generalized test variables for the quantile θ under the condition that $\sigma_1^2 \leq \sigma_2^2$. In this section, we will compare the performances of all those proposed test procedures for various combinations of sample sizes in terms of size and power.

To compute the size and power, we have generated 10,000 random samples from each normal population with a common mean and ordered variances, using the Monte Carlo simulation method. In order to compute the generalized test statistics, the inner loop is repeated 5000 times. Though several choices of parameters and sample sizes have been considered in the simulation study, in Tables 3 and 4, we have presented the size and power for a few specific choices of parameters and sample sizes. Throughout the simulation study, we have taken $\eta = 1.96$. For computing the size and power, we have taken $\theta_0 = 1.96$. Observe that all the tests are location invariant; hence, its size only depends on $\rho = \sigma_1/\sigma_2 > 0$ through σ_2. The values of ρ have been varied from 0 to 1 by fixing $\sigma_1 = 1$, so that the condition $\sigma_1^2 \leq \sigma_2^2$ is satisfied.

In Table 3, we present the size of all the eight test procedures for the specific sample sizes. In Table 3, the first column presents the choice of ρ. Corresponding to one value of ρ, there are eight values from the second column onward that present sizes of tests in the given order of sample sizes. In each cell, the size values will be read vertically downward. It is to be noted that for equal sample sizes, that is, when $n_1 = n_2$ as the estimators $\hat{\mu}_{\text{GD}} = \hat{\mu}_{\text{KS}}$, $\hat{\mu}_{\text{MK}} = \hat{\mu}_{\text{TK}}$, $\hat{\mu}^a_{\text{GD}} = \hat{\mu}^a_{\text{KS}}$ and $\hat{\mu}^a_{\text{MK}} = \hat{\mu}^a_{\text{TK}}$, so their corresponding tests, as well as sizes and powers, are also equal. For the unequal sample sizes that is when $n_1 \neq n_2$, $\mu^a_{\text{GD}} = \hat{\mu}_{\text{GD}}$, $\mu^a_{\text{KS}} = \hat{\mu}_{\text{KS}}$, $\mu^a_{\text{MK}} = \hat{\mu}_{\text{MK}}$ and $\mu^a_{\text{TK}} = \hat{\mu}_{\text{TK}}$, so similarly their corresponding tests, sizes and powers are also equal. The maximum bound for the simulation error has been seen up to 0.003 to attain a high level of accuracy. The following observations have been made from our simulation study and also from Tables 3 and 4, which we write in the forms of a remark.

Remark 2
1. It has been observed that all the tests attain the nominal level within 20% of the specified level of significance $\alpha = 0.05$. All the tests are qualified for further power comparison.
2. The powers of all the tests have been computed by varying θ from θ_0, through $\mu = 0.4, 0.6, 0.8, 1, 2$ and fixing $\sigma_1 = \sigma_2 = 1$.

Table 1 p-values for all the proposed tests

Method	T_{GD}	T_{KS}	T_{MK}	T_{TK}	T^a_{GD}	T^a_{KS}	T^a_{MK}	T^a_{TK}
p-value	0.0036	0.0036	0.0034	0.0034	0.0038	0.0038	0.0037	0.0037

Table 2 p-values for all the proposed tests

Method	T_{GD}	T_{KS}	T_{MK}	T_{TK}	T^a_{GD}	T^a_{KS}	T^a_{MK}	T^a_{TK}
p-value	0.7048	0.6972	0.7432	0.7492	0.7048	0.6972	0.7432	0.7492

3. It is further observed that as the difference between θ and θ_0 increases, the powers of all the tests increase up to 1. Moreover, when the sample size increases, the powers of all the tests increase.
4. In the case of equal sample sizes, the test based on the improved estimator μ^a_{GD} and μ^a_{KS} has the best performance in terms of power, whereas all other tests are competing with each other. However, in the case of unequal sample sizes, none of the tests dominates others; that is, all the tests compete well with each other.

5 Concluding Remarks with Real-life Examples

In this article, we have derived several generalized test procedures to test the hypothesis regarding the quantile θ of the first normal population among two normal populations with a common mean and order restricted variances. Utilizing some estimators proposed by Jena et al. [7], the generalized variable as well as test statistics have been constructed to test a hypothesis regarding the quantile. All the proposed test procedures have been compared in terms of their sizes and powers numerically using the Monte Carlo simulation method.

From our simulation study, we have concluded that all the proposed tests attain the nominal level within 20% of the level of significance. It is also concluded that the test based on the estimators μ^a_{GD} and μ^a_{KS} (i.e., T^a_{GD} and T^a_{KS}) have the best performance in terms of power for equal sample sizes, whereas for unequal sample sizes, all the tests compete with each other. We hope that the current research work will enlighten the inference on the quantiles, which have many real-world applications. Below, we discuss two examples, which will illustrate the methods of tests proposed in this paper.

Example 1 We consider the two datasets of equal sample size 10 as given in Jena et al. [7] which have a common mean and ordered variances $\sigma_1^2 \leq \sigma_2^2$. The sufficient statistics for this data is $\bar{x}_1 = 25.36$, $\bar{x}_2 = 25.04$, $s_1^2 = 57.69$ and $s_2^2 = 36.46$. It has been noted that $s_1^2 > s_2^2$ for this data. Suppose, one is interested to test hypothesis $H_0 : \theta = 40$ against the alternative $H_1 : \theta \neq 40$ at the level of significance $\alpha = 0.05$. Table 1 presents the the p-values of all the proposed tests to test this hypothesis.

Table 3 Sizes of the proposed tests for the sample sizes (5, 5), (12, 12), (25, 25), (40, 40), (5, 10), (10, 5), (12, 20), (20, 12) and $\alpha = 0.05$

ρ	T_{GD}	T_{KS}	T_{MK}	T_{TK}	T_{GD}^a	T_{KS}^a	T_{MK}^a	T_{TK}^a
0.2	0.0452	0.0452	0.0472	0.0472	0.0452	0.0452	0.0472	0.0472
	0.0500	0.0500	0.0516	0.0516	0.0500	0.0500	0.0516	0.0516
	0.0512	0.0512	0.0524	0.0524	0.0512	0.0512	0.0524	0.0524
	0.0560	0.0560	0.0500	0.0500	0.0560	0.0560	0.0500	0.0500
	0.0368	0.0412	0.0428	0.0420	0.0368	0.0412	0.0428	0.0420
	0.0352	0.0356	0.0400	0.0400	0.0352	0.0356	0.0400	0.0400
	0.0420	0.0412	0.0452	0.0444	0.0420	0.0412	0.0452	0.0444
	0.0480	0.0476	0.0472	0.0472	0.0480	0.0476	0.0472	0.0472
0.4	0.0384	0.0384	0.0428	0.0428	0.0384	0.0384	0.0428	0.0428
	0.0448	0.0448	0.0456	0.0456	0.0448	0.0448	0.0456	0.0456
	0.0448	0.0448	0.0472	0.0472	0.0448	0.0448	0.0472	0.0472
	0.0516	0.0516	0.0544	0.0544	0.0516	0.0516	0.0544	0.0544
	0.0400	0.0412	0.0404	0.0404	0.0400	0.0412	0.0404	0.0404
	0.0456	0.0464	0.0468	0.0468	0.0456	0.0464	0.0468	0.0468
	0.0496	0.0496	0.0452	0.0444	0.0496	0.0496	0.0452	0.0444
	0.0528	0.0528	0.0560	0.0560	0.0528	0.0528	0.0560	0.0560
0.6	0.0420	0.0420	0.0424	0.0424	0.0436	0.0436	0.0436	0.0436
	0.0496	0.0496	0.0484	0.0484	0.0504	0.0504	0.0484	0.0484
	0.0448	0.0448	0.0432	0.0432	0.0448	0.0448	0.0432	0.0432
	0.0468	0.0468	0.0472	0.0472	0.0468	0.0468	0.0472	0.0472
	0.0440	0.0444	0.0420	0.0416	0.0440	0.0444	0.0420	0.0416
	0.0468	0.0484	0.0480	0.0480	0.0468	0.0484	0.0480	0.0480
	0.0400	0.0408	0.0396	0.0396	0.0400	0.0408	0.0396	0.0396
	0.0524	0.0520	0.0532	0.0512	0.0524	0.0520	0.0532	0.0512
0.8	0.0456	0.0456	0.0464	0.0464	0.0500	0.0500	0.0492	0.0492
	0.0404	0.0404	0.0408	0.0408	0.0456	0.0456	0.0448	0.0448
	0.0444	0.0444	0.0432	0.0432	0.0440	0.0440	0.0432	0.0432
	0.0496	0.0496	0.0492	0.0492	0.0500	0.0500	0.0492	0.0492
	0.0412	0.0408	0.0412	0.0408	0.0412	0.0408	0.0412	0.0408
	0.0488	0.0508	0.0516	0.0504	0.0488	0.0508	0.0516	0.0504
	0.0432	0.0448	0.0412	0.0404	0.0432	0.0448	0.0412	0.0404
	0.0516	0.0536	0.0488	0.0484	0.0516	0.0536	0.0488	0.0484
1.0	0.0372	0.0372	0.0372	0.0372	0.0424	0.0424	0.0400	0.0400
	0.0380	0.0380	0.0385	0.0385	0.0432	0.0432	0.0464	0.0464
	0.0488	0.0488	0.0480	0.0480	0.0496	0.0496	0.0484	0.0484
	0.0468	0.0468	0.0488	0.0488	0.0484	0.0484	0.0480	0.0480
	0.0452	0.0436	0.0436	0.0428	0.0452	0.0436	0.0436	0.0428
	0.0552	0.0552	0.0560	0.0560	0.0552	0.0552	0.0560	0.0560
	0.0376	0.0380	0.0348	0.0348	0.0376	0.0380	0.0348	0.0348
	0.0504	0.0500	0.0492	0.0492	0.0504	0.0500	0.0492	0.0492

Table 4 Powers of the proposed tests for the sample sizes (5, 5), (12, 12), (25, 25), (40, 40), (5, 10), (12, 20), (10, 5), (20, 12) and $\alpha = 0.05$

θ	T_{GD}	T_{KS}	T_{MK}	T_{TK}	T^a_{GD}	T^a_{KS}	T^a_{MK}	T^a_{TK}
2.36	0.1072	0.1072	0.1156	0.1156	0.1300	0.1300	0.1292	0.1292
	0.1488	0.1488	0.1552	0.1552	0.1612	0.1612	0.1604	0.1604
	0.3044	0.3044	0.3044	0.3044	0.3048	0.3048	0.3036	0.3036
	0.4432	0.4432	0.4424	0.4424	0.4464	0.4464	0.4444	0.4444
	0.1056	0.1048	0.1064	0.1064	0.1056	0.1048	0.1064	0.1064
	0.1820	0.1828	0.1824	0.1828	0.1820	0.1828	0.1824	0.1828
	0.1176	0.1200	0.1116	0.1128	0.1176	0.1200	0.1116	0.1128
	0.2364	0.2376	0.2332	0.2340	0.2364	0.2376	0.2332	0.2340
2.56	0.2224	0.2224	0.2316	0.2316	0.2428	0.2428	0.2432	0.2432
	0.3308	0.3308	0.3380	0.3380	0.3508	0.3508	0.3480	0.3480
	0.6068	0.6068	0.6076	0.6076	0.6100	0.6100	0.6072	0.6072
	0.7820	0.7820	0.7808	0.7808	0.7832	0.7832	0.7812	0.7812
	0.1876	0.1864	0.1852	0.1832	0.1876	0.1864	0.1852	0.1832
	0.3512	0.3484	0.3520	0.3504	0.3512	0.3484	0.3520	0.3504
	0.2388	0.2428	0.2328	0.2348	0.2388	0.2428	0.2328	0.2348
	0.4776	0.4800	0.4696	0.4720	0.4776	0.4800	0.4696	0.4720
2.76	0.4136	0.4136	0.4200	0.4200	0.4336	0.4336	0.4296	0.4296
	0.5444	0.5444	0.5496	0.5496	0.5608	0.5608	0.5552	0.5552
	0.8564	0.8564	0.8552	0.8552	0.8564	0.8564	0.8552	0.8552
	0.9612	0.9612	0.9612	0.9612	0.9612	0.9612	0.9612	0.9612
	0.2904	0.2856	0.2884	0.2872	0.2904	0.2856	0.2884	0.2872
	0.5812	0.5816	0.5792	0.5784	0.5812	0.5816	0.5792	0.5784
	0.3992	0.4040	0.3732	0.3816	0.3992	0.4040	0.3732	0.3816
	0.7368	0.7388	0.7272	0.7292	0.7368	0.7388	0.7272	0.7292
2.96	0.6140	0.6140	0.6224	0.6224	0.6312	0.6312	0.6304	0.6304
	0.7380	0.7380	0.7376	0.7376	0.7440	0.7440	0.7412	0.7412
	0.9688	0.9688	0.9684	0.9684	0.9688	0.9688	0.9684	0.9684
	0.9984	0.9984	0.9984	0.9984	0.9984	0.9984	0.9984	0.9984
	0.4432	0.4388	0.4376	0.4328	0.4432	0.4388	0.4376	0.4328
	0.8012	0.8020	0.7996	0.7992	0.8012	0.8020	0.7996	0.7992
	0.6212	0.6204	0.6020	0.6056	0.6212	0.6204	0.6020	0.6056
	0.9084	0.9084	0.9100	0.9108	0.9084	0.9084	0.9100	0.9108
3.96	0.9988	0.9988	1.0000	1.0000	0.9988	0.9988	1.0000	1.0000
	1.0000	1.0000	1.0000	1.0000	1.0000	1.0000	1.0000	1.0000
	1.0000	1.0000	1.0000	1.0000	1.0000	1.0000	1.0000	1.0000
	1.0000	1.0000	1.0000	1.0000	1.0000	1.0000	1.0000	1.0000
	0.9760	0.9816	0.9784	0.9788	0.9760	0.9816	0.9784	0.9788
	1.0000	1.0000	1.0000	1.0000	1.0000	1.0000	1.0000	1.0000
	0.9988	0.9988	0.9980	0.9980	0.9988	0.9988	0.9980	0.9980
	1.0000	1.0000	1.0000	1.0000	1.0000	1.0000	1.0000	1.0000

Since all the p-values are less than 0.05, the null hypothesis H_0 is rejected with significance level 0.05.

Example 2 Rohatgi and Saleh [19] (p. 515) took up an example for a two-sample problem on the mean life of bulbs (in hours). The mean life of the first sample of nine light bulbs is ($\bar{x}_1$) 1309 h and standard deviation 420 h. The second sample of 16 bulbs from other population has a mean ($\bar{x}_2$) 1205 h and a standard deviation 390 h. Equality of means has been tested by two-sample t-test. Further, we apply the F-test and observed that the variances are ordered, that is, $\sigma_1^2 \leq \sigma_2^2$. Therefore, these datasets are considered for our model. It is our interest to test hypothesis $H_0 : \theta = 2040$ against the alternative $H_1 : \theta \neq 2040$ at significance level 0.05. and present the p-values in Table 2.

We can conclude from the above p-values that hypothesis H_0 cannot be rejected using all the test procedures at level 0.05 (Tables 3 and 4).

Acknowledgements The authors would like to express their sincere thanks to the two anonymous reviewers for their constructive comments, which have helped in improving the presentation of this article.

References

1. Brewster JF, Zidek JV (1974) Improving on equivariant estimators. Ann Stat 2(1):21–38
2. Chang CH, Pal N (2008) Testing on the common mean of several normal distributions. Comput Stat Data Anal 53(2):321–333
3. Cohen A, Sackrowitz HB (1984) Testing hypotheses about the common mean of normal distributions. J Stat Plann Infer 9(2):207–227
4. Elfessi A, Pal N (1992) A note on the common mean of two normal populations with order restricted variances. Commun Stat-Theory Methods 21(11):3177–3184
5. Graybill FA, Deal RB (1959) Combining unbiased estimators. Biometrics 15(4):543–550
6. Guo H, Krishnamoorthy K (2005) Comparison between two quantiles: the normal and exponential cases. Commun Stat-Simul Comput 34(2):243–252
7. Jena AK, Tripathy MR, Pal N (2019) Alternative estimation of the common mean of two normal populations with order restricted variances. REVSTAT. https://www.ine.pt/revstat/pdf/AlternativeEstimation.pdf
8. Keating JP, Tripathi RC (1985) Estimation of percentiles. Encycl Stat Sci 6:668–674
9. Khatri CG, Shah KR (1974) Estimation of location parameters from two linear models under normality. Commun Stat-Theory Methods 3(7):647–663
10. Khatun H, Tripathy MR, Pal N (2020) Hypothesis testing and interval estimation for quantiles of two normal populations with a common mean. Commun Stat-Theory Methods. https://doi.org/10.1080/03610926.2020.1845735
11. Krishnamoorthy K, Lu Y (2003) Inferences on the common mean of several normal populations based on the generalized variable method. Biometrics 59(2):237–247
12. Kumar S, Tripathy MR (2011) Estimating quantiles of normal populations with a common mean. Commun Stat-Theory Methods 40(15):2719–2736
13. Li X, Tian L, Wang J, Muindi JR (2012) Comparison of quantiles for several normal populations. Comput Stat Data Anal 56(6):2129–2138
14. Lin SH, Lee JC (2005) Generalized inferences on the common mean of several normal populations. J Stat Plann Infer 134(2):568–582

15. Misra N, van der Meulen EC (1997) On estimation of the common mean of k($\geq$ 2) normal populations with order restricted variances. Stat Probab Lett 36(3):261–267
16. Moore B, Krishnamoorthy K (1997) Combining independent normal sample means by weighting with their standard errors. J Stat Comput Simul 58(2):145–153
17. Nagamani N, Tripathy MR (2020) Improved estimation of quantiles of two normal populations with common mean and ordered variances. Commun Stat-Theory Methods 49(19):4669–4692
18. Pal N, Lim WK, Ling CH (2007) A computational approach to statistical inferences. J Appl Probab Stat 2(1):13–35
19. Rohatgi VK, Saleh AKME (2017) An introduction to probability and statistics, 2nd edn. Wiley
20. Rukhin AL (1983) A class of minimax estimators of a normal quantile. Stat Probab Lett 1(5):217–221
21. Saleh AKME (1981) Estimating quantiles of exponential distribution. Stat Relat Top 9:145–51
22. Tripathy MR, Kumar S (2010) Estimating a common mean of two normal populations. J Stat Theory Appl 9(2):197–215
23. Tripathy MR, Kumar S, Jena AK (2017) Estimating quantiles of several normal populations with a common mean. Commun Stat-Theory Methods 46(11):5656–5671
24. Tsui KW, Weerahandi S (1989) Generalized p-values in significance testing of hypotheses in the presence of nuisance parameters. J Am Stat Assoc 84(406):602–607
25. Zidek JV (1969) Inadmissibility of the best invariant estimator of extreme quantiles of the normal law under squared error loss. Ann Math Stat 40(5):1801–1808

An Approach to Experimental Data-Based Mathematical Modelling for a Green Roof

Pramod Belkhode, Pranita Belkhode, and Kanchan Borkar

Abstract Planted roof or green roof is an environmentally friendly approach to overcome climatic challenges and environmental issues created in the urban area. Literature reviews highlight that for many years planted roof building has been beneficial to the society as it reduces the roof temperature. So it is very important to study various geometrical parameters of the plant roof to optimize the dimensional parameters by means of independent and dependent variables to generate an exact mathematical model between different geometrics, and to develop the mathematical model based on the experimental data recorded during the experimentation. The experimental outputs, such as heat flux through planted roof, are obtained by varying the inputs such as solar radiation, velocity, humidity and various temperatures recorded at different layers of planted roof. This correlation is formulated using the approach suggested by H. Schenck Jr. in his book 'Theories of Engineering Experimentation'. Thus, the factors influencing the performance of the planted roof activity have been identified, so as to optimize the performance of the heat flow through planted roof.

Keywords Planted roof · Mathematical model · Solar radiation · Optimization · Dimensional parameters · Green roof · Global warning

1 Introduction

Due to global warming, the average indoor and outdoor temperature has increased. Mechanical air-conditioning systems are used for comfort in India. Energy consumption in conventional domestic housing in India is high, and the equipment is highly

P. Belkhode (✉)
Laxminarayan Institute of Technology, Nagpur, India
e-mail: pramodb@rediffmail.com

P. Belkhode
Lady Amritbai Daga College for Women, Nagpur, India

K. Borkar
Military Engineering Services, Nagpur, India

S. S. Ray et al. (eds.), *Applied Analysis, Computation and Mathematical Modelling in Engineering*, Lecture Notes in Electrical Engineering 897,
https://doi.org/10.1007/978-981-19-1824-7_16

energy intensive. Planted roofs are now an established technology for improvement of thermal efficiency or heat reduction process of buildings.

A detailed literature review about the thermal performance of planted roofs, planted roof materials and construction, feasibility of planted roofs in arid regions, the effect of planted roof in energy efficiency of the buildings, etc., has been done. Paper demonstrates the formulation of experimental data-based model and the influence of the individual pie terms based on its indices to optimize the performance of planted roof so as to control the heat flow through planted roof. The mathematical model investigates the effectiveness of green roof in terms of energy efficiency on small building depending on the heat flux. The formulated mathematical equation of the green roof highlights the most effective pie terms so that accordingly the care of that pie term will be taken to obtain best performance of planted roof [1].

2 Working Principal of Planted Roof

Green roof consists of different layers such as

Planted Roof Layer is as follows:

(1) Plants (chosen depending on unique qualities).
(2) Mixture of soil, minerals and nutrients.
(3) Retention medium to hold the water.
(4) Drainage layer that sometimes has a built in reservoir.
(5) Water proofing membrane with a root barrier.
(6) Original roof structural material (Concrete slab)

Figure 1 shows the different layers of the planted roof under the extensive roofing with thin medium layer for plantation of different plants and vegetation [2–4].

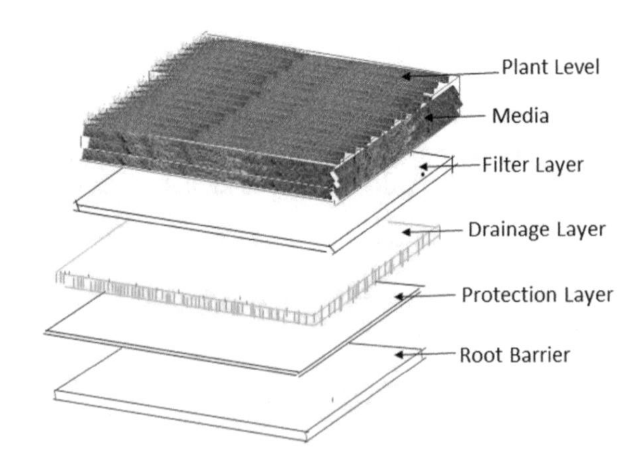

Fig. 1 Different layers of green roofing

3 Model Formulation

Model formulation of any activities related to field or experimental investigation plays an important role in the analysis. The recorded observations were studied by following the analysis steps of formulation as given below.

- Recognize various inputs and corresponding outputs related to the system.
- As it is difficult to handle many inputs in the equation, dimensionless analysis is used to minimize the number of inputs under the group of suitable pie.
- Once the pie terms are selected, it is necessary to fix the range.
- The data which is not within the range is rejected.
- Final stage is to correlate the data by formulating the mathematical model.

The independent variable or parameters involved in the experimental or field data are identified as the Theories of Experimentation. The inputs and output involved in the green roof is identify to effects of these variables.

4 Variables of Planted Roof

The input variables of the green roof and corresponding output are given in Table 1.

5 Dimensionless Pie Terms

5.1 Formation of Pie Terms

All the input terms involved in the experimental set-up are grouped to the individual pie term so that the effect of each pie term, i.e. the group of variable, can be studied effectively. The group of these variables is given in Table 2.

5.2 Formulation of Experimental Data-Based Model

All the variables described in Table 1 are grouped as given in Table 2. Further model formulation is done to identify the curve fitting constant and the indices of each pie term. The total nine pies are P_1, P_2, P_3, P_4, P_5, P_6, P_7, P_8 and P_9 with dependent pie term $\pi_{D.}$ The approximate mathematical model is presented as follow [5],

$$(z_1) = k * [(P_1)^{a_1} * (P_2)^{b_1} * (P_3)^{c_1} * (P_4)^{d_1} * (P_5)^{e_1} * (P_6)^{f_1} * (P_7)^{g_1} * (P_8)^{h_1} * (P_9)^{i_1}] \quad (1)$$

Table 1 Planted roof–variable

S. No	Description of variables	Type of variable	Symbol	Dimension
1	Heat flux	Dependent	Q_f	$M^1L^0T^{-3}$
2	Heat transferred into the room	Independent	Q_{tr}	M^1L^0T
3	Evaporation rate	Independent	R_{ep}	$M^1L^{-2}T^{-1}$
4	Transpiration rate	Independent	R_{tp}	$M^1L^{-2}T^{-1}$
5	Heat generation by respiration	Independent	Q_{rp}	M^1L^0T
6	Temperature	Independent	T_{sr}	$M^0L^0T^0$
7	Wind velocity at surrounding	Independent	W_{vo}	$M^0L^1T^{-1}$
8	Wind velocity inside room	Independent	W_{vi}	$M^0L^1T^{-1}$
9	Humidity	Independent	H_u	$M^0L^0T^0$
10	Ambient temperature	Independent	T_a	$M^0L^0T^0$
11	Nutrient concentration in plant	Independent	N_c	$M^0L^0T^0$
12	Water content of plant tissue	Independent	W_c	$M^0L^0T^0$
13	Plant type	Independent	P_t	$M^0L^0T^0$
14	Humidity of plant	Independent	H	$M^0L^0T^0$
15	Leaf index of plant	Independent	L_i	$M^0L^0T^0$
16	Leaf area	Independent	L_a	$M^0L^2T^0$
17	Leaf area ratio	Independent	L_{ar}	$M^{-1}L^2T^0$
18	Areal density of plant	Independent	A_d	$M^1L^{-3}T^0$
19	Plant height	Independent	P_h	$M^0L^1T^0$
20	Heat storage of plant	Independent	H_s	$M^1L^2T^{-2}$
21	Solar storage converted by photosynthesis	Independent	S_e	$M^1L^2T^{-2}$
22	Plant temperature	Independent	P_{tc}	$M^0L^0T^0$
23	Thickness layer of soil	Independent	T_s	$M^0L^1T^0$
24	Soil temperature	Independent	S_t	$M^0L^{-2}T^{-2}$
25	Thermal conductivity of soil	Independent	T_{cs}	$M^1L^1T^{-2}$
26	Areal density of soil	Independent	D_e	$M^1L^{-3}T^0$
27	Sp. heat of soil	Independent	S_{hs}	$M^0L^{-2}T^{-2}$
28	Vol. heat capacity	Independent	V_{hc}	$M^1L^{-1}T^{-2}$
29	Latent heat of soil	Independent	L_s	$M^0L^{-2}T^{-2}$
30	Surface area of soil	Independent	S_a	$M^0L^2T^0$
31	Evaporation rate	Independent	E_r	$M^1L^{-2}T^{-1}$
32	Velocity	Independent	V	$M^0L^1T^{-1}$
33	Density	Independent	D_e	$M^1L^{-3}T^0$
34	Temperature	Independent	T	$M^0L^0T^0$
35	Evaporation rate	Independent	E_r	$M^1L^{-2}T^{-1}$
36	Area of flow of water	Independent	A_w	$M^0L^2T^0$

(continued)

Table 1 (continued)

S. No	Description of variables	Type of variable	Symbol	Dimension
37	Rate of flow of water	Independent	F_w	$M^1L^0T^{-1}$
38	Heat storage	Independent	H_s	$M^1L^2T^{-2}$
39	Sp. Heat of concrete	Independent	S_p	$M^0L^{-2}T^{-2}$
40	Bulk density	Independent	B_d	$M^1L^{-3}T^0$
41	Thickness of concrete	Independent	T	$M^0L^1T^0$
42	Area of concrete	Independent	A_r	$M^0L^2T^0$

Table 2 Grouped independent pie terms

S. No	Independent dimensionless ratios	Nature of basic physical quantities
01	$\pi_1 = [(Q\text{rp}/\text{Qtr})(R_{\text{ep}}/R_{\text{tp}})$	Solar radiation
02	$\pi_2 = [W_{\text{vo}}/W_{\text{vi}}]$	Wind velocity
03	$\pi_3 = [H_{\text{u}}]$	Relative humidity
04	$\pi 4 = [T_a]$	Ambient temperature
05	$\pi_5 = [(N_c/Wc)(P_t\text{H}/L_i)(L_a L_{\text{ar}} A_d/P_h)(H_s/S_e)(P_{\text{tc}})]$	Temperature gradient across plant layer
06	$\pi_6 = [(S_{\text{hs}}A_t/V_{\text{hc}})(T_s S_t T_{\text{cs}} D_s/L_s S_a E_r)$	Temperature Gradient across Soil Layer
07	$\pi_7 = [E_r T/VD_e)$	Temperature gradient across retention layer
08	$\pi_8 = [(TErAw/F_w)$	Temperature gradient across drainage layer
09	$\pi_9 = [H_s/S_p B_d TA_r]$	Temperature gradient across concrete layer
10	π_D	Heat flux through planted roof

Equation 1 shows the exponential form of model formed between the pie terms. In the above equation, dependent variable z_1 represents heat flux through planted roof. Independent variable π_1 represents solar radiation data, π_2 represents wind velocity data, π_3 represents relative humidity, π_4 represents ambient temperature, π_5 represents roof plant layer data, π_6 represents roof soil layer data, π_7 represents roof retention layer data, π_8 represents roof drainage layer data and π_9 represents specification concrete layer data. To determine the values of k, a_1, b_1, c_1, $d_{1,}$ $e_{1,}$ $f_{1,}$. $g_{1,}$ h_1 and i_1 to arrive at the regression hyper plane, the above equations are presented as follows. Equation 2 is formed by taking the log on both sides of Eq. 1.

$$\text{Log}\, z_1 = \log k + a_1 \log P_1 + b_1 \log P_2 + c_1 \log P_3 + d_1 \log P_4 + e_1 \log P_5 + f_1 \log P_6 + g_1 \log P_7 + h_1 \log P_8 + i_1 \log P_9 \quad (2)$$

Let, $Z_1 = \log z_1$, $K = \log k_1$, $A = \log P_1$, $B = \log P_2$, $C = \log P_3$, $D = \log P_4$, $E = \log P_5$, $F = \log P_{6,}$, $G = \log P_7$, $H = \log P_8$, and $I = \log P_9$.

Substituting the above terms in Eq. 2 to simplify the equation to find out the unknown.

$$Z_1 = K + a_1 A + b_1 B + c_1 C + d_1 + e_1 E + f_1 F + g_1 G + h_1 H + i_1 I \quad (3)$$

As in the experimental set-up, the total nine pie terms with curve fitting constant K can be evaluated by formulating the total ten equations. This is done by taking the summation of each pie term as shown in Eq. 4.

$$\sum Z_1 = nK + a_1 \sum A + b_1 \sum B + c_1 \sum C + d_1 \sum D + e_1 \sum E + f_1 \sum F + g_1 \sum G + h_1 \sum H + i_1 \sum I \quad (4)$$

In the above equations, n represents the number of reading and A, B, C, D, E, F, G, H and I represent the independent p_i terms $P_1, P_2, P_3, P_4, P_5, P_6, P_7, P_8$ and P_9 while Z represents dependent p_i term. All the equations are represented in the matrix form as shown below.

$$[Z] = [W] * [X]$$

All the equations are arranged in the matrix form as presented above. The matrix W is a 10×10 matrix with the multipliers of k, $a_1, b_1, c_1, d_1, e_1, f_1, g_1, h_1$ and i_1.

Matrix

$$Z_1\ x \begin{bmatrix} n \\ A \\ B \\ C \\ D \\ E \\ F \\ G \\ H \\ I \end{bmatrix} = \begin{bmatrix} n & A & B & C & D & E & F & G & H & I \\ A & A^2 & BA & CA & DA & EA & FA & GA & HA & IA \\ B & AB & B^2 & CB & DB & EB & FB & GB & HB & IB \\ C & AC & BC & C^2 & DC & EC & FC & GC & HC & IC \\ D & AD & BD & CD & D^2 & ED & FD & GD & HD & ID \\ E & AE & BE & CE & DE & E^2 & FE & GE & HE & IE \\ F & AF & BF & CF & DF & EF & F^2 & GF & HF & IF \\ G & AG & BG & CG & DG & EG & FG & G^2 & HG & IG \\ H & AH & BH & CH & DH & EH & FH & GH & H^2 & IH \\ I & AI & BI & CI & DI & EI & FI & GI & HI & I^2 \end{bmatrix} x \begin{bmatrix} k \\ a_1 \\ b_1 \\ c_1 \\ d_1 \\ e_1 \\ f_1 \\ g_1 \\ h_1 \\ i_1 \end{bmatrix}$$

The unknown of the above matrix is evaluated with the help of MATLAB software and substituted in the exponential mathematical model to form the final mathematical model of green roof.

5.3 Proposed Form of Model for Dependent Variables of Planted Roof

The proposed form of models for dependent variables of heat flux of planted roof is as under.

$$Q = (Z_1) = 3.25 * \left[(P_1)^{2.5}(P_2)^{-0.4}(P_3)^{0.33}(P_4)^{1.7}(P_5)^{1.32}(P_6)^{-2.5} (P_7)^{-0.5}(P_8)^{0.3}(P_9)^{-3.7}\right]$$

In the above equations, (Z_1) is relating to response variable for heat flux of planted roof. As the soil depth increases, it reduces the temperature in the room, at the same time thicker roof would increase the dead weight upon the concrete roof. Around 20–30 cm thickness of soil gives the best performance. Heat transfer is controlled by leaf cover area. With the increase in leaf area index, heat transfer reduces. The experimental investigation shows that the room air temperature of green roofs is always lower than normal roof throughout the experimentation. The average difference can be predicted by varying the experimental data, i.e. pie terms of independent variables. Thus, the model can be utilized as a tool to find the optimum parameter and shows that green roof is one through which energy consumption impacts the residential building.

6 Results and Conclusions

The indices of the mathematical models of green roof show the effect of the individual pie term so that the index with high value or low value will indicate the dominating nature over the output heat flux of the green roof.

Curve fitting constant k and the indices of each pie term show the influence of the causes on the effects. The curve fitting constant k is the collectively combined effect of the entire extra variable which is not included in the experimentation. If the value of k is 1, it means the model is perfect. If it is too low, the causes are overestimated; if it is too high, the causes are underestimated. This would decide when to repeat the investigation again or to refine the approach in subsequent attempts. The magnitude of exponents of the causes indicates the degree of influences of those causes on the specific response. The results indicate that the planted roof can greatly affect the room temperature profile. Planted roofs are potentially good for climates in terms of energy and cost.

The value of the curve fitting constant in this model for (Z_1) is 3.25. This collectively represents the combined effect of all extraneous variables such as soil properties, concrete materials and plant types. Further, as it is positive, this indicates that these causes have an increasing influence on heat flux through planted roof.

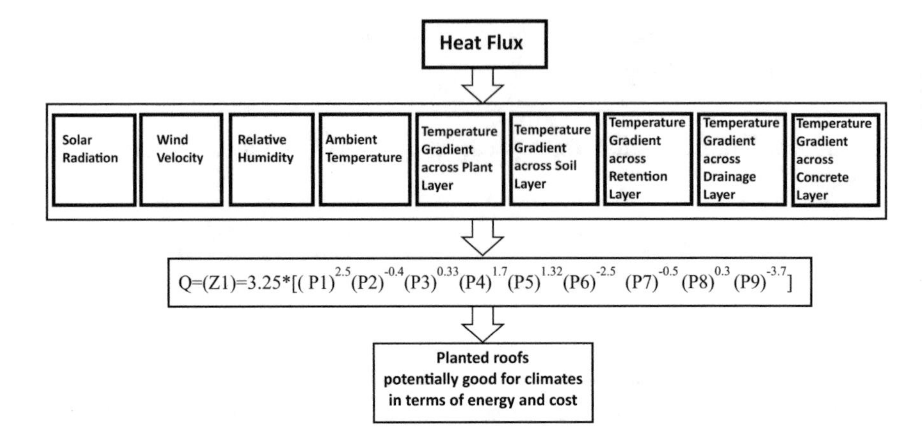

Flow chart of performance of planted roof

1. The absolute index of π_1 is the highest, viz. 2.5. Thus, the term related to the specification of the solar radiation involved the most influencing π term in this model. The value of this index is positive indicating that the heat flux through planted roof (Z_1) is directly proportional to term related to the specification of heat generation by respiration and evaporation rate. Heat flux increases with the increase in heat generation by respiration and evaporation rate.
2. The absolute index of π_3 is the lowest, viz. 0.3. Thus, the term related is the effect influencing π term in this model. The value of this index is positive indicating that the heat flux through planted roof (Z_1) is directly proportional to the term related to temperature gradient across drainage layer [π_8]. The heat flux through planted roof increases as [π_8] increases on the effect of predetermining parameters such as temperature, evaporation rate, area of flow of water and rate of flow of water. Suggestions regarding selection of appropriate area of flow of water will reduce the heat flux.
3. The sequence of influence of another independent π term present in this model is π_4, π_5, π_3, π_8, π_2, π_6 and π_9 having absolute indices as 1.7, 1.32, 0.33, 0.3, −0.4, −2.5 and −3.7 in the order, respectively.

References

1. Gowthami L, Vijayabhaskar V (2019) Green roofing: an eco-friendly approach in urban areas. Int J Chem Stud 7(3):5049–5061
2. MacDonagh LP (2005) Benefits of green roofs. Implications 4(8):1–6
3. Vijayaraghavan K, Raja FD (2015) Pilot-scale evaluation of green roofs with Sargassum biomass as an additive to improve runoff quality. Ecol Eng 75:70–78
4. Rao SS (1994) Optimization theory and application. Wiley Eastern Ltd
5. Belkhode PN (2017) Mathematical modelling of liner piston maintenance activity using field data to minimize overhauling time and human energy consumption. J Inst Eng (India) Ser C 1–9. Springer Publication

6. Schenck Jr H (1961) Theories of engineering experimentation. Mc-Graw Hill
7. Gibbs J, Luckett K, Jost V, Morgan S, Yanand T, Retzlaff W (2006) Evaluating performance of a green roof system with different growing mediums, sedumspecies and fertilizer treatments. In: Proceedings of Midwest regional green roof symposium, Edwardsville, Ill, June 30 2006
8. Belkhode PN, Vidyasagar V (2014) Mathematical model for face drilling in underground mining operation. IJERST Int J Eng Res Sci Technol 3(2)

Modal Analysis of a Thermally Loaded Functionally Graded Rotor System Using ANSYS

Waseem Shameer, Abhishek Mishra, and Prabhakar Sathujoda

Abstract To overcome the demerits of the traditional composite materials, such as debonding due to the high residual stress at inter-laminar layers and delamination of layers at higher temperature gradients, functionally graded materials (FGMs) have been developed. The present study deals with the modal analysis of a Jeffcott FG rotor system, consisting of an FG shaft mounted on linear bearings at the ends. The shaft is functionally graded which is made up of a mixture of stainless steel (SS) and zirconium dioxide (ZrO_2), where the volume fraction of metal (SS) decreases towards the outer radius and ceramic (ZrO_2) volume fraction increases. The material gradation is applied following the exponential gradation law, whereas the thermal gradients across the radius of the FG shaft are achieved through the exponential temperature distribution method (ETD). 3D finite element modelling and the modal analysis of the FG rotor system have been carried out using ANSYS software with suitable validations to determine the natural and whirl frequencies. A Python code was developed to generate the functionally graded temperature-dependent material properties of the shaft. The influence of material gradation and temperature gradients on the rotor-bearing system's natural and whirl frequencies are studied.

Keywords Functionally graded material · Rotor-bearing system · Exponential law · Exponential temperature distribution

W. Shameer (✉) · A. Mishra · P. Sathujoda
Department of Mechanical and Aerospace Engineering, Bennett University, Greater Noida, India
e-mail: ws5738@bennett.edu.in

A. Mishra
e-mail: am4265@bennett.edu.in

P. Sathujoda
e-mail: prabhakar.sathujoda@bennett.edu.in

S. S. Ray et al. (eds.), *Applied Analysis, Computation and Mathematical Modelling in Engineering*, Lecture Notes in Electrical Engineering 897,
https://doi.org/10.1007/978-981-19-1824-7_17

1 Introduction

When designing a rotor system, the study of rotor-dynamic characteristics such as critical speeds and the vibrational response is vital to avoid failure. In rotor systems, the use of composites stems from the quest for new materials that would enhance the integrity of the structure. Composite materials found common use in the industries due to their obvious additional benefits such as higher stiffness–weight ratio and increased specific strength, but the applications quickly reduced due to delamination and debonding at higher temperatures combined with their brittle nature and low toughness. Efforts to mitigate these problems without sacrificing rotor efficiency have therefore led to the establishment of an advanced class of composites, functionally graded materials (FGM). An FGM is an advanced composite obtained by the inhomogeneous mixing of various material phases. The constituent elements used for the processing of FGM are typically metals and ceramics, ceramic delivers superior heat tolerance and metal ensure structural integrity. Based on different material laws, the volume fractions of the constituent materials are gradually varied. The first applications of functionally graded materials were accomplished at the National Aerospace Laboratory of Japan in 1984 [1].

The following subsections present some of the relevant works in rotor-bearing systems and their dynamic analysis. Nelson and Mcvaugh [2] proposed a dynamic mathematical model for rotor systems in both rotational and fixed frames of reference. Zorzi and Nelson [3] research focus on the stability analysis of rotors with damping modelled using linear finite elements. The works of [4] discuss cracked shafts, their vibrational behaviour, and the detection of said cracks. Sankar [5], using exponential law, has provided an elasticity solution for functionally graded beams. Gayen and Roy [6] discuss the modal analysis of a spinning FG shaft system, utilizing beam elements built on the Timoshenko beam theory. Sathujoda [7] analyses vibrational properties of thermally graded FG beams. Obalareddy et al. [8] have modelled FG beams for dynamic analysis.

Much of the preceding work on functionally graded systems was carried out utilizing beam elements. It is essential to analyse functionally graded materials using 3D elements as they find more and more purpose in the industry, which will help in modelling the problem more accurately. Most of the studies available on the FG rotor involves gradation of power-law and methods of nonlinear temperature distribution. The current work aims to investigate the natural and whirl frequencies of an exponentially graded FG rotor-bearing system subjected to temperature gradients using ANSYS. To distribute the temperature-dependent material properties, the exponential temperature distribution method is used. The material properties will be affected by the temperature gradients, those effects are noted, and changes this will cause to the natural and whirl frequencies are determined. The FE analysis was conducted by adjusting the FG shaft's inner metallic temperature (Tm), similar work was done by Bose [9] using beam elements.

2 Material Properties and Temperature Distribution

The microstructure of an FGM is visualized in Fig. 1. In FG shafts, the gradation in material properties varies radially, with the inner core comprising entirely of metal and the outer core of ceramic. Here the volume fraction of metal decreases along the radius while that of ceramic increases. The Voigt model [10] gives us the position-dependent material properties P_i for every level as given in Eq. 1.

$$P_i = P_c V_c + P_m V_m \tag{1}$$

Here, represents the material properties, and represents the volume fraction. c refers to ceramic while m refers to metal. At any given point, the sum of and always equals one. The FG rotor-bearing system in the current work has been modelled using exponential law. Other than position-dependent changes, there are temperature-dependent changes as well, as discussed in Touloukian [11] in Eq. 2. Here $_{-1}$, $_0$, $_1$, $_2$, $_3$ are material-dependent temperature constants as given in Table 1, where E is the modulus of elasticity, K is the thermal conductivity, and υ is the Poisson's ratio. The values of the coefficients are obtained from the works of Reddy et al. [10].

$$P(T) = P_0\left(P_{-1}T^{-1} + 1 + P_1T^1 + P_2T^2 + P_3T^3\right) \tag{2}$$

In the present work, a stainless steel-zirconium dioxide (SS-ZrO_2) FGM is modelled as the shaft in the rotor system. Using the exponential temperature distribution (ETD) method, the thermomechanical responses of an FGM could be modelled. For shafts, the temperature at any given radial distance could be calculated by solving the

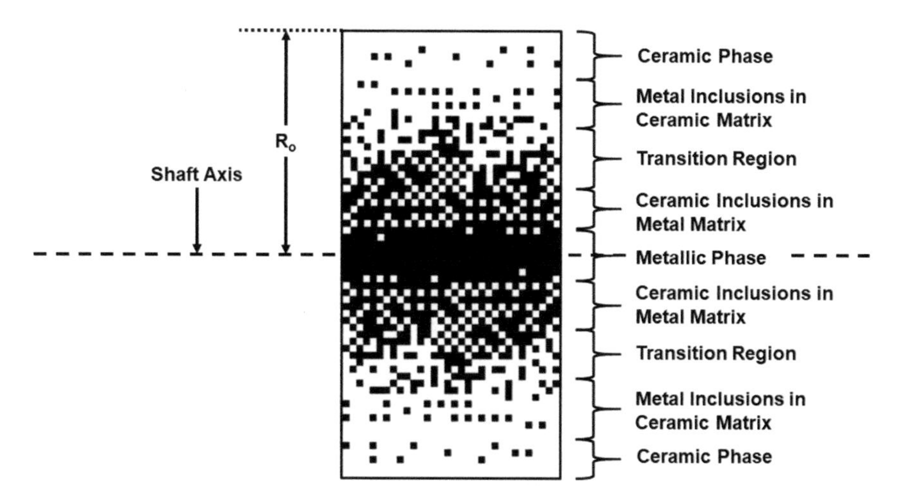

Fig. 1 Cross-sectional view of the graded microstructure of an FG shaft

Table 1 Temperature coefficients

Material properties		P_{-1}	P_0	P_1	P_2	P_3
Stainless steel	E	2.0104×10^{11}	0	-2.794×10^{-4}	3.998×10^{-9}	0
	K	15.379	0	-1.264×10^{-3}	2.092×10^{-6}	-7.223×10^{-10}
	v	0.3262	0	-2.002×10^{-4}	3.797×10^{-7}	0
ZrO_2	E	2.4427×10^{11}	0	-1.371×10^{-3}	1.214×10^{-6}	-3.681×10^{-10}
	K	1.7000	0	1.276×10^{-4}	6.648×10^{-8}	0
	v	0.2882	0	1.133×10^{-4}	0	0

1-D Fourier heat conduction equation with zero heat generation as expressed in Eq. 3. Here, K is the thermal conductivity, r is the radius of the shaft, and T is the temperature.

$$\frac{\mathrm{d}}{\mathrm{d}r}\left[rK(r)\frac{\mathrm{d}T}{\mathrm{d}r}\right] = 0 \tag{3}$$

2.1 Material Property Gradation Using Exponential Law with ETD

The exponential law for position-dependent material properties for a cylindrical shaft of an FG rotor-bearing system is expressed in Eq. 4 [12].

$$P(r) = P_m \cdot e^{\lambda(r-R_i)}; \quad \lambda = \left[\frac{\ln\left(\frac{P_c}{P_m}\right)}{R_o - R_i}\right]; \quad R_i \le r \le R_o \tag{4}$$

Here, () represents the material property as a function of radius. c and m are used to denote the position at the ceramic and metal-rich areas, correspondingly. All material properties could be obtained using this law, other than Poisson's ratio, which is taken as constant. Within the exponential law, the radial temperature distribution is given by ETD. To get the radial temperature distribution of the shaft, Eq. 5 [1] is to be used.

$$T(r) = A + B.e^{-\lambda(r-R_i)}; \; R_i \le r \le R_o$$
$$A = T_m - \frac{(T_c - T_m)}{e^{-\lambda(R_o-R_i)} - 1}; \; B = \frac{(T_c - T_m)}{e^{-\lambda(R_o-R_i)} - 1}; \; \lambda = \left[\frac{\ln\left(\frac{P_c}{P_m}\right)}{R_o - R_i}\right] \tag{5}$$

2.2 Modelling in Present Work

The present work considers an SS-ZrO_2 FG rotor. Based on a convergence study during validation with literature, the shaft is modelled with twenty elements radially to assign graded material properties to twenty layers. This division is, however, the minimum required for a smooth gradation of material properties, more could be added based on the available computational capability. A Python script was written to generate the temperature-dependent exponentially graded material properties in the form of ANSYS macros, containing the modulus of elasticity, density, and Poisson's ratio. In ANSYS, the user can provide the dimensions of the rotor, layer thickness, and can assign different material properties for each layer. The layer-wise material assignment plot of the FG shaft is shown in Fig. 2.

For a more accurate material assignment, a layer-averaged material assignment is followed. The material property function is taken to vary as exponential law and is averaged over the inner radius to the outer radius of the layer. This allows for a more uniform gradation in material properties than directly applying the mid-radius property. This layer-wise averaging is expressed in Eq. 6. Here r_i and r_o are the inner and outer radius of the layer, respectively.

$$P_{avg} = \frac{\int_{r_i}^{r_o}\left[P_m \cdot e^{-\lambda(r-R_i)}\right]dr}{\int_{r_i}^{r_o} dr};\ \lambda = \left[\frac{\ln\left(\frac{P_c}{P_m}\right)}{R_o - R_i}\right] \tag{6}$$

Fig. 2 Shaft modelled using ANSYS-material gradation in SS-ZrO2

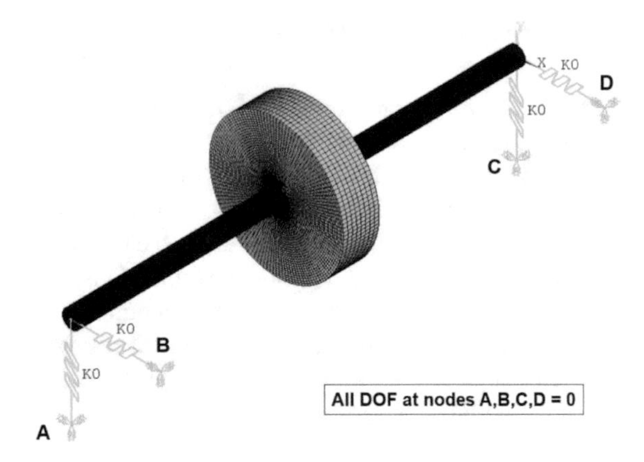

Fig. 3 Meshed finite element model of a rotor-bearing system

3 Finite Element Modelling Using ANSYS

The finite element modelling of the rotor-bearing system is performed using ANSYS software. A single-disc rotor-bearing system (Fig. 3) has been modelled to study its natural and whirl frequencies. The shaft is 0.5 m long and has a radius of 0.01 m. The disc is mounted at midspan, with a thickness of 0.04 m and an outer radius of 0.075 m. The shaft is supported on an isotropic bearing at both ends with no internal damping effect. The stiffness of the bearings is 10^5 N/m with a damping coefficient of 100 Ns/mm. The disc and the shaft are modelled using the SOLID185 element. The isotropic bearings are modelled using the COMBIN-14 element. The model can be meshed with a map mesh option with the shaft being partitioned into twenty different layers. The graded material properties could now be applied. The disc is completely made of stainless steel, whereas the shaft is exponentially graded. To model the isotropic bearings, one end of the COMBIN-14 element is attached to the shaft centre and the other end is restrained in all degrees of freedom. More details about the element types are described in the ANSYS manual [].

4 Validations

The validation of the material properties and the FE model needs to be done before moving on to the natural and whirl frequency investigations of an exponentially graded rotor-bearing system. The temperature-dependent exponential gradation of material properties and the natural frequencies of an FG rotor-bearing system is verified for validation. The findings indicate considerable validation of the literature evidence. The FE model is also verified for the natural frequencies of the rotor system.

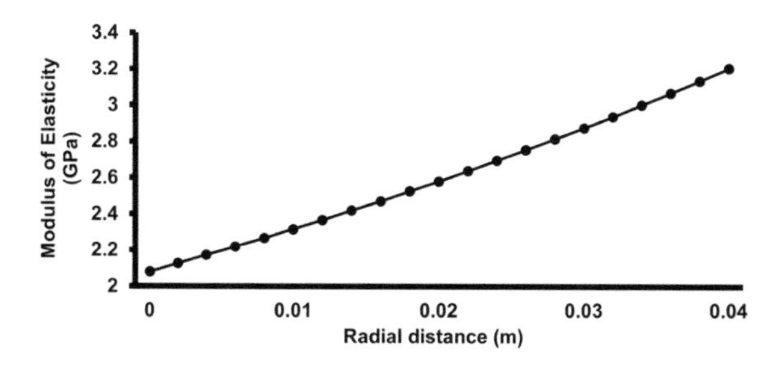

Fig. 4 4. Radial effect of exponential law on the modulus of elasticity of an SS-Al_2O_3 shaft at room temperature

Table 2 Validation of the model with natural frequencies

	Natural frequency		
Modes	Present work	Bose et al. [7]	Error (%)
1st	150.11	151.40	0.85
2nd	622.04	636.95	2.34

4.1 *Verification of Exponentially Graded Material Properties*

A Python script was written to generate the temperature-dependent exponentially graded material properties. To verify this code, the FG shaft in [13] was replicated. It is an SS-Al_2O_3 FG shaft with a radius of 40mm at room temperature. The code-generated radial variation of the modulus of elasticity has been plotted (Fig. 4), and it demonstrates significant validation with the literature.

4.2 *Verification of the FE Model*

The model has been validated by replicating the study in [9] of the modal analysis of a non-rotating FG rotor. The rotor-bearing system is at room temperature, and the first and second natural frequencies have been used to verify the model. The percentage of error is coming out to 0.85% for the first natural frequency and 2.34% for the second (Table 2).

5 Results

The natural and whirl frequency analysis of the developed FG rotor system has been carried out to investigate the effect of temperature and material gradation.

5.1 Natural Frequency Analysis of Non-Rotating Exponentially Graded FG Rotor-Bearing System

The variation in fundamental frequencies of a non-rotating exponentially graded FG rotor-bearing system due to the application of temperature gradients is analysed. Here, the inner metal core is subjected to four different temperatures (Tm) varying from 300 to 600 K. Four different temperature gradients are applied across the radius between zero and 600K to investigate the thermal effects (Fig. 5).

The first three natural frequencies of the FG rotor subjected to temperature gradients are given in Table 3. The three-mode shapes generated from ANSYS are shown in Fig. 6. When the first natural frequencies are plotted against the different temperature gradients applied (Fig. 6), it shows that the natural frequencies of the rotor-bearing system drop as the temperature gradient is increased. It could also be observed from the same plot that at higher temperatures at the inner core, the natural frequencies still drop even without changing the temperature gradient. This behaviour is anticipated as FG materials become flexible at higher temperatures, thus reducing the total stiffness of the FG rotor, subsequently allowing the natural frequencies to decline.

The percentage drop ($\omega_D\%$) in natural frequencies at different temperature gradients is plotted to further understand the effect of temperature gradients (Fig. 7). The percentage drop ($\omega_D\%$) is calculated using Eq. 6. Here Tm is the inner metal core temperature and ΔT is the change in temperature between the inner core and the outermost layer.

$$\omega_D\% = \frac{\omega_{Tm,\Delta T=0} - \omega_{Tm,\Delta T=k}}{\omega_{Tm,\Delta T=0}} \times 100\% \tag{7}$$

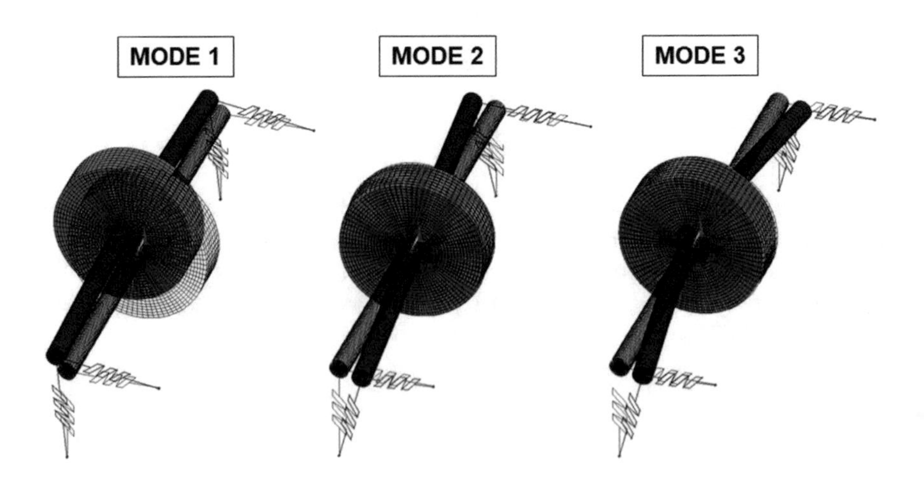

Fig. 5 Mode shapes from ANSYS

Table 3 Natural frequencies (rad/s) of an exponentially graded FG rotor subjected to thermal gradients (ΔT)

Modes	Tm (K)	$\Delta T = 0$ K	$\Delta T = 200$ K	$\Delta T = 400$ K	$\Delta T = 600$ K
1st	300	150.11	148.19	146.62	145.41
	400	148.52	146.81	145.48	144.49
	500	147.06	145.6	144.5	143.68
	600	145.78	144.56	143.66	142.9
2nd	300	622.04	619.62	617.58	615.98
	400	620.07	617.86	616.1	614.77
	500	618.21	616.27	614.8	613.68
	600	616.54	614.89	613.66	612.63
3rd	300	1792	1726.6	1678.3	1644.2
	400	1737.4	1684	1646	1619.6
	500	1691.5	1649.3	1619.9	1598.9
	600	1654.4	1621.3	1598.4	1580

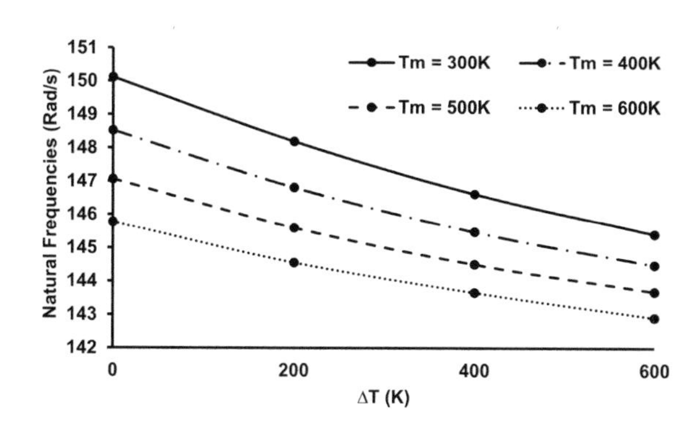

Fig. 6 Variation in the fundamental frequencies of the FG shaft with (ΔT) at different (Tm)

Here $\omega_{Tm,\Delta T=0}$ refer to the natural frequency at a certain Tm and $\Delta T = 0$. $\omega_{Tm,\Delta T=k}$ represents the natural frequency at the same Tm at different thermal gradients studied in present studies (k = 200, 400, 500, 600).

It is observed that as the inner core temperature rises, the $\omega_D\%$ or the percentage drop in natural frequencies also decreases. This leads us to the conclusion that the operating temperatures of FG rotors must be carefully monitored to avoid a significant reduction in natural frequencies and stiffness.

5.2 *Whirl Frequency Analysis of Rotating Exponentially Graded FG Rotor-Bearing System*

To observe the effect of temperature gradients on the whirl frequency response of an exponentially graded FG rotor system, the whirl frequency analysis is performed.

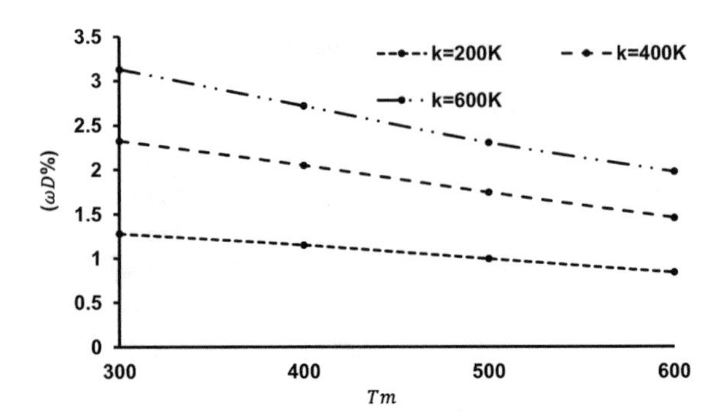

Fig. 7 Influence of ETD on the dropping rate of natural frequencies at different temperature gradients

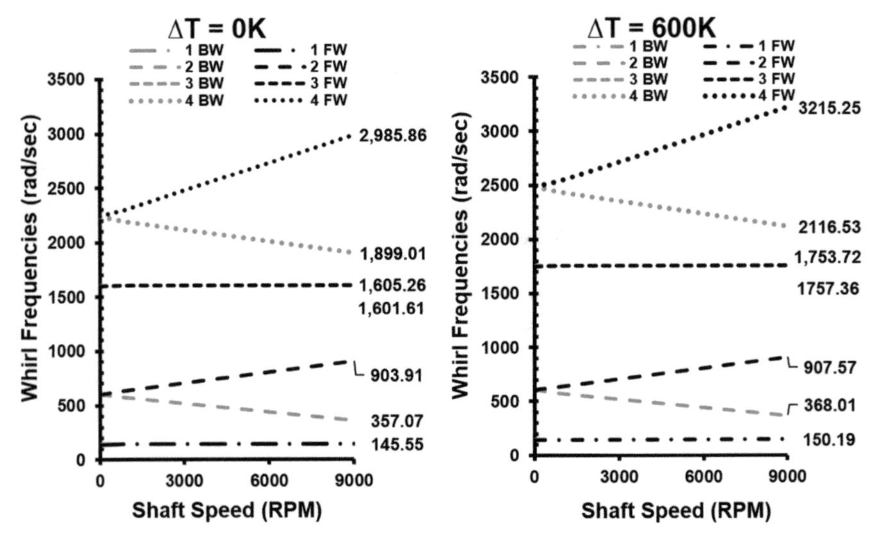

Fig. 8 Campbell diagrams of exponentially graded FG shaft under different thermal environments

The rotational speed of the shaft varying from zero to 9000 RPM to cover at least the first two natural frequency modes. The Campbell diagrams are generated to better visualize the results in Fig. 8. The frequencies that occur in pairs at zero speeds detach at higher rotating velocities due to the gyroscopic effects of the rotor system. The first four whirl frequencies, both forward and backward, are visualized in the plot. Due to a drop in stiffness, the general behaviour of decreasing whirl frequencies with an increase in temperature gradients is observed. The separation between the forward and backward whirl frequencies of the first and third modes is not visible in the plot.

The separation of the first whirl frequencies is an important parameter that establishes the effect of ETD on the whirl frequencies of the FG rotor-bearing system

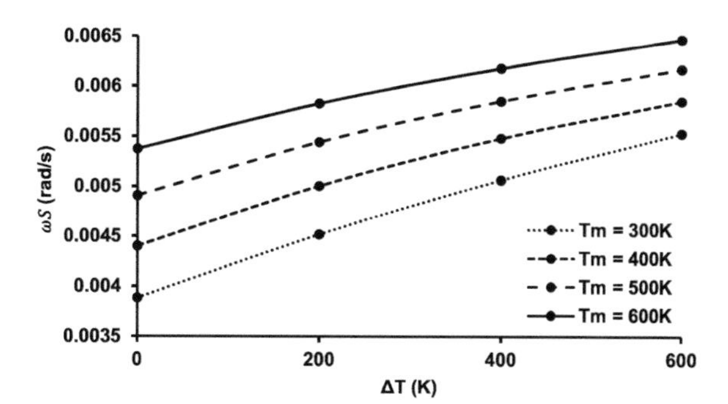

Fig. 9 Effect of ETD on the separation of whirl frequencies

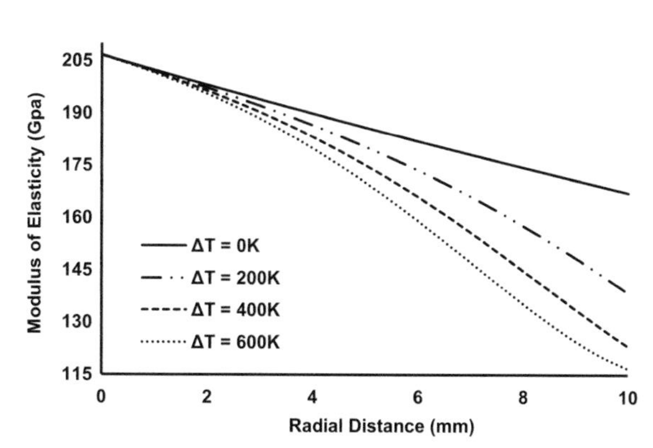

Fig. 10 Variation in modulus of elasticity radially under thermal gradients for an exponentially graded shaft with inner core temperature of 300 K

is expressed in Eq. 7. Fig. 9 illustrates the variation of ω_S with an increase in thermal gradients. The plot also shows the separation between shafts at different inner core temperatures. It is observed that with a rise in thermal gradients, the separation between the first whirl frequencies also increases. It is also observed that the shafts with higher inner core temperatures have a higher separation in their first natural frequencies. This is due to the change in modulus of elasticity at different thermal gradients as shown in Fig. 10.

$$\omega_S = \omega_{FW} - \omega_{BW} \tag{8}$$

6 Conclusions

Using ANSYS, the effect of thermal gradients on the natural and whirl frequencies of an exponentially graded FG rotor is analysed. An FG rotor of constituents, stainless steel, and zirconia is considered. Exponential law has been used for material gradation and ETD, for temperature distribution. Now, a Python code is developed to achieve this multi-layer gradation of material properties. These graded material properties are applied to the model by creating ANSYS macros with the help of Python. The material properties are validated by comparing the gradation in material properties with that of [13]. The model is validated by doing a natural frequency analysis in ANSYS and comparing the results with that of [9]. Natural and whirl frequency analysis of the FG rotor-bearing system is carried out, and the effects of the exponential distribution of material properties and ETD have been investigated. The variation in the fundamental frequencies of the FG shaft with the temperature gradient (ΔT) at different inner core temperatures (Tm) is plotted; it is observed that the natural frequencies decrease with an increase in temperature gradient or the inner core temperature. Campbell diagrams are created after the whirl frequency analysis. The observation is that, with a higher inner core temperature or a higher thermal gradient, the separation in the first natural frequency increases.

References

1. Gayen D, Tiwari R, Chakraborty D (2019) Static and dynamic analyses of cracked functionally graded structural components: a review. Compos Part B: Eng 173:106982–106982
2. Nelson HD, Mcvaugh JM (1976) The dynamics of rotor-bearing systems using finite elements. J Eng Ind 98(2):593–600
3. Zorzi E, Nelson HD (1977) Finite element simulation of rotor-bearing systems with internal damping. J Eng Power 99:71–76
4. Prabhakar S, Sekhar AS, Mohanty AR (2001) Detection and monitoring of cracks using mechanical impedance of rotor-bearing system. J Acoust Soc Am 110(5):2351–2359
5. Sankar BV (2001) An elasticity solution for functionally graded beams. Compos Sci Technol 61(5):689–696
6. Reddy JN, Chin CD (1998) Thermomechanical analysis of functionally graded cylinders and plates. J Thermal Stresses 21(6):593–626
7. Sathujoda P, Batchu A, Obalareddy B, Canale G, Maligno A, Citarella R (2020) Free vibration analysis of a thermally loaded porous functionally graded rotor-bearing system. Appl Sci 10(22)
8. Obalareddy B, Batchu A, Sathujoda P, Delhi I (2021) Modelling of futuristic exponentially graded rotor bearing system for dynamic analysis. In: Proceedings of the international conference on futuristic technologies
9. Bose A, Sathujoda P (2021)
10. Reddy JN (1998) Thermo-mechanical behavior of functionally graded materials. AFOSR Grant F49620–95- 1–0342. D.C., August, Washington
11. (1967) Thermophysical properties of high-temperature solid materials. In: TouloukianYS (eds) Air force materials laboratory, Macmillan
12. Afsar AM, Go J (2010) Finite element analysis of thermoelastic field in a rotating FGM circular disk. Appl Math Modell 34(11):3309–3320
13. Bose A, Sathujoda P (2020) Effect of thermal gradient on vibration characteristics of a functionally graded shaft system. Mathe Modell Eng Prob 7(2):212–222

Numerical Solution of Laplace and Poisson Equations for Regular and Irregular Domain Using Five-Point Formula

Malabika Adak

Abstract In many areas of science and engineering, to determine the steady-state temperature, potential distribution, electricity, gravitation, Laplace and Poisson elliptic partial differential equation is required to solve. It is difficult to obtain an analytical solution of most of the partial differential equations that arise in mathematical models of physical phenomena. So, five-point finite difference method (FDM) is used to solve the two-dimensional Laplace and Poisson equations on regular (square) and irregular (triangular) region. To solve partial differential equation, specific boundary conditions are required. In this study, Dirichlet and Robin boundary conditions are considered for solving the system of equations at each iteration. When the function itself is specified, the boundary is called Dirichlet boundary. In some problems, a linear combination of function and its normal derivative is specified on the boundary, called Robin boundary. The obtained numerical results are compared with analytical solution. The study objective is to check the accuracy of FDM for the numerical solutions of square and triangular bodies of 2D Laplace and Poisson equations.

Keywords Finite difference method · Dirichlet boundary condition · Robin boundary condition · Laplace equation · Poisson equation

1 Introduction

Most of the engineering problems can be modeled mathematically which is governed by partial differential equations. Poisson/Laplace equation describes the behavior of electric, steady-state temperature and fluid potentials, respectively. In partial differential equations, the equation is always associated with a particular type of boundary conditions. It is difficult to obtain an analytical solution of initial-boundary value problems. There exist several numerical methods for the approximate solution of

M. Adak (✉)
Department of Applied Mathematics and Humanities, Yeshwantrao Chavan College of Engineering, Nagpur, India
e-mail: malabikaadak@yahoo.co.in

S. S. Ray et al. (eds.), *Applied Analysis, Computation and Mathematical Modelling in Engineering*, Lecture Notes in Electrical Engineering 897,
https://doi.org/10.1007/978-981-19-1824-7_18

elliptic partial differential equation. A popular, simple and powerful technique is the finite difference method which is mostly preferable for irregular and regular domains. Potential distribution is related to the charge density which is calculated by Poisson's equation and in a charge-free region of space it appears Laplace equation. We prefer to calculate the potential distribution and steady-state temperature distribution of this problem. Ames [6], Burden and Faires [8], Clive [9], Cooper [10], Mortan and Mayer [15], Golub and Ortega [12], and Adak [1–5] provide modern introduction of partial differential equation using finite difference method with convergence study of numerical methods. Patil and Prasad [17], Morales et al. [14], Li et al. [13], Dhumaland kiwne [11], and Ubaidullah and Muhammad [18] discussed approximate solution of two-dimensional Laplace equation with Dirichlet conditions. Solution of two-dimensional Poisson equation with Dirichlet boundary condition is presented by Benyam and Purnachandra [7], and Pandey and Jaboob [16].

From the literature survey, one can observe that maximum problems of Laplace and Poisson equation with Dirichlet boundary condition are solved. Main objective of this study is to solve Dirichlet initial-boundary (well-posed) problems and Robin initial-boundary (ill-posed) problems in the square (regular) and triangular (irregular) domain using five-point finite difference method.

2 Problem Formulation

2D steady-state heat conduction equation is Laplace equation given by

$$\frac{\partial^2 T}{\partial x^2} + \frac{\partial^2 T}{\partial y^2} = 0 \quad 0 \leq x, y \leq 1 \tag{1}$$

where $T(x, y)$ is the steady-state temperature distribution in the square domain.

The Poisson equation (potential distribution) is given by

$$\frac{\partial^2 T}{\partial x^2} + \frac{\partial^2 T}{\partial y^2} = -f(x, y) \quad 0 \leq x, y \leq 1 \tag{2}$$

where $T(x, y)$ is potential distribution in the square domain with the Dirichlet boundary condition $T(x, y) = f(x, y)$ on S and the Robin boundary (Mixed boundary) condition $\frac{\partial T}{\partial n} + T(x, y) = g(x, y)$ in square domain.

Equations (1) and (2) are considered with Dirichlet boundary condition in an irregular (triangular) domain.

3 Solution of Finite Difference Approximation

In finite difference approximation, derivative is replaced by finite differences, so that differential equation becomes finite difference equation. Finite difference solution depends on the following steps:

(i) The solution domain is discretized using horizontal and vertical lines into small rectangles, and intersection points are called nodes or grids.
(ii) The equivalent finite difference equations give the solutions at the grid points.
(iii) Using the initial-boundary condition, difference equation produces the system of simultaneous equation.
(iv) To solve the system of equation, use Gauss–Seidel iteration method.

In the region on XY plane, rectangles or meshes h and k along x and y direction, respectively, such that $x = \text{ih}$, $i = 0, 1, 2\ldots$ and $y = \text{jh}$, $j = 0, 1, 2\ldots$ are considered.

Derivatives in Laplace/Poisson equation are replaced by Taylor's series expansion

$$\frac{\partial^2 T}{\partial x^2} = \frac{T_{i-1,j} - 2T_{i,j} + T_{i+1,j}}{h^2} + O\left(h^2\right) \tag{3}$$

$$\frac{\partial^2 T}{\partial y^2} = \frac{T_{i,j-1} - 2T_{i,j} + T_{i,j+1}}{k^2} + O\left(k^2\right) \tag{4}$$

The terms $O\left(h^2\right)$ and $O\left(k^2\right)$ denote the order of local truncation error and are also known as the order of method. After neglecting the truncation, error and simplifying, we obtain the following difference equations. Using Eqs. (3) and (4) in Eqs. (1) and (2), we obtain the five-point finite difference formulae

$$T_{i,j} = \frac{1}{4}\left(T_{i+1,j} + T_{i-1,j} + T_{i,j+1} + T_{i,j-1}\right) \tag{5}$$

$$T_{i,j} = \frac{1}{4}\left(T_{i+1,j} + T_{i-1,j} + T_{i,j+1} + T_{i,j-1}\right) + h^2 f\left(x_i, y_j\right) \tag{6}$$

This shows that value of T at any point of the region R is the average of the surrounding points in the five-point stencil of the following Fig. 1.

The temperature or potential on the four sides of a region is given, at all the internal points the temperature/potential is considered according to boundary value. Region is divided into a finite number of rectangular elements. Here, we consider mesh size $h = (1/4)$ in x direction and $k = (1/4)$ in y direction will split the region into 16 rectangular elements. The interior nodes lie inside the region, and exterior nodes lie on the boundary. The nodes of the region are numbered and shown in Fig. 2.

Here, our problem is to find numerical value of a function $T(x, y)$ at the interior node points of the region provided that the Laplace/Poisson equation and the boundary condition given in Eqs. (1) and (2) are satisfied.

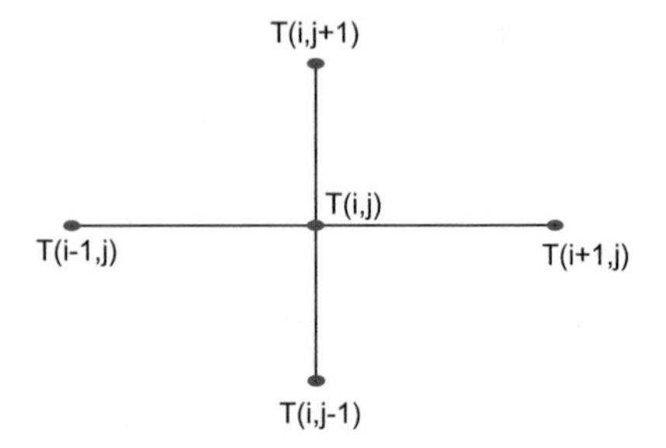

Fig. 1 Temperature at any point average of surrounding points

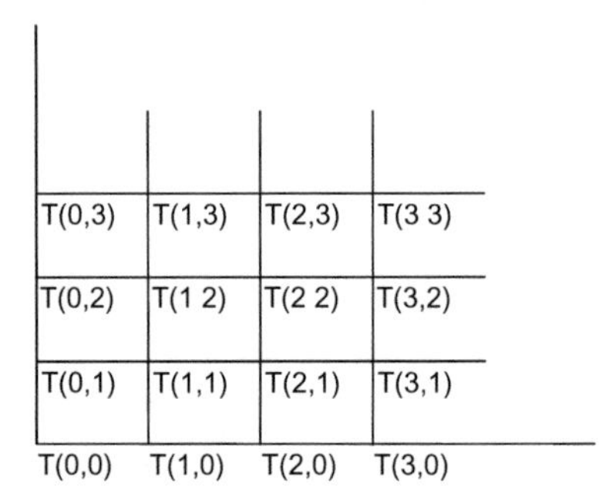

Fig. 2 Rectangular elements with node numbers

This study used the Gauss–Seidel iterative method for solving the system of equations.

The stopping criterion for the iteration is

$$\varepsilon_{\text{ij}} = \left| \frac{T_{i,j\text{present}} - T_{i,j\text{previous}}}{T_{i,j\text{present}}} \right| < 10^{-n}$$

3.1 Analytical Solution

For analytical solution, method of separation of variables is used to reduce partial differential equation to ordinary differential equations. In separation of variable method, the solution of Laplace equation is considered as a product of two individual functions as.

$T(x, y) = X(x)Y(y)$. The exact solution of Laplace equation is

$$X(x) = A\cos(\alpha x) + B\sin(\alpha x) \tag{7}$$

4 Numerical Experiments

4.1 Regular Domain

Example 1 Consider Laplace Equation with **Dirichlet** boundary condition in a **square (regular)** domain defined in Fig. 3 to find the steady-state temperature distribution.

Using five-point finite difference formula and Leibmann's iterative method, following equations will give the results at interior nodes.

$$T_1^{(n+1)} = \frac{1}{4}\left[0 + 0 + T_2^{(n)} + T_4^{(n)}\right]$$

$$T_2^{(n+1)} = \frac{1}{4}\left[0 + 0 + T_3^{(n)} + T_1^{(n+1)}\right]$$

$$T_3^{(n+1)} = \frac{1}{4}\left[0 + 1 + T_2^{(n+1)} + T_4^{(n)}\right]$$

$$T_4^{(n+1)} = \frac{1}{4}\left[0 + 1 + T_1^{(n+1)} + T_3^{(n+1)}\right]$$

Putting $n = 0, 1, 2, 3, \ldots.$ we obtain the temperature distribution at T_1, T_2, T_3, T_4.

The numerical solution of Gauss–Seidel iteration has been given in Table 1.

8th iteration is showing the approximate solution of Laplace equation with Dirichlet condition in square domain. Numerical results are compared with analytical results which calculated from Eq. (7) using separation of variable method. Table 2 has shown the well agreement of exact and numerical solution obtained by FDM. The results of Table1 and Fig. 4 showed the prediction of steady-state temperature distribution in regular square domain with Dirichlet conditions.

Example 2 Consider Poisson equation $\frac{\partial^2 T}{\partial x^2} + \frac{\partial^2 T}{\partial y^2} = -3(x + y), 0 \leq x, y \leq 1$ with **Dirichlet** boundary condition in a **square** domain defined in Fig. 5 to find the potential distribution.

Using Eq. (6), we obtain the approximate solution at each node. The Leibmann's iterative process for numerical solution has been given in Table 3.

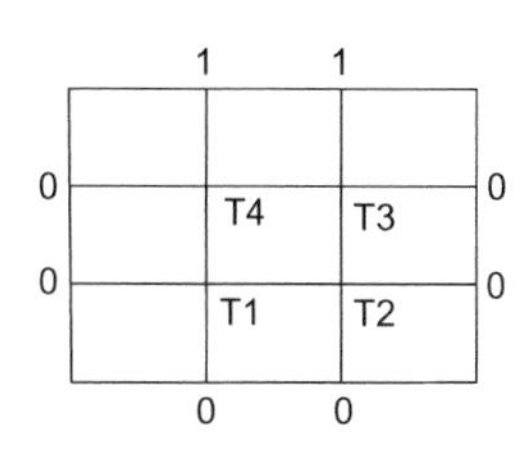

Fig. 3 Dirichlet boundary condition on the four sides of a square domain for Laplace equation

Table 1 Leibmann's iterative process for Laplace Equation with **Dirichlet** boundary condition in a **square** domain

Iterations	T_1	T_2	T_3	T_4
1	0.25	0.3125	0.5625	0.46875
2	0.21875	0.17187	0.42187	0.39844
3	0.14844	0.13672	0.38672	0.38086
4	0.13086	0.12793	0.37793	0.37646
5	0.12646	0.12573	0.37573	0.37537
6	0.125275	0.12525	0.375155	0.3751075
7	0.125275	0.125061	0.375015	0.3750327
8	0.1250706	0.125028	0.375015	0.3750214

Table 2 Comparison with exact solution and numerical solution

Nodes	Numerical solution	Exact solution	Error
T_1	0.1250706	0.0893405	0.0357301
T_2	0.125028	0.0894536	0.0355744
T_3	0.375015	0.3206756	0.0543394
T_4	0.3750214	0.3206732	0.0543482

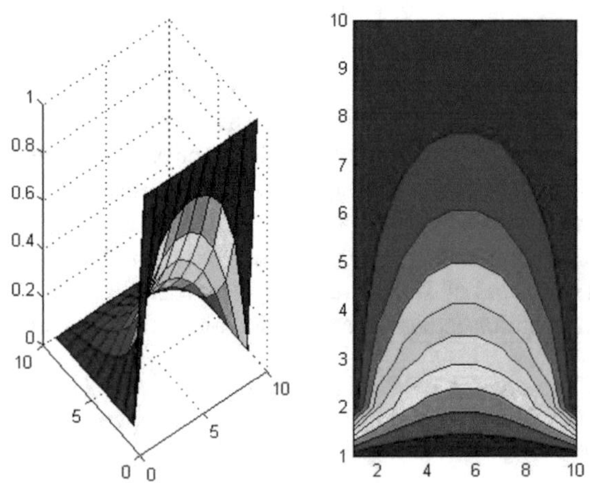

Fig. 4 Contour plot of solving Laplace equation with Dirichlet condition

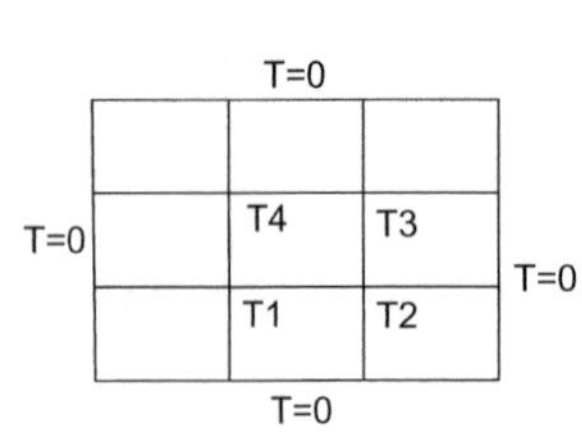

Fig. 5 Dirichlet boundary conditions on the four sides of a square domain for Poisson equation

Table 3 Leibmann's iterative process for Poisson equation with **Dirichlet** boundary condition in a **square** domain

Iterations	T_1	T_2	T_3	T_4
1	1.5	2.625	3.65625	3.539
2	3.041	3.92431	4.8658	4.2267
3	3.53775	4.35088	5.14439	4.4205
4	3.6928	4.4593	5.21995	4.4788
5	3.73437	4.48858	5.24169	4.494013
6	3.745648	4.49683	5.24674	4.49809
7	3.74873	4.49886	5.24924	4.49949
8	3.74958	4.499707	5.24992	4.499846
9	3.74988	4.49757	5.24918	4.49302

Nine iterations are necessary to reach a solution of Poisson's equation with Dirichlet condition in square domain. Figure 6 has shown the contour plot of electric potential distribution.

Example 3 Consider Laplace equation with **Robin boundary** conditions in a **square** domain as follows and defined in Fig. 7.

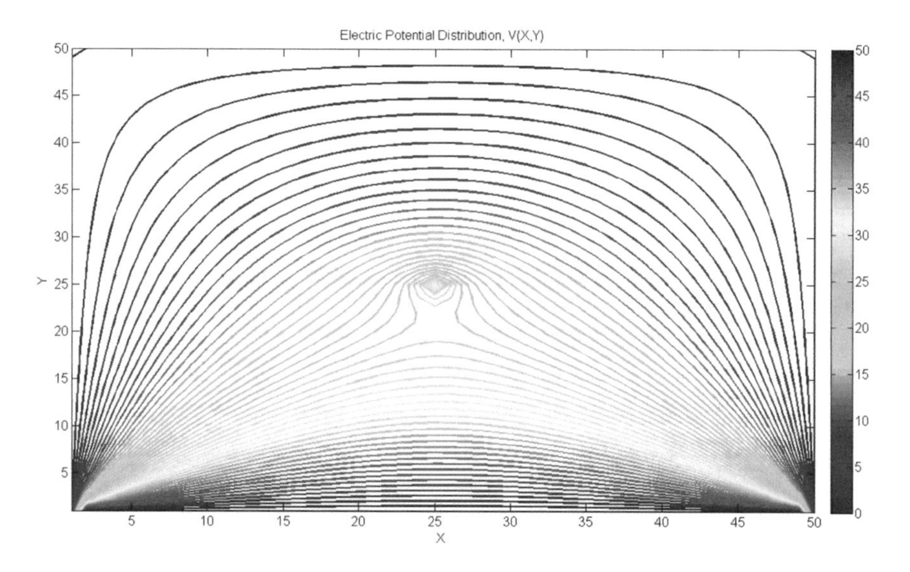

Fig. 6 Contour plot of electric potential distribution

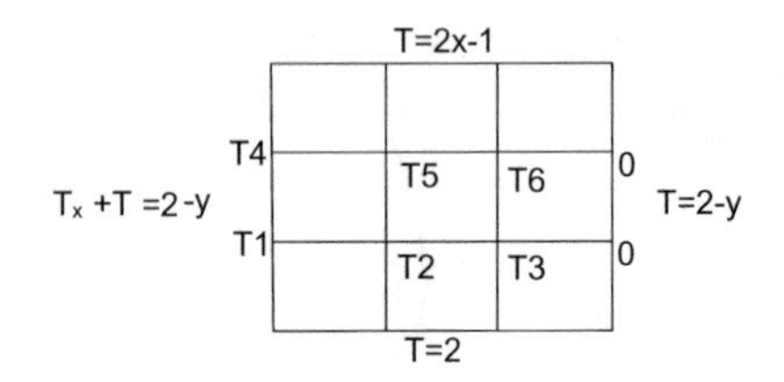

Fig. 7 Robin boundary condition on the left side of a square domain for Laplace equation

$$\begin{aligned}
&\nabla^2 T = 0, && 0 \le x, y \le 1\\
&T = 2x, && y = 0\\
&T = 2x - 1, && y = 1\\
&T_x + T = 2 - y, && x = 0\\
&T = 2 - y, && x = 1
\end{aligned}$$

At T_1 and T_4, mixed boundary condition is defined. So, at T_1, consider boundary condition

$$T_x + T = 2 - y$$

Apply forward difference approximation for derivative in condition, then. $\frac{T_2 - T_1}{h} + T_1 = 2 - y$ Assume $h = 1/3$.
$T_1 = \frac{1}{2}\left(3T_2 - \frac{5}{3}\right)$, similarly $T_4 = \frac{1}{2}\left(3T_5 - \frac{4}{3}\right)$.
At T_2, T_3, T_5, T_6, Gauss–Seidel iteration formulae are

$$T_2^{(n+1)} = \frac{2}{5}\left(T_3^{(n)} + T_5^{(n)} - \frac{1}{6}\right) \quad T_3^{(n+1)} = \frac{1}{4}\left(T_2^{(n+1)} + T_6^{(n)} + 3\right)$$

$$T_5^{(n+1)} = \frac{2}{5}\left(T_2^{(n+1)} + T_6^{(n)} - 1\right) \quad T_6^{(n+1)} = \frac{1}{4}\left(T_3^{(n+1)} + T_5^{(n+1)} + \frac{5}{3}\right).$$

The Leibmann's iterative process for numerical solution is given in Table 4.

Nine iterations are necessary to get a solution of Dirichlet's equation with Robin condition in a square domain.

4.2 Irregular Domain

Example 4: Laplace equation with Dirichlet boundary condition in a **triangular (irregular)** domain with vertices $(0, 0)$, $(7, 0)$, $(0, 7)$ is shown in Fig. 8. Assume mesh size 2.

Five-point finite difference formula with Gauss–Seidel iteration is

$$T_1^{(n+1)} = \frac{1}{4}\left(T_2^{(n)} + T_3^{(n)}\right) \quad T_2^{(n+1)} = \frac{1}{4}\left(T_1^{(n+1)} + 8\right)$$

Table 4 Leibmann's iterative process for Laplace equation with **Robin** boundary condition in a **square** domain

Iterations	T_2	T_3	T_5	T_6
1	0.466687	1.19999	0.319871	0.7966405
2	0.5412768	1.084479	0.1351669	0.721586
3	0.421178	1.035693	0.0571056	0.6898739
4	0.37045	1.015081	0.0241312	0.6764775
5	0.349016	1.006373	0.0101975	0.67081764
6	0.32725	0.983240	0.0098431	0.65320425
7	0.30456	0.952452	0.0094061	0.6332156
8	0.30748	0.950471	0.0094043	0.6321457
9	0.30732	0.950456	0.0094033	0.6320852

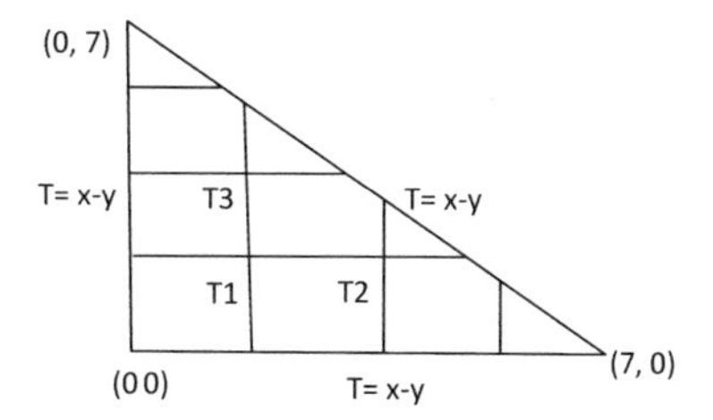

Fig. 8 Dirichlet boundary condition on the four sides of a triangular domain for Laplace equation

$$T_3^{(n+1)} = \frac{1}{4}\left(T_1^{(n+1)}\right)$$

Table 5 shows the Leibmann's iteration process for numerical solution of Laplace equation in irregular domain.

Six iterations are necessary to reach a solution of Laplace equation with Dirichlet condition in triangular domain.

Table 5 Leibmann's iterative process for Laplace equation with **Dirichlet** boundary condition in a **triangular** domain

Iterations	T_1	T_2	T_3
1	0.75000	2.187506	0.1875
2	0.593750	2.1484375	0.1484375
3	0.5742187	2.143555	0.1435547
4	0.571777	2.142944	0.14294425
5	0.571472	2.142868	0.14286806
6	0.571470	2.142839	0.14286800

5 Conclusion

This study focused on various domains to obtain solutions of Laplace and Poisson equations with Dirichlet and Robin boundary conditions numerically. Five-point finite difference method (FDM) is used for predicting temperature and potential distribution. In this study, four problems are considered in various domains with simple and mixed boundary conditions. The table results have shown the agreement of numerical and exact solution. The results of tables were the steady-state temperature distribution and potential distribution in regular and irregular domain with Dirichlet and Robin conditions. Contour plots are drawn using MATLAB code. Contour plots look like a very nice presentation of temperature and potential distribution. Finite difference method is superior in both competence and accuracy. This technique can be successfully implemented for temperature and potential distribution in more complicated geometries in future work.

References

1. Adak M, Mandal NR (2010) Numerical and experimental study of mitigation of welding distortion. Appl Math Model 34:146–158
2. Adak M, SoaresGuedes C (2014) Effects of different restraints on the weld-induced residual deformations and stresses in a steel plate. Int J Adv Manuf Technol 71:699–710
3. Adak M, SoaresGuedes C (2016) Residual deflections and stresses in a thick T joint plate structure. J Appl Mech Eng 5:6
4. Adak M (2020) Comparison of explicit and implicit finite difference schemes on diffusion equation. In: Mathematical modeling and computational tools, pp 227–238
5. Adak M, Mandal A (2020) Numerical solution for predicting soil or rock temperature with different thermal diffusivity. Solid State Technol 63(4):1938–1947
6. Ames FW (1992) Numerical methods for partial differential equations, 3rd edn. Academic Press Inc, Boston
7. Banyam M, Purnachandra RK (2015) Numerical solution of a two dimensional equation with dirichlet boundary conditions. Am J Appl Math 3(6):297–304
8. Burden RL, Faires JD (1997) Numerical analysis, 6th edn. Brooks/Cole Publishing Co., New York
9. Fletcher CAJ (1988) Computational techniques for fluid dynamics. Springer-Verlag, Berlin
10. Cooper J (1998) Introduction to partial differential equations with Matlab. Birkhäuser, Boston
11. Dhumal ML, Kiwne SB (2014) Finite difference method for laplace equation. Int J Stat Math 9:11–13
12. Golub G, Ortega JM (1993) Scientific computing: an introduction with parallel computing. Academic Press Inc., Boston
13. Li ZC, Zhang LP, Wei Y, Lee MG, Chaiang JY (2015) Boundary method for dirichlet problems of laplace equation in elliptic domains with elliptic holes. Eng Anal Boundary Elem 61
14. Morales M, Rodolfo DRA, Herrera WJ (2015) Solutions of laplace's equation with simple boundary condition and their applications for capacitors with multiple symmetries. J Electro stat 78:31–45
15. Morton KW, Mayers DF (1994) Numerical solution of partial differential equations: an introduction. Cambridge University Press, Cambridge, England
16. Pandey PK, Jaboob SSA (2018) A finite difference method for a numerical solution of elliptic boundary value problems. Appl Math Nonlinear Sci 3(1):311–320

17. Patil PV, Prasad SVR (2013) Numerical solution of two dimensional laplace equation with dirichlet boundary conditions. IOSR J Math 6:66–75
18. Ubaidullah MSC (2016) Finite difference method with dirichlet problems of 2D laplace's equation in elliptic domain. Pakistan J Eng Technol Sci 6(2):136–144

Selecting Ultrasound Machine Using ELECTRE Method

Seema Gupta Bhol, Jnyana Ranjan Mohanty, and Prasant Kumar Pattnaik

Abstract Disease diagnosis provides crucial information and serves as primary guide for treatment. Ultrasound machines are largely used by medical experts for this purpose. There are several models of ultrasound machines available in market with variety of features. Thus, selection of good ultrasound machine comes under the category of multi-criteria decision making. This paper focuses on use of Electre method to select best ultrasound machine fulfilling user's criteria. Six criteria were considered to select best model among five alternatives. The method can be applied to any number of characteristics and alternatives.

Keywords Multi-criteria decision making · Electre · Ultrasound · Outranking · Concordance · Discordance

1 Introduction

Ultrasound is a type of sound wave with a typical wavelength of more than 20,000 Hz. It lies in the range of frequency which human beings cannot hear. Medical ultrasound, an ultrasound technique, is used in clinical setting for image-based diagnostic procedures. Ultrasound machines are not only successfully helping medical doctors in examining medical conditions of a patient but also assisting in making treatment plan. Various organs of human body are scanned with the help of images produced by ultrasound machine. To high light the body structure, high-frequency sound waves are sent by ultrasound machine. A smart computing device is attached that receives the waves and convert them into an image. An ultrasound machine is characterized by variety of factors like application areas, transducer types, imaging modes, number of probes, image resolution, etc. Multiple criteria should be evaluated before selecting a best ultrasound machine. Multi-criteria decision-making problem offers variety of tools for solving decision-making problems involving consideration of

S. G. Bhol (✉) · J. R. Mohanty · P. K. Pattnaik
Kalinga Institute of Industrial Technology, Bhubaneswar, India
e-mail: seemaguptabhol@gmail.com

S. S. Ray et al. (eds.), *Applied Analysis, Computation and Mathematical Modelling in Engineering*, Lecture Notes in Electrical Engineering 897,
https://doi.org/10.1007/978-981-19-1824-7_19

multiple features. Multi-criteria decision-making approach of decision making is well equipped with variety of tools like AHP, Promethee, Topsis, Electre, etc.

Electre is one of the powerful tools that overcomes the problems associated with traditional decision-making aids [1]. Moreover, complex situations where decision making involves multiple criteria those include not only quantitative data but also qualitative data along with involvement of more than one decision maker are efficiently handled using Electre method. A model using Electre is proposed for handling collaboration between, buyers, suppliers and business partners [2]. Electre method is used in selecting best location for agriculture based industry [3]. To reduce environmental pollution by manufacturing industries, selection of suitable cutting fluid is made using Electre method [4]. Electre along with AHP is applied in evaluation of cyber security metrics [5]. Electre method is used in supply chain management in the selection of best supplier in flexible packing industry [6]. Desirable and consistent supplier for supplying fabric in clothing industry is made with help of Electre [7].

A model using Electre is proposed for handling collaboration between, buyers, suppliers and business partners [8]. To reduce environmental pollution by manufacturing industries, selection of suitable cutting fluid is made using Electre method [9]. Sustainability assessment of manufacturing industries having more safety requirements and large customer base is successfully carried out using Electre and Promethee method [10]. For in interrupted mobile communication, Electre method is applied by reducing the handoff failures [11]. For designing a new approach that includes stake holders and new technologies for implementation of Water Framework Directive, Electre and AHP methods were used [12]. Prioritization of requirement of customers banking, finance and investment sector [13], Electre method is applied in design and assessment of public transport system for big cities [14].

Recent trends show large use of multi-criteria decision-making methods in healthcare applications. This technique can assist decision makers (usually medical doctors) in health sector by empowering them in selecting best option among available set of alternatives. A report from Emerging Good Practices Task Force described the key steps and provided an overview of the principal methods involved in multi-criteria decision making [15]. In the research conducted by Dursun [16], the authors used both ordered weighted averaging (OWA) operator and fuzzy Topsis for efficient health waste management. Dehe [17] used evidential reasoning (ER) as well as analytical hierarchy process (AHP) to aid decision process for selecting best location for healthcare infrastructure. Multi-criteria decision making is also used by Carmen [18] in asset management in healthcare organizations by assisting in selecting best set of maintenance policies. Charles [19] applied multi-criteria decision making to design a best healthcare plan that meets requirements of both insurer and payer. Amini [20] successfully used AHP and Topsis to rank health care with regards to the execution of the family health program. Multi-criteria decision making is used in health technology assessment (HTA) to obtain best value for money [21] and management of health care is improved by evaluation of quality of services [22].

In this present paper, the use of Electre method is described with a help of a case study in selecting best ultrasound machine and meeting the specified requirements.

A brief introduction of Electre method is followed by its implementation in a case study.

2 ELECTRE (Elimination and Choice Translating Algorithm)

B. Roy proposed Electre Elimination Et Choix Traduisant la Realité (Elimination Et Choice Translating Reality) method in 1960s. A multi-criteria decision-making problem (MCDM) is described by:

(i) $A_i, I = 1, \ldots, m$, which denotes the alternatives.
(ii) $g_j, j = 1, \ldots, n$, which denotes the criteria.
(iii) $w_j, j = 1, \ldots, n$ also $\sum_{j=1}^{n} w_j$, which denotes the criteria.

The objective is to find the best alternative among available alternatives, where the performance matrix ($m \times n$) possess value for every alternative corresponding to every criterion. Also, the weights of the criteria are given in a weight matrix.

Let the defined criteria be like $g_{ij} = 1, 2 \ldots, n$ where A corresponds to the set of alternatives. Any two alternatives a and b in set of alternatives A have either of following relations:

- $a\ P\ b$ Here, the preference is given to a than b $g(a) > g(b)$
- $a\ I\ b$ Here a remains not-so-different to b $g(a) > g(b)$
- $a\ R\ b$ (it remains not feasible to compare a and b)

  where $a,b \in A$,
  $g(a)$: value of alternative a for criteria g
  $g(b)$: value of alternative b for criteria g

Outranking relation is computed to describe the importance of one criterion over other. There are two sorts of comparisons required for the computation of outranking relation as given herewith, i.e., concordance and discordance.

- The decision maker is equipped by concordance test to check whether the alternative a can at least be good alike b.
- The intention of discordance test falls under the criteria in which the performance of a is worse compared to b. When there is a failure in this result, the high opposition can be said as vetoing the concordance test.
- The outranking relation of a S b can be finalized as true only when both concordance and discordance tests are passed.

2.1 Electre Method Steps

Step 1: Decision matrix formation

Here, the rows in decision matrix ($m \times n$) correspond to m alternatives whereas the criteria is denoted by n columns.

Step 2: Assignment of weights to criteria

Each of the criteria is assigned with weights of importance. Criteria with highest weight are most important to decision maker.

Step 3: Computation of concordance and discordance sets

Decision matrix data is compared for all the pairs of alternatives in terms of every criterion. Further, both concordance and discordance sets are analyzed. For every alternative pair a and b ($a, b = 1, 2, \ldots m$), the set of criteria is portioned into two following subsets:

- Concordance set, C: It consists of all the criteria due to which the alternative a is given preference against the other alternative i.e., b

$$C(a, b) = \{j : g_j(a) \geq g_j(b)\} \tag{1}$$

  where g_j (a) corresponds to alternative a's weight in terms of jth criteria.

  Thus, C (a, b) denotes a set of criteria in which the alternative a is either better or equal to that of the alternative b.

- Discordance set, D: It is compliment of concordance set C (a, b). D contains all criteria for which alternative a is worse than alternative b.

$$D(a, b) = \{j : g_j(a) < g_j(b)\} \tag{2}$$

  where $g_j(a)$ is weight of alternative a with respect to jth criteria.

Step 4: Concordance matrix computation

When criteria value weights, for elements present in concordance set, are measured, it results in concordance matrix.

$$C(a, b) = \sum_{j \in C(a,b)} w_j \tag{3}$$

where W_j is weight of the criteria g_j for which alternative a is better than or equal to alternative b.

Step 5: Discordance matrix computation

In general, the discordance matrix is calculated by dividing the values of discordance set members with the total value of whole set.

$$D(a, b) = \frac{\max_{j \in D(a,b)} |g_j(a) - g_j(b)|}{\max_j |g_j(a) - g_j(b)|} \tag{4}$$

Step 6: Computation of outranking relationship

In this step, the average values of both concordance and discordance are considered. In case of concordance matrix, any C (a, b) value which is higher than or equal to C average is considered to be 1, whereas in case of discordance matrix, any value less than or equal to D average is specified to be 0.

Step 7: Formation of net concordance and discordance matrix

In this step, the calculation of net concordance and discordance values is performed in order to rank among the set of available alternatives.

$$C_a = \sum_{k=1,k\neq a}^{m} C(a,k) - \sum_{k=1,k\neq a}^{m} C(k,a) \tag{5}$$

$$D_a = \sum_{k=1,k\neq a}^{m} D(a,k) - \sum_{k=1,k\neq a}^{m} D(k,a) \tag{6}$$

3 Application of ELECTRE Method in Selecting Best Ultrasound Machine

Five midrange price category models are considered from five leading brands, suitable for conducting ultrasound exams for patients across all ages and body types and offer fast and reliable diagnosis. Affinity 30 (Philips), Logiq P7 (GE), HS 50 (Samsung), DC 70 (mindray), S -1000 (Siemens) are the alternatives under consideration. Selection of ultrasound machine depends upon many factors like brand reputation, automatic OB measure, touch screen size, number of transducer ports, image quality and contrast imaging as set of evaluating criteria. We referred various websites [23, 24] and product manuals to get comparative data for the above mentioned models. The performance matrix for set of alternatives and criteria is shown in Table 1.

Assignment of weights to criteria is an important step in getting best result using Electre method. A questionnaire was prepared and sent to 15 healthcare providers (doctors and technicians performing ultrasounds). However, only eight of them responded. The weighted average of their criteria weights was computed. The criteria

Table 1 Performance matrix

Model name	Brand reputation 1	Automatic OB measure 2	Touch screen size 3	No of transducer ports 4	Image quality 5	Contrast imaging 6
Affinity 30	High	No	12	4	High	No
Logiq P7	High	Yes	10.4	3	Mid	Yes
HS 50	Mid	Yes	10.1	4	Mid	Yes
DC70	Low	Yes	10.4	4	Mid	Yes
S-1000	High	Yes	7	3	High	Yes

with highest weight are considered as more important. The associated weight matrix is shown in Table 2.

Pair-wise comparisons were performed for all elements in set of alternatives A and concordance sets are determined (Table 3).

Next, we compute concordance index for each alternative pair by adding weights for the criteria included in the corresponding concordance set. The concordance indexes are shown in Table 4.

Discordance computations will not be performed as the data in our preference matrix does not allow it. Next, outranking relationships will be computed by defining dominance relationship between pair of alternatives. Higher values of concordance

Table 2 Criteria weights

Criteria	Weights
Brand reputation	0.25
Automatic OB measure	0.20
Touch screen size	0.15
No. of transducer ports	0.10
Image quality	0.20
Contrast imaging	0.10

Table 3 Concordance set

C (A, L)	1, 3, 4, 5
C (A, H)	1, 3, 4, 5
C (A, D)	1, 3, 4, 5
C (A, S)	1, 3, 4, 5
C (L, A)	1, 2, , 6
C (L, H)	1, 2, 3, 5, 6
C (L, D)	1, 2, 3, 5, 6
C (L, S)	1, 2, 3, 4, 6
C (H, A)	2, 4, 6
C (H, L)	2, 4, 5, 6
C (H, D)	1, 2, 4, 5, 6
C (H, S)	2, 3, 4, 6
C (D, A)	2, 4, 6
C (D, L)	2, 3, 4, 5, 6
C (D, H)	2, 3, 4, 5, 6
C (D, S)	2, 3, 4, 6
C (S, A)	1, 2, 5, 6
C (S, L)	1, 2, 4, 5, 6
C (S.H)	1, 2, 5, 6
C (S, D)	1, 2, 5, 6

Table 4 Concordance indexes

	Affinity 30	Logiq P7	HS 50	DC 70	S-1000
Affinity 30	–	0.70	0.70	0.70	0.70
Logiq P7	0.55	–	0.9	0.9	0.80
HS 50	0.40	0.6	–	0.85	0.55
DC70	0.40	0.75	0.75	–	0.55
S–1000	0.75	0.85	0.75	0.75	–

index C (a, b) show that alternative a is dominant over alternative b. The method defines that a outranks b when C $(a, b) >$ Average (C).

Average $C = 0.695$.

The outranking relationship between various alternatives is defined in Table 5.

Ranking among various alternatives will be evaluated with the help of net concordance and discordance matrix. Now, ranking will be determined by computing advantages. The net concordance is computed by following equation number 6. The net concordance is shown in Table 6.

After sorting net concordance in increasing order, it is determined that alternative Affinity 30 is most desirable followed by S-1000, Logiq P7, DC 70, HS 50 (Fig. 1).

Table 5 Outranking relationships

	Affinity 30	Logiq P7	HS 50	DC 70	S-1000
Affinity 30	–	1	1	1	1
Logiq P7		–	1	1	1
HS 50			–	1	
DC70		1	1	–	
S-1000	1	1	1	1	–

Table 6 Net concordance

	C (a, b)	C (b, a)	Net C
Affinity 30	4	1	3
Logiq P7	3	3	0
HS 50	1	4	−3
DC70	2	4	−2
S-1000	4	2	2

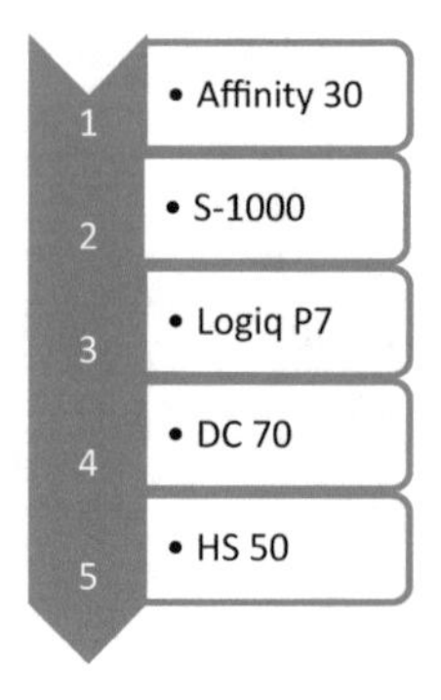

Fig. 1 Ranking of alternatives

4 Result Analysis and Discussions

Medical decisions related to both diagnostic and possible treatment very important as they have direct impact over life of individual. Many times these decisions become more complex if there are contradicting objectives. Thus, the well-structured approach, that can take care of nity gritties of complex decision-making problem, has a potential to improve the quality of decision making, up to a large extent. The Electre method proves to be a strong tool for decision-making problems involving both quantitative and qualitative attributes. The method starts with computation of concordance and discordance sets followed by computation of net concordance and out ranking relation. In the present paper, we have not considered discordance because the type of data we have does not support discordance. Finally, net concordance was computed, and final ranking of alternatives is obtained. Affinity30 is selected as best alternative and HS 50 as worst alternative. It is also observed that change in assignment of weights by decision maker can alter the ranking.

The problem of selecting best ultrasound machine is also modeled and evaluated with the help of AHP. Same set of criteria and alternatives was taken. Electre method asks experts to provide weights for various criteria while AHP method has a detailed procedure to calculate criteria weights by making pair-wise comparisons of criteria. The AHP method has a mechanism for checking consistency of decision makers but Electre method has no such provisions. It has been noticed in the study, for same set of consistent decision makers, Electre method and AHP method provided the same ranking.

5 Conclusions

Decision making in healthcare sector is complex in nature as a best decision should find a fine balance between various parameters from various stake holders. Judgments are not replaced by multi-criteria decision-making tools. It is actually involved in the identification, collection and structuring of the information for people who make judgments and to support the deliberative process. Multi-criteria decision making provides transparent and consistent decisions. The present paper explained the working of Electre method in detail. Each of the steps that lead to ranking of relations is explained with the help of a case study.

Like any other method, the Electre method too has limitations. The main drawback is that if the criteria and criteria weights are not selected carefully, Electre method will yield non optimal results. Hence, with the selection of correct criteria and criteria weights the Electre method results in strong decision making aide for healthcare experts. The paper suggested easy to use and reliable tool for selecting best ultrasound machines that suits ones requirements.

This paper carries methodological importance as it explains in detail how set of available alternatives can be analyzed to suggest a ranking method for alternatives that can be ordered in the range of best to worst. Electre method was validated in a selection of ultrasound machine that includes only six parameters. It is evident from the literature that this is a scalable method though the current research work missed to validate it. The future research is intended to validate the method involving large-scale requirements.

References

1. Figueira JR, Greco S, Roy B, Słowinski R (2013) An overview of ELECTRE methods and their recent extensions. J Multi-Criteria Decis Anal 20(1–2):61–85
2. Guarnieri P, Hatakeyama K (2016) The process of decision-making regarding partnerships including a multicriteria method. J Adv Manuf Syst 15(3):101–131
3. Bhol SG, Mohanty JR, Pattnaik PK (2020) Selecting location for agro-based industry using ELECTRE III method. In: IoT and WSN applications for modern agricultural advancements: emerging research and opportunities. IGI Global USA, pp 99–121
4. Jayant A, Chaudhary N (2018) A decision-making framework model of cutting fluid selection for green manufacturing: a synthesis of 3 mcdm approaches. ICAET-2018, Sangrur
5. Bhol SG, Mohanty JR, Pattnaik PK (2020) Cyber security metrics evaluation using multi-criteria decision-making approach. In: Smart intelligent computing and applications. Proceedings of the third international conference on smart computing and informatics, vol 2, Springer Nature Singapore, pp 665–67
6. Cristea C, Cristea M (2017) A multi-criteria decision-making approach for Supplier selection in the flexible packaging Industry. MATEC Web Conf 94:06002
7. Galinska B, Bielecki M (2017) Multiple criteria evaluation of suppliers in company operating in clothing industry. In: 17th international scimitar conference business logistics in modern management. Osijek, Coratia
8. Majdiiman (2013) Comparative evaluation of promethee and ELECTRE with application to sustainability assessment
9. Preethi GA, Chandrasekar C (2015) Seamless handoff using ELECTRE III and promethee methods. Int J Comput Appl 126(13):32–38
10. Bruen M (2007) Systems analysis—a new paradigm and decision support tools for the water framework directive. Hydrol Earth Syst Sci Discuss 4
11. Arul MS, Suganya G (2016) Multi-criteria decision making using ELECTRE. Circ Syst 07:1008–1020
12. Solecka K, Zak J (2014) Integration of the urban public transportation system with the application of traffic simulation. Transp Res Procedia 3

13. Kevin M, Maarten I, Thokala P, Baltussen R, Meindert B, Zoltan K, Thomas L, Filip M, Stuart P, Watkins J, Devlin N (2016) Multiple criteria decision analysis for health care decision making—emerging good practices: report 2 of the ISPOR MCDA emerging good practices task force. Value Health 19(2):125–137
14. Dursun M, Ertugrul KE, Melis AK (2011) A Fuzzy MCDM Approach for health-care waste management. World Acad Sci, Eng Technol Int J Ind Manuf Eng 5(1)
15. Dehe B, Bamford D (2015) Development, test and comparison of two multiple criteria decision analysis (MCDA) models: a case of healthcare infrastructure location. Expert Syst Appl 42:6717–6727
16. María CC, Andres G (2016) A multi criteria decision making approach applied to improving maintenance policies in healthcare organizations. BMC Medical Inf Decis Making 16–47
17. Charles H, Smith H, Roland W (2000) On designing health care plans and systems from the multiple criteria decision making (MCDM) perspective. In: Research and practice in multiple criteria decision making, Springer-Verlag Berlin Heidelberg
18. Amini F, Rezaeenour J (2016) Ranking healthcare centers using fuzzy analytic hierarchy process and TOPSIS: Iranian experience. Int J Appl Oper Res 6(1):25–39
19. Aris A, Panos K (2017) Multiple criteria decision analysis (MCDA) for evaluating new medicines in health technology assessment and beyond: the advance value framework. Soc Sci Med 188:137–156
20. Meltem M, Gulfem T, Bahar S (2017) Multi-criteria decision making techniques for healthcare service quality evaluation: a literature review. Sigma J Eng Nat Sci 35(3):501–512
21. Mindray Homepage. http://www.mindray.com/en/category/Medical_Imaging_System.html. Last accessed 5 Feb 2020
22. Philips Homepage. http://www.philips.co.in/healthcare/solutions/ultrasound. Last accessed 5 Feb 2020
23. Samsung Homepage. http://www.samsunghealthcare.com/en/products/UltrasoundSystem. Last accessed 5 Feb 2020
24. Siemens Homepage. http://www.healthcare.siemens.co.in/MedicalImaging. Last accessed 5 Feb 2020

A Fatigue Crack Path Analysis in Rail Weldment Under Mixed Mode Loading Condition: A Computational Simulation

Prakash Kumar Sen, Mahesh Bhiwapurkar, and S. P. Harsha

Abstract Most of the usual AT butt-welded rail fatigue incidents consist of either weld web fractures or rolling surface fractures. In the preliminary phase, a crack spreads parallel to the rail rolling surface, but within a slant surface area, it continues to spread after it. The object of this study is to examine the processes included for initiation along with expansion of a crack on the web of the rail weldment in order to anticipate the direction of fracture crack and secondary, the intervals of weld inspections. With this viewpoint, the finite element study for the expected cracking was performed to measure the brief history of stress intensity factors, K_{I} and K_{II} at various tips of cracking tip throughout the transit of the usual loadings caused by the train. In observations, it has been shown that basic flaw appears to follow a direction where K_{II} is close to the maximum possible value and there is a tiny superimposed K_{I} traction. Around the crack edge, mixed mode conditions are characterized, and stress intensity factors are measured using the method of crack opening displacement. The direction of crack development is estimated by the criterion of minimum strain energy density. Fractographic studies have verified that crack propagation mostly dominated by type of mode II and then only after the final divergence of the crack mode I occur. Computational simulations and experimental findings made by RDSO on the three-dimensional growth of fatigue crack are compared here. An acceptable relationship is revealed between the direction of numerical and experimental crack growth, both for the three-dimensional element by corner cracks and through the elliptical crack thickness.

Keywords Alumino thermit (AT) weld · Crack · FEM · Stress intensity factor

P. K. Sen (✉) · M. Bhiwapurkar
Mechanical Engineering Department, O.P. Jindal University, Raigarh, India
e-mail: prakashkumarsen@gmail.com

S. P. Harsha
Department of Mechanical and Industrial Engineering, IIT Roorkee, Roorkee, India

S. S. Ray et al. (eds.), *Applied Analysis, Computation and Mathematical Modelling in Engineering*, Lecture Notes in Electrical Engineering 897,
https://doi.org/10.1007/978-981-19-1824-7_20

1 Introduction

Rails are manufactured in finite lengths and therefore connected end to end to provide a flat surface for trains to run over. The usual way of connecting rails is to use metal fish plates to bolt them together, resulting in jointed track. Trains passing over jointed tracks make a clattering sound due to the small gaps between the rails. Jointed track does not have the same ride quality as welded rails and is thus less suitable for high-speed trains unless it is well-maintained. However, because of the lower cost of installing and maintaining, jointed track has been used in railways on lower speed lines and sidings. In Indian railways, many accidents of passengers carrying trains are due to rail fractures and weld failures, resulting loss of life of passengers. In the view of this scenario, as a measure to reduce rail fractures and weld failures, and to reduce rail accidents on Indian railways, specification of 52 and 60 kg, rails and process of welding should be revised with objective to have track with zero rail fractures.

1.1 Mix Mode Loading and Crack Growth

In many situations, the development of fatigue cracks leads to failure of not only railway structures, but also multiple crack propagation in machine and technical components. Defects or fractures, which originally become stable as a result of service loading, are the source of these damages. In several cases, such cracking is exposed to complex loading conditions, referred to as mixed mode loadings, which are caused by the component's geometry or external loading.

Figure 1 shows a rail specimen with fractures that are normally exposed to plane mixed mode loading. The stress intensity factors (SIFs), K_{I} and K_{II}, which illustrate the same phase of the applied stresses $\sigma(t)$ in the case of quasi-static load, describe the local cyclic loading at the tip of the crack.

Combined load consideration on cracks could be seen in the case of three fundamental fracture modes as shown in Fig. 2 which occurs partially or completely in

Fig. 1 Rail weld crack under a mix mode situation

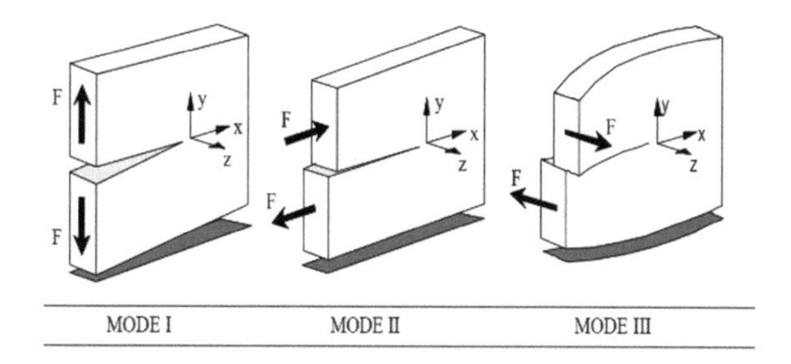

Fig. 2 Three modes of fracture

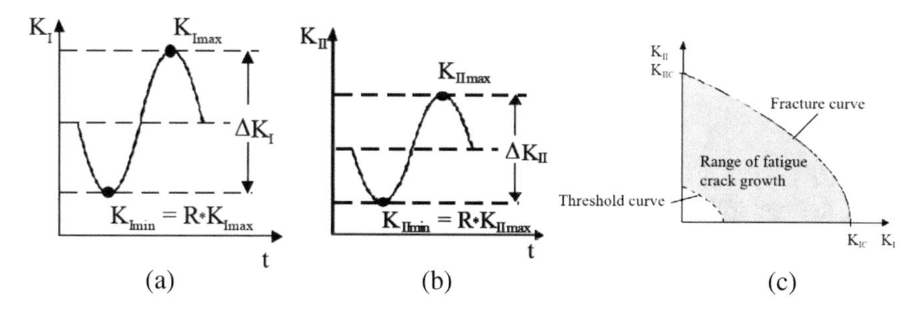

Fig. 3 **a** and **b** Stress intensity factors (SIF) K_{I} and K_{II} for a crack in mix mode. **c** $K_{\text{I}}/K_{\text{II}}$-diagram with the range of fatigue crack growth

combination due to any externally load or crack direction. This indicates that the structure's loading develops an unsymmetrical, single stress site near the cracking face. As a result, the fracture changes shape to the point where not just an opening, but also a linear shifting of two cracks surfaces can be detected. Then as a result, the stress distribution around the crack is now described by that of the stress intensity factor.

As illustrated in Fig. 3a and b, the fatigue crack development is regulated by the cyclic stress intensity factors, K_{I} and K_{II}, as well as the cyclic relative stress intensity factor ΔK_V, which was established by [1, 2].

$$\Delta K_v = \frac{\Delta K_{\text{I}}}{2} + \frac{1}{2}\sqrt{\Delta K_{\text{I}}^2 + 6\Delta K_{\text{II}}^2} \quad (1)$$

If ΔK_V reaches the mode I threshold value Kth, crack initiation and growth are possible until ΔK_V exceeds the fracture toughness value $\Delta K_C = (1 - R)\, K_{\text{IC}}$ [2], and crack growth is stable and managed. In a plane mix mode type of situation as shown in Fig. 3c depicts the ranging of fatigue crack initiation.

The SIFs K_{I}, K_{II} and/or K_{III} play an important role. Cracks those are kinked or dispersed, several cracks which are generating from the notch positions or crack points in either welded or adherent joints on a rail, and cracks in composites are the examples of crack complications that arises in reality. Static, dynamic, or thermic

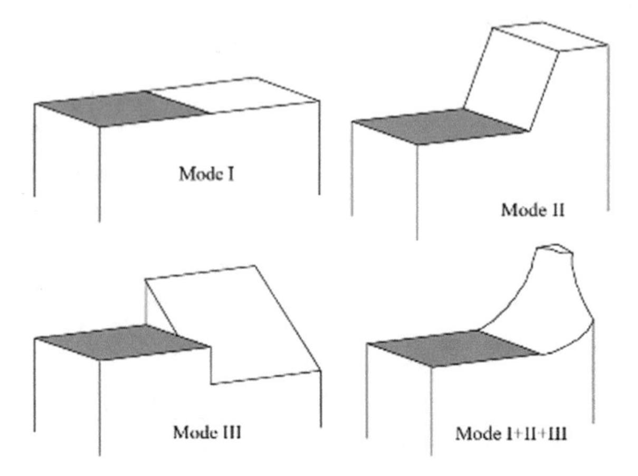

Fig. 4 Various crack growth types due to mixed mode loading

loads, the multiplicity of loads, thermal and fatigue stress, as well as changes in loading or its allocation, are root cause to form cracks.

The superposition of modes I and II at a crack is known under plane mixed mode loading, while the superposition of all three modes I, II, and III is known as spatial mixed mode loading. Any changes in the structure's loading condition may result in a change in the crack orientation and path. Because in an isotropic material, a "long crack" (from a fracture mechanical point of view) can only grow straight on under pure mode I loading, a change in the mixed mode portions causes the crack growth direction to change that can be seen on Fig. 4. Mode II loading causes the crack to kink, while mode III causes the crack front to twist. However, the superposition of modes I, II, and III results in a unique crack surface, which is frequently observed in action.

1.2 Introduction and Failure Description for Rail Fracture and Weld Failures

Rail steels are prone to fatigue failures, which can result in rail fractures and unexpected downtime, and other safety problems in railway operations [3, 4]. Maya-Johnson [5] examined at the rate of fatigue fracture propagation in two common rail steels. Masoudi Nejad et al. [6] used Franc3D to investigate the SIFs of rolling contact fatigue fractures. According to Yang Liu et al. [7], there are few works on establishing SIFs for particular surface fracture examples for three-dimensional cracks on the transverse plane of a thermite weld cross section. The additional stresses produced by the effect of moving loads on the rail joint reduce the life of rails and welds. Due to fatigue stress in the rail ends and weld, they are especially battered and harmed, and the possibilities of rail/weld fracture in the joints are very high. Rail/weld fractures develop when a small crack grows into a large one, and they

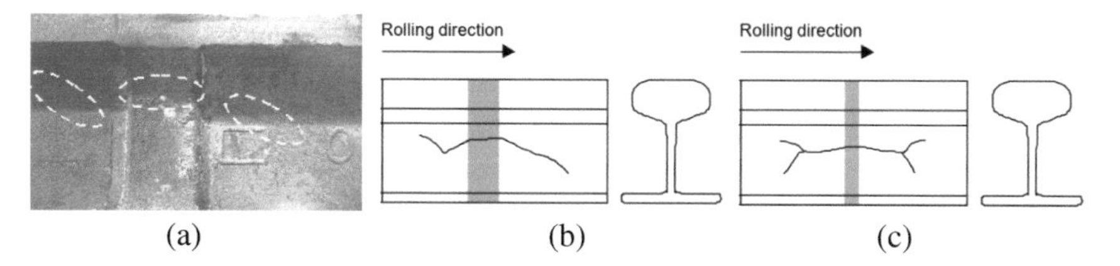

Fig. 5 **a** Crack failures starting from welded joints. **b** The two kink crack and its schematic drawing. **c** A schematic drawing of the other distinctive crack with branches

occur during the season when the mean temperature difference in the daytime are vary significantly, causing the tracks to expand or contract. Lack of maintenance and installation will also result in a small crack that leads to rail splitting over time. This causes a track fracture, resulting the bogie to fall off the rail track, disrupting all of the coaches following it. The most common cause of rail failure, according to the Indian Railways' Ultrasonic Flaw Detection Manual, is fatigue fracture, which is caused by material imperfections or crack development during operation [8].

The most common causes of rail fractures in operations are either crack on the surface of rail weld with the wheel running on its surface or, quite often, cracking in weld connections. Some typical crack failures that start with welded joints are depicted in Fig. 5a. The cracking direction is defined by its first part in which the defect forms approximately parallel to plane and a second part in which it reveals two kink corners. Figure 5b represents the cracking develops toward the rail running surface in the direction of head corner (in a rolling way context) and stretches toward its foot at the trail tip; on the other hand, Fig. 5c shows the pattern of a parallel to the plane crack having two branches of the crack propagation in the lead and trail tip corners.

Differences observed in crack direction indicated that cracks grow in mode II before switching to mode I at the kink or branch point in these failures. Any cracks caused by rolling contact fatigue (RCF) due to wheel–rail interaction have been stated in literature [9, 10]. A per these conditions, mode II spreading has been studied, and it has been shown that high compressive stresses due to contact strengthen this crack propagation mode [8]. The considered failures are however situated in a position on the web where compressive stresses induced would not be too severe due to significant distance from the contact stress field.

A descending crack branch or kink, like surface crack formation caused by rolling contact fatigue, can cause the rail/weld rupture and catastrophic failure [11], so it is important to know the requirements that regulate crack formation. However, if mode II spreading is verified in the first portion of the cracking, it will be necessary to examine the standards that used to determine the location of welds along the rail.

2 Materials and Methodology

In order to investigate fatigue and fracture of railway structures, finite element method is used to analyze the model and the wheel/rail interaction, to determine the crack initiation position. The analysis of contact stresses and wheel/rail interaction is carried out with ANSYS FEA software for this purpose. Prominent profiles according to the Indian Railways UIC60 rail with AT weld configuration and regular wheel profile as shown in Fig. 6 were used to geometrically model the wheel and rail having dimensions as given in Tables 1 and 2. For wheel/rail touching, the friction coefficient is 0.29, and it is constant in both directions [4]. Eight-node three-dimensional components of style solid 45 is used for model mesh. However, apart from that, contact 173 and target 170 are the contact elements for the wheel and rail, respectively, as shown in the CAD assembly in Fig. 8.

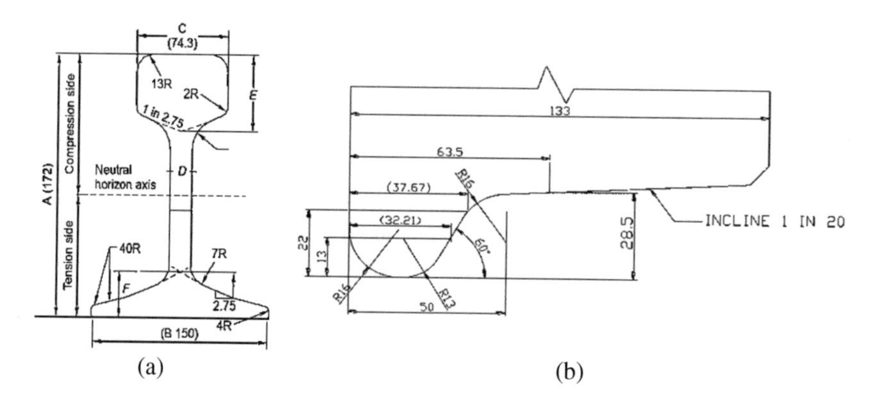

Fig. 6 **a** UIC 60 rail profile. **b** Wheel profile

Table 1 UIC rail dimensions

Rail section	Wt/Meter (kg)	Area of the section (mm^2)	Dimensions (mm)					
60 kg (UIC)	60.34	7686	A	B	C	D	E	F
			172.0	150.0	74.3	16.5	51.0	31.5

Table 2 Wheel parameter

Wheel parameter	Range (in mm)
Diameter	950 to 1150
Rim width	100 to 150
Rim thickness	40 to 80

2.1 *Material Properties of Rail/Weld and Wheel with Axle Under Material Selection*

Rail/weld, wheel, and axel mechanical properties are adapted from Yuan-qing Wang, Hui Zhou [12] with bi-axial (linear) kinematic hardened elastio-plastio material model. The properties of used elastic material values are given in Tables 3 and 4.

Figure 7 is a flow chart used for this work which represents the parameters applied in ANSYS structural analysis for preprocessing and then automatic meshing of assembly. The model assembly is composed of a rail length of 800 mm between

Table 3 Mechanical properties of assembly

Part name	Modulus of elasticity (GPa)	Modulus of plasticity (GPa)	Yield stress (MPa)	Poisson's ratio
Rail	206.9	22.7	483	0.295
Wheel	205	22.7	640	0.3
Axle	205	–	–	0.3

Table 4 Weld properties

Mechanical property	Value
Poisons ratio	0.3
Young's modulus (GPa)	207 GPa
Ultimate tensile strength (MPa)	996.7 MPa
Yield strength	675.7 MPa
Percentage reduction of area	4.22
Percentage elongation	3.09

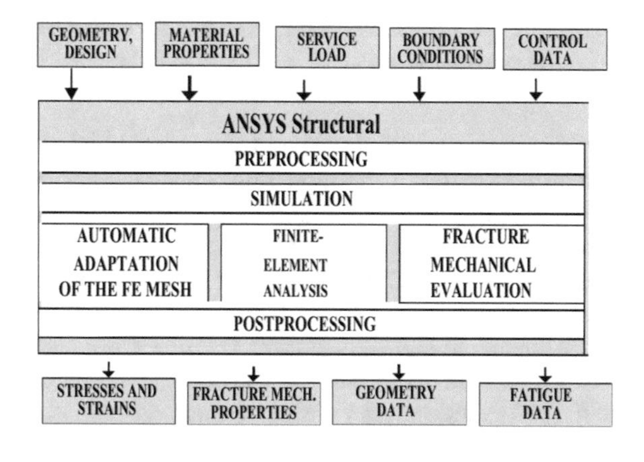

Fig. 7 Flow chart for fatigue crack growth simulation in FEM

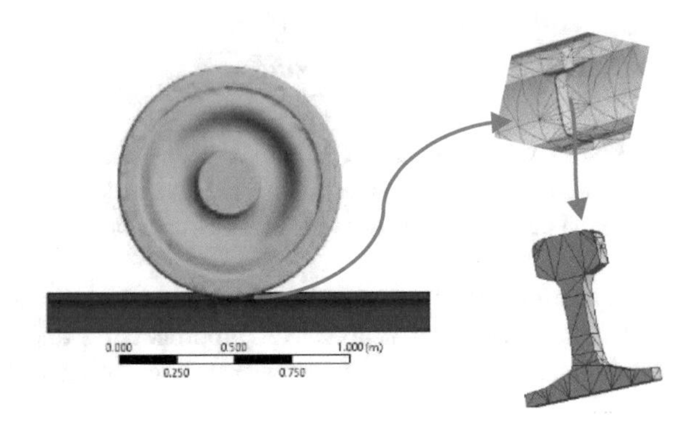

Fig. 8 CAD model of rail wheel and weld

the two sleepers and a wheel diameter of 1150 mm. Boundary conditions are used to retain material integrity as a fixed rail from the bottom and ends, are as follows.

(a) Rail foot surface area is completely fixed with the condition of the translation and rotation end as null in all three directions.
(b) End conditions of symmetry have been applied to the assembly.
(c) The movement is allowed to visualize the result of the force on the model only in the *y*-direction.

A fracture tool is adopted which is preexists in ANSYS for fracture crack analysis to identify the values of SIFs and required fatigue data. Figure 8 shows the CAD model of rail/weld/wheel assembly model with meshed weld which is under axle load of 22.5 ton in vertical direction.

3 Results and Discussion

3.1 Finite Element Analysis for Crack Direction and Its Propagation

According to many field experiments, semi-elliptical cracking has been the most common fault in rail transportation. The crack position is determined as described in the previous part, and a semi-elliptical cracking is being used for modeling and simulation through fracture tool. The first crack has a diameter of one millimeter and is measured in millimeters. Quite apart from that, cracks are examined in two directions: primarily in the direction of loading and secondary in the reverse direction of loading as in shown in Fig. 7. The first crack has a diameter of one millimeter and is measured in millimeters. The first one is related with the mass applied vertical to the rail cross section by train, and the other is associated with rotational inertia force created by wheel–rail rolling contact in this study. Crack planes are subjected to the secondary force, also called as rolling friction, which allows them to slip over one another. Now elements are specified, and the CAD model is meshed. The stress analysis is conducted using finite element method (FEM) software. Calculations for stress intensity factors (SIFs) have been done for all three kinds of fracture conditions. The crack propagation path is identified by SIFs. The crack growth line is measured using maximum shear stress (τ_{max}) criteria, and the crack will spread one step subsequently when the growth line is determined.

The element is redefined and prepared for resolving after the first process in crack initiation. Stress strength parameters and calculations from sources [13, 14] are used to measure a weld's fatigue life. The updated Paris model also accounts for crack development as well as fatigue crack closing.

For different stages of crack propagation, multiple FEM simulations were conducted. Figure 9b depicts three virtual crack tip locations for the weld fracture. The findings for the leading tip can be seen in Figs. 10, 11 and 12. In both cases, it shows the angle formed by potential kinking and the crack's current direction as $\alpha = 0$, shown in Fig. 9a, which indicates that there is no deviation. It can be seen in the first and second steps in the Figs. 10 and 11, and a maximum intensity of K_{II} and a minimal K_I is associated with propagation. The actual propagation direction, in particular, belongs to a K_{II} value close to the limit that is "assisted" by a smaller K_I. As the crack approaches to kinking, Fig. 12 indicates the results from the stress level variables. As it can be shown, that the actual propagation angle correlates directly to

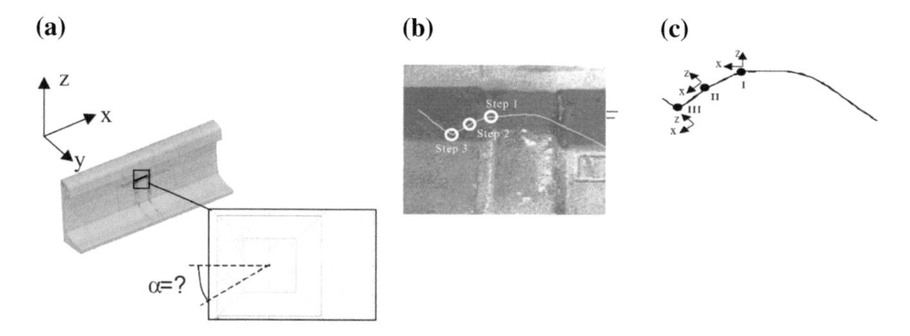

Fig. 9 **a** Lead tip analysis points. **b** Coordinate axis system convention. **c** Kink angle "α" convention

Fig. 10 SIF variation with different wheel load position

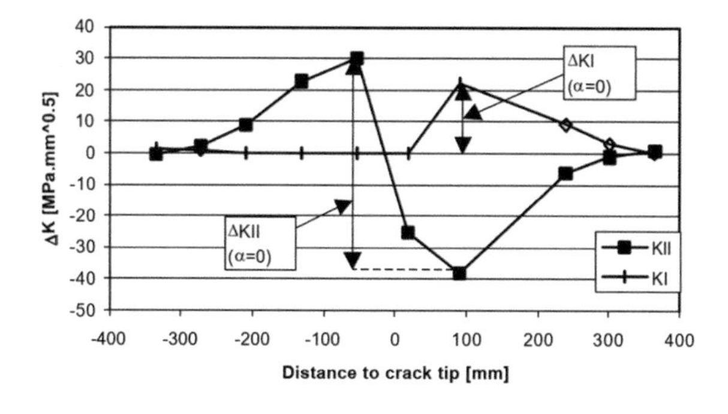

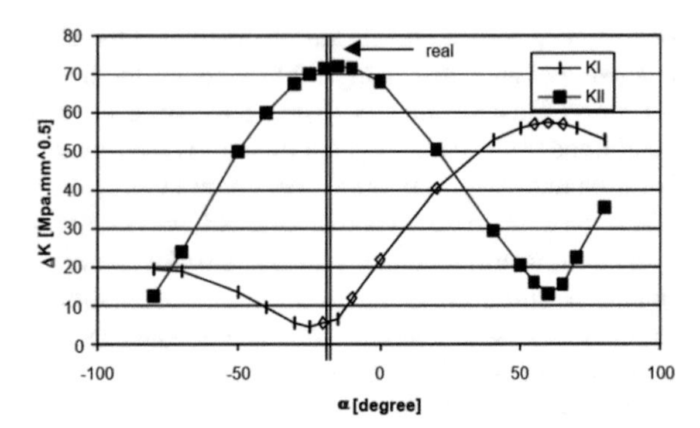

Fig. 11 SIF K_I and K_{II} variation with kinking angle for leading tip at the first step

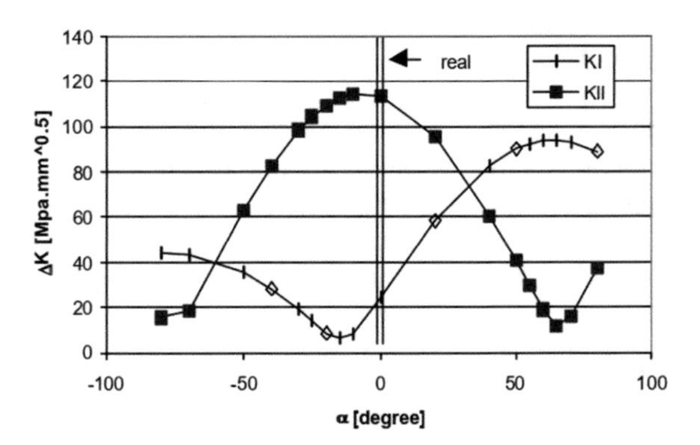

Fig. 12 K_I and K_{II} variations along with kinking angle at step 2 on the leading edge

the maximum mode I, meaning that kinking is necessary to a transition from mode II to mode type I crack development. Different outcomes were determined for the trail tip corner.

3.2 *RDSO Test on At Weld and Fractographic Study*

Fatigue testing of 60 kg (90 uts) A.T. Welded Rail Joint has been performed on rail track panel fatigue testing equipment having an arrangement shown Fig. 13, to identify the crack pattern under fatigue loading. According to scheme, three weld samples, made with one meter long new rail pieces, having overall length of 2.0 m have been prepared at thermit portion plant, RDSO Lucknow, and send to track laboratory. Weld geometry of two samples out of three has been ground on foot, to the level of rail, and one has been left intact. Each sample has to be placed on test frame as shown in Fig. 13, having span of 1.5 m, and is to be subjected to dynamic loading for stress range of tensile 20 kg/sq mm to compressive 4 kg/sq mm (these are the stress on the bottom of rail foot). All four samples have undergone through downward lodging of 219.351 kN and upward lodging of 43.879 kN with a frequency

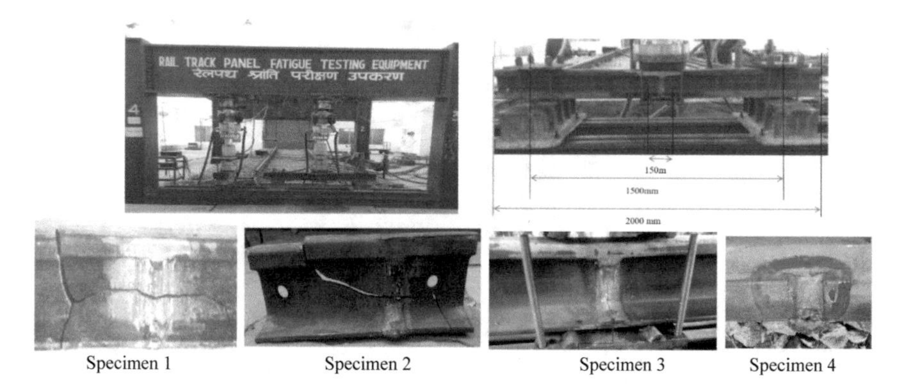

Fig. 13 Rail track fatigue equipment and the crack outcomes after the experiment on four specimens

of 9 Hz. The weld test results shown in Fig. 13 under specimens 1, 2, 3, and 4 and data stated in Table 5 represent specific type of crack pattern similar to discussed earlier.

SEM studies were performed in order to better explain the crack process. According to Wong [11], the crack surfaces are weakened by scratching due to frictional contact and opposite faces, frictional wear, respectively in almost all of the samples studied. As a result, neither the crack initiation point nor the fatigue inclusion could be identified in testing. As a matter of fact, as shown in Fig. 14a, the surface was sliced along its symmetry axis. The specimens are examined using a scanning electron microscope (SEM) to see the key properties of the crack direction after scrubbing the inner face.

Table 5 Weld specimen results

S No.	Specimen No.	Crack develops and broken on No of cycles	Pattern of crack
1	1	506,137 cycles	Mode II to mode I
2	2	2 million cycles	Mode II
3	3	912,294 cycles	Mode II
4	4	1,623,410 cycles	Mode II to mode I

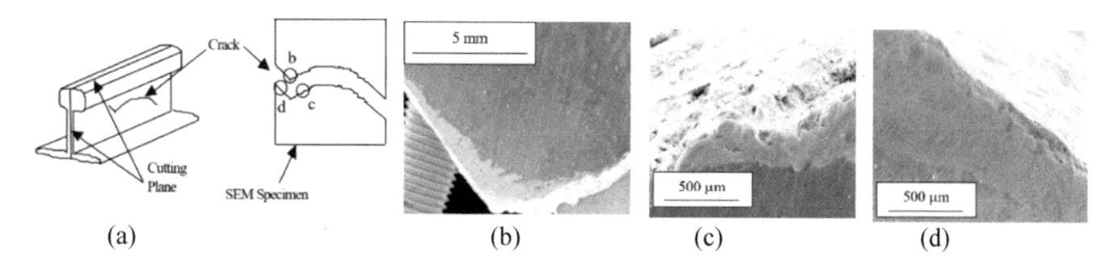

Fig. 14 Kink area SEM analysis at the leading position: **a** View of the crack on rail weld. **b** Kink view. **c** View of fracture on the surface just before kink starts. **d** View of the fracture surface just after the kink

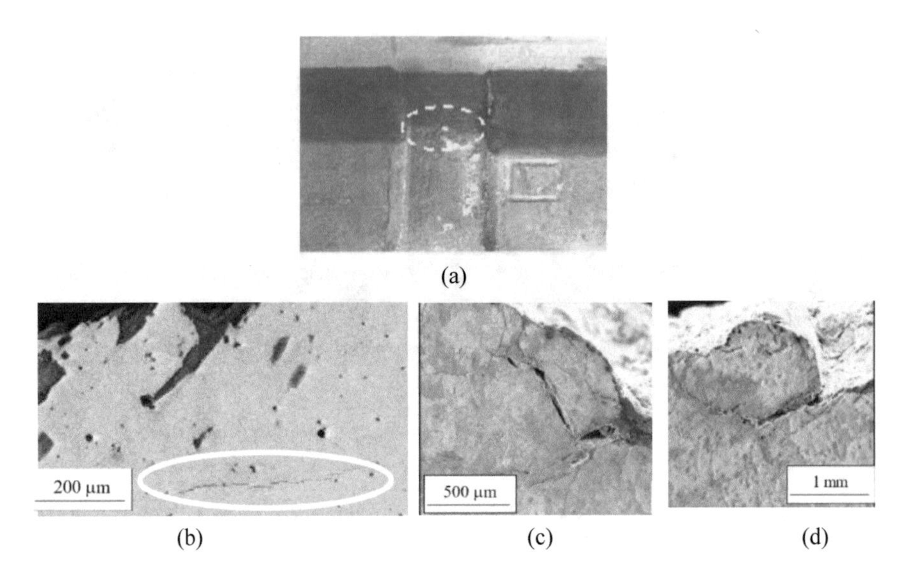

Fig. 15 Weldment fracture SEM analysis: **a** Probed area. **b** and **c** Fracture plane damage due to mode II. **d** Another cracking pattern having a stepping kind of view

Figure 14b shows the crack development path on pre and post kinking. Before kinking, the fracture surface has a highly damage pattern as shown in Fig. 14c, but the damage rapidly decreases after the kinking as shown in Fig. 12d. Mode II has been related to a higher degree of elastic–plastic degradation and short splitting, according to literature [12, 15]. This result appears to support the main assumption: a change in cracking transmission mode through beginning mode II to the last mode I is related to a change in defect layout. Figure 15 shows the SEM images of the center portion of the fracture (a region of nucleation). Again, there is a considerable amount of damages associated with the outbreak of crack. A second layer with a pattern like a step-wise shape is seen in Fig. 15d, which is a specific form of the damage and can be seen in type II of fracture mode diffusion [13, 16].

3.3 Analytical Validation by Equations

A 3D-FEM analyses were performed using the commercial finite element program ANSYS R3 [14] to corroborate the notion of a first mode II propagation followed by branching or kinking transition to mode I. The crack surfaces contact was introduced using the master-slave surface approach and 20 node solid components. The finite element mesh for the first stage is shown in Fig. 9. The relative crack opening displacement U_z and sliding displacement U_x of crack nodes were used to calculate K_I and K_{II} values at the fracture tip using well-known formulas [17]:

$$K_{\rm I} = \frac{2\pi G \Delta U_Z}{(3 - 4\mathcal{V})\sqrt{2\pi L}} \text{ as } L \to 0 \tag{2}$$

$$K_{\rm II} = \frac{2\pi G \Delta U_x}{(3 - 4\mathcal{V})\sqrt{2\pi L}} \text{ as } L \to 0 \tag{3}$$

Here, G is the shear modulus, $\mathcal{V}$ is the Poisson's ratio, and L is the fracture tip distance. Figures 10 and 11 depict the variance of $K_{\rm I}$ and $K_{\rm II}$ for the leading crack tip at various wheel locations. [12] obtains the local $K_{\rm I}$ and $K_{\rm II}$ for a potential kinking angle.

$$K_{\rm I}(\alpha) = C_{11}K_{\rm I} + C_{12}K_{\rm II} \tag{4}$$

$$K_{\rm II}(\alpha) = C_{21}K_{\rm I} + C_{22}K_{\rm II} \tag{5}$$

$K_{\rm I}(\alpha)$ and $K_{\rm II}(\alpha)$ are local stress intensity factors at the kink's tip, while $K_{\rm I}$ and $K_{\rm II}$ are stress intensity factors for the main fracture, as determined by FEM analysis using Eqs 2 and 3. The coefficients $C_{\rm ij}$ are calculated as follows:

$$C_{11} = \frac{3}{4}\cos\frac{\alpha}{2} + \frac{1}{4}\cos\frac{3\alpha}{2}; \tag{6.1}$$

$$C_{12} = -\frac{3}{4}\left(\sin\frac{\alpha}{2} + \sin\frac{3\alpha}{2}\right) \tag{6.2}$$

$$C_{21} = \frac{1}{4}\left(\sin\frac{\alpha}{2} + \sin\frac{3\alpha}{2}\right); \tag{7.1}$$

$$C_{22} = \frac{1}{4}\cos\frac{\alpha}{2} + \frac{3}{4}\cos\frac{3\alpha}{2} \tag{7.2}$$

Two primary values are produced for each angle and each simulated step (in Fig. 10, simply the case of α=0, which signifies no kinking) $K_{\rm II}$ and $K_{\rm I}$. Crack closure phenomena are ignored in the last one for $K_{\rm I}$.

4 Conclusions and Remarks for Future WorK

The propagation direction of crack growth within rail weldment has been investigated. Due to variation in cracking pattern which is manifested by kinking or fracturing the SEM studies and finite element analysis confirmed that this type of crack propagates in mode II at first, before migrating to type I mode propagation. The effect of the weldment's location on the rail on $\Delta K_{\rm II}$ variance is critical since mode II is the initiating mode of propagation. In all of the cases, discussed here the weldment is considered always in the mid position of two parallel laying sleepers. In addition, in

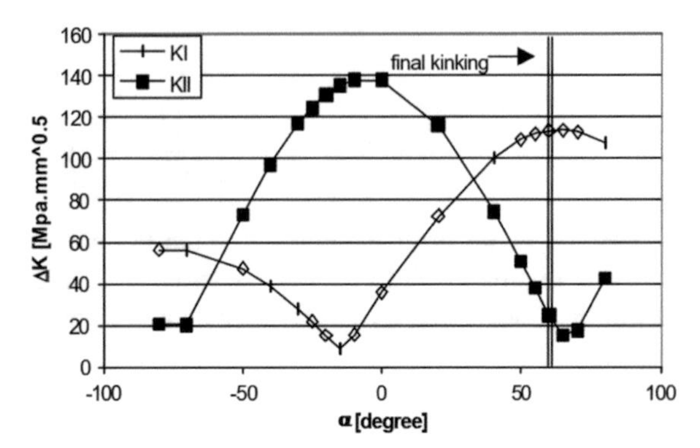

Fig. 16 K_I and K_{II} verses kink angle variations: Lead tip on the third step (Its final kink angle which maximizes K_I)

order to determine the maintenance cycles for weldment on railway lines, it should be important to analyze fracture growth speed that can be done on the basis of this study (Fig. 16).

Acknowledgements This study was done on the student edition of ANSYS under the academic license.

References

1. Kondo K, Yoroizaka K, Sato Y (1996) Fatigue crack growth behavior of surface crack in rails. Wear 191:199–203
2. Ishida M, Abe N (1996) Wear 191:65–71
3. Bonniot T, Doquet V, Mai SH (2018) Mixed mode II and III fatigue crack growth in a rail steel. Int J Fatigue 115:42–52
4. Jiang WJ, Liu C, He CG, Guo J, Wang WJ, Liu QY (2017) Investigation on impact wear and damage mechanism of railway rail weld joint and rail materials. Wear 376–377:1938–1946
5. Maya-Johnson S, Ramirez AJ, Toro A (2015) Fatigue crack growth rate of two pearlitic rail steels. Eng Fract Mech 138:63–72
6. Masoudi NR, Khalil F, Mahmoud S, Majid M (2017) Stress intensity factors evaluation for rolling contact fatigue cracks in rails. Tribol Trans 60(4):645–652
7. Liu Y, Tsang KS, Hoh HJ, Shi X, Pang JHL, Structural fatigue investigation of transverse surface crack growth in rail steels and thermite welds subjected to in-plane and out-of-plane loading, doi.org/https://doi.org/10.1016/j.engstruct.2019.110076
8. Indian Railways' Ultrasonic Flaw Detection Manual
9. Akama M, Mori T (2002) Wear 253:35–41
10. Jung-Kyu K, Chul-Su K (2002) Mater Sci Eng, A 338:191–201
11. Melin S (1986) Int J Fract 30:103–114
12. Wong SL, Bold PE, Brown MW, Allen RJ (1996) Wear 191:45–53
13. NASA (2000) Fatigue crack growth computer program NASGRO version 3.0–reference manual. JSC-22267B, NASA, Lyndon B. Johnson Space Center, Texas
14. ANSYS User's Manual 2019R3
15. Forman RG, Kearney VE, Engel RM (1967) Numerical analysis of crack propagation in cyclic-loaded structures. J Basic Eng 89(3):459–463

16. Anderson TL (1994) Fracture mechanics, fundamentals and applications, 2nd edn. CRC Press, Boca Raton, USA
17. Anderson TL (1995) Fracture mechanics. CRC Press, Boca Raton

Nonexistence of Solutions for a Higher-Order Wave Equation with Delay and Variable-Exponents

Erhan Pişkin and Hazal Yüksekkaya

Abstract In this article, we deal with a higher-order wave equation with delay term and variable exponents. Under suitable conditions, we prove the nonexistence of solutions in a finite time. Generally, the problems with variable exponents arise in many branches in sciences such as nonlinear elasticity theory, electrorheological fluids and image processing. Time delays often appear in many practical problems such as thermal, biological, chemical, physical and economic phenomena.

Keywords Delay term · Higher-order wave equation · Nonexistence of solutions · Variable exponent

Mathematics Subject Classification 35B44 · 35L05 · 35L55

1 Introduction

In this paper, we study the higher-order wave equation with delay term and variable exponents as follows:

$$\begin{cases} u_{tt} + (-\Delta)^m u + \mu_1 u_t (x,t) |u_t|^{q(x)-2} (x,t) & \\ +\mu_2 u_t (x,t-\tau) |u_t|^{q(x)-2} (x,t-\tau) = bu |u|^{p(x)-2} & \text{in } \Omega \times (0,\infty), \\ u(x,t) = 0 & \text{in } \partial\Omega \times [0,\infty), \\ u(x,0) = u_0(x), \ u_t(x,0) = u_1(x) & \text{in } \Omega, \\ u_t(x,t-\tau) = f_0(x,t-\tau) & \text{in } \Omega \times (0,\tau), \\ D^\alpha u(x,t) = 0, \quad |\alpha| \le m-1 & \text{in } x \in \partial\Omega, \end{cases} \tag{1}$$

E. Pişkin · H. Yüksekkaya (✉)
Department of Mathematics, Dicle University, Diyarbakir, Turkey
e-mail: hazally.kaya@gmail.com

E. Pişkin
e-mail: episkin@dicle.edu.tr

S. S. Ray et al. (eds.), *Applied Analysis, Computation and Mathematical Modelling in Engineering*, Lecture Notes in Electrical Engineering 897,
https://doi.org/10.1007/978-981-19-1824-7_21

where Ω is a bounded domain with smooth boundary $\partial\Omega$ in R^n, $n \geq 1$. $\tau > 0$ is a time delay term, $b \geq 0$ is a constant, μ_1 is a positive constant and μ_2 is a real number. $A = (-\Delta)^m u$, $m \geq 1$ is a natural number, Δ is the Laplace operator and $\alpha = (\alpha_1, \alpha_2, \cdots, \alpha_n)$, $|\alpha| = \sum_{i=1}^n |\alpha_i|$, $D^\alpha = \prod_{i=1}^n \frac{\partial^{\alpha_i}}{\partial x_i^{\alpha_i}}$. u_0, u_1, f_0 are the initial data functions to be specified later.

The variable exponents $q(\cdot)$ and $p(\cdot)$ are given as measurable functions on $\overline{\Omega}$ satisfy:

$$2 \leq q^- \leq q(x) \leq q^+ \leq p^- \leq p(x) \leq p^+ \leq p^* \tag{2}$$

where

$$q^- = ess\inf_{x\in\Omega} q(x),\ q^+ = ess\sup_{x\in\Omega} q(x),$$
$$p^- = ess\inf_{x\in\Omega} p(x),\ p^+ = ess\sup_{x\in\Omega} p(x),$$

and

$$p^* = \begin{cases} 2 < p < \infty, & \text{if } n \leq 2m \\ 2 < p \leq \frac{2m}{n-2m}, & \text{if } n > 2m. \end{cases}$$

The problems with variable exponents arise in many branches in sciences such as electrorheological fluids, nonlinear elasticity theory and image processing [8, 9]. Time delay often appears in many practical problems such as thermal, biological, chemical, physical and economic phenomena [18].

Without delay term ($\mu_2 u_t(x, t-\tau) |u_t|^{q(x)-2}(x, t-\tau)$), the Eq. (1) turns into as follows

$$u_{tt} + (-\Delta)^m u + u_t |u_t|^{q(x)-2} = u |u|^{p(x)-2}. \tag{3}$$

Pişkin [22] obtained the nonexistence of solutions for $E(0) < 0$ of the Eq. (3).

Without delay term and when $m = 1$, the Eq. (1) becomes in the form as follows

$$u_{tt} - \Delta u + u_t |u_t|^{q(x)-2} = u |u|^{p(x)-2}. \tag{4}$$

Messaoudi et al. [17] considered the existence of weak local solution by Faedo–Galerkin method. The authors also established the blow up result with $E(0) < 0$ for this equation. Then, Ghegal et al. [11] proved a global result and obtained the stability result by using an integral inequality due to Komornik to the Eq. (4).

Without the delay term, when $m = 1$ and $q(x)$, $p(x)$ are constants, problem (1) becomes in the form as follows

$$u_{tt} - \Delta u + u_t |u_t|^{q-2} = u |u|^{p-2}. \tag{5}$$

When the damping term $|u_t|^{m-2} u_t$ is absence, Ball [7] proved that the source term $u|u|^{p-2}$ ensures finite time nonexistence results for $E(0) < 0$ in this equation, in 1977. Haraux and Zuazua [12] established that the damping term provides global existence, in the absence of the source term. When $m = 2$, Levine [15] proved a finite time nonexistence result with $E(0) < 0$. Then, Georgiev and Todorova [10], in the case $m > 2$, improved Levine's result, in 1994. They obtained that, when $m \geq p$, for arbitrary initial data, the global solution exists. Also, they showed that, if $p > m$, solutions with sufficiently $E(0) < 0$ blow up in finite time. Then, Messaoudi [19] extended the result of Georgiev and Todorova in 2001. He obtained blow up of solutions in a finite time with $E(0) < 0$.

In [5], Antontsev et al. studied the following Petrovsky equation with variable exponents

$$u_{tt} + \Delta^2 u - \Delta u_t + |u_t|^{p(x)-2} u_t = |u|^{q(x)-2} u. \tag{6}$$

They obtained the local existence and nonexistence result of (6).

When $m = 1$, Messaoudi and Kafini [20] looked into the equation as follows

$$u_{tt} - \Delta u + \mu_1 u_t |u_t|^{q(x)-2} + \mu_2 u_t (x, t-\tau) |u_t|^{q(x)-2} (x, t-\tau) = bu |u|^{p(x)-2}. \tag{7}$$

They established the global nonexistence and decay results for equation (7).

In recent years, some other authors investigated the related equations (see [2, 6, 13, 14, 16, 23–37, 39]).

Motivated by the above studies, in this work, we study the nonexistence results for the problem (1) under appropriate conditions. There is no research, to our best knowledge, related to higher-order wave equations with delay term and variable exponents; hence, our paper is generalization of the previous ones. Our main aim is to get the nonexistence results of the Eq. (1) with delay term and variable exponents.

The plan of this paper is as follows. Firstly, in Sect. 2, the definition of the variable exponent Sobolev $W^{1,p(\cdot)}(\Omega)$ and Lebesgue $L^{p(\cdot)}(\Omega)$ spaces are introduced. In Sect. 3, we establish the nonexistence of solutions for $E(0) < 0$.

2 Preliminaries

In this part, we state the results related to Lebesgue $L^{p(\cdot)}(\Omega)$ and Sobolev $W^{1,p(\cdot)}(\Omega)$ spaces with variable exponents (see [4, 9]).

Assume that $p : \Omega \to [1, \infty)$ be a measurable function. We define the variable exponent Lebesgue space with a variable exponent $p(\cdot)$

$$L^{p(\cdot)}(\Omega) = \left\{ u : \Omega \to R;\ \text{measurable in } \Omega\ : \int_\Omega |u|^{p(\cdot)}\, dx < \infty \right\},$$

with a Luxemburg-type norm

$$\|u\|_{p(\cdot)} = \inf \left\{ \lambda > 0 : \int_\Omega \left| \frac{u}{\lambda} \right|^{p(x)} dx \leq 1 \right\}.$$

Equipped with this norm, $L^{p(\cdot)}(\Omega)$ is a Banach space (see [9]).

Next, we define the variable exponent Sobolev space $W^{1,p(\cdot)}(\Omega)$ as following:

$$W^{1,p(\cdot)}(\Omega) = \left\{ u \in L^{p(\cdot)}(\Omega) : \nabla u \text{ exists and } |\nabla u| \in L^{p(\cdot)}(\Omega) \right\}.$$

Variable exponent Sobolev space with the norm

$$\|u\|_{1,p(\cdot)} = \|u\|_{p(\cdot)} + \|\nabla u\|_{p(\cdot)}$$

is a Banach space. $W_0^{1,p(\cdot)}(\Omega)$ is the space which is defined as the closure of $C_0^\infty(\Omega)$ in $W^{1,p(\cdot)}(\Omega)$. For $u \in W_0^{1,p(\cdot)}(\Omega)$, we can define an equivalent norm

$$\|u\|_{1,p(\cdot)} = \|\nabla u\|_{p(\cdot)}.$$

The dual of $W_0^{1,p(\cdot)}(\Omega)$ is defined as $W_0^{-1,p'(\cdot)}(\Omega)$, similar to Sobolev spaces, where $\frac{1}{p(\cdot)} + \frac{1}{p'(\cdot)} = 1$.

Meanwhile, we define Sobolev space $H^m(\Omega)$, such that

$$H_0^m(\Omega) = \left\{ u \in H^m(\Omega) : \frac{\partial^i u}{\partial v^i} = 0,\ i = 0, 1, \cdots, m-1 \right\},$$

where v is the unit outward normal on $\partial\Omega$ and $\frac{\partial^i u}{\partial v^i}$ shows the i-order normal derivation of u.

As usual, $(.,.)$ and $\|.\|_{p(\cdot)}$ show the inner product in the space $L^2(\Omega)$ and the norm of the space $L^{p(\cdot)}(\Omega)$, respectively. For brevity, we denote $\|.\|_2 = \|.\|_{L^2(\Omega)}$ and $\|.\|_{p(\cdot)} = \|.\|_{L^{p(\cdot)}(\Omega)}$ for $p > 1$.

We also suppose that:

$$|q(x) - q(y)| \leq -\frac{A}{\log|x-y|} \text{ and } |p(x) - p(y)| \leq -\frac{B}{\log|x-y|} \text{ for all } x, y \in \Omega, \tag{8}$$

$A, B > 0$ and $0 < \delta < 1$ with $|x-y| < \delta$. (log-Hölder condition)

Lemma 1 *[1, 38] (Sobolev-Poincarė inequality) Suppose that $p(\cdot)$ satisfies (2). Then, $H_0^m(\Omega) \hookrightarrow L^{p(\cdot)}(\Omega)$ and*

$$\|u\|_{p(\cdot)} \leq c \left\|D^m u\right\|_{p(\cdot)} \textit{ for all } u \in H_0^m(\Omega),$$

where $c = c\left(p^-, p^+, |\Omega|\right) > 0$.

Lemma 2 *[3] If $p^+ < \infty$ and $p : \Omega \to [1, \infty)$ is a measurable function, then $C_0^\infty(\Omega)$ is dense in $L^{p(\cdot)}(\Omega)$.*

Lemma 3 *[3] (Hölder' inequality) Let $p, q, s \geq 1$ be measurable functions defined on Ω and*

$$\frac{1}{s(y)} = \frac{1}{p(y)} + \frac{1}{q(y)}, \text{ for a.e. } y \in \Omega,$$

satisfies. If $f \in L^{p(\cdot)}(\Omega)$ *and* $g \in L^{q(\cdot)}(\Omega)$, *then* $fg \in L^{s(\cdot)}(\Omega)$ *and*

$$\|fg\|_{s(\cdot)} \leq 2\|f\|_{p(\cdot)} \|g\|_{q(\cdot)}.$$

Lemma 4 *[3] (Unit ball property) Suppose that* $p \geq 1$, *then*

$$\|f\|_{p(\cdot)} \leq 1 \text{ if and only if } \varrho_{p(\cdot)}(f) \leq 1,$$

where

$$\varrho_{p(\cdot)}(f) = \int_{\Omega} |f(x)|^{p(x)} dx.$$

Lemma 5 *[4] If* $p \geq 1$ *is a measurable function on* Ω, *then*

$$\min\left\{\|u\|_{p(\cdot)}^{p^-}, \|u\|_{p(\cdot)}^{p^+}\right\} \leq \varrho_{p(\cdot)}(u) \leq \max\left\{\|u\|_{p(\cdot)}^{p^-}, \|u\|_{p(\cdot)}^{p^+}\right\},$$

for any $u \in L^{p(\cdot)}(\Omega)$ *and for a.e.* $x \in \Omega$.

3 Nonexistence of Solutions

In this part, for the case $b > 0$, we establish the nonexistence of solutions for the problem (1). Now, as in the work of [21], we introduce the new function

$$z(x, \rho, t) = u_t(x, t - \tau\rho), \ x \in \Omega, \ \rho \in (0, 1), \ t > 0,$$

which implies that

$$\tau z_t(x, \rho, t) + z_\rho(x, \rho, t) = 0, \ x \in \Omega, \ \rho \in (0, 1), \ t > 0.$$

Consequently, problem (1) can be transformed as follows:

$$\begin{cases} u_{tt} + (-\Delta)^m u + \mu_1 u_t(x, t) |u_t(x, t)|^{q(x)-2} & \\ +\mu_2 z(x, 1, t) |z(x, 1, t)|^{q(x)-2} = bu|u|^{p(x)-2} & \text{in } \Omega \times (0, \infty), \\ \tau z_t(x, \rho, t) + z_\rho(x, \rho, t) = 0 & \text{in } \Omega \times (0, 1) \times (0, \infty), \\ z(x, \rho, 0) = f_0(x, -\rho\tau) & \text{in } \Omega \times (0, 1), \\ u(x, t) = 0 & \text{on } \partial\Omega \times [0, \infty), \\ u(x, 0) = u_0(x), \ u_t(x, 0) = u_1(x) & \text{in } \Omega. \\ D^\alpha u(x, t) = 0, \ |\alpha| \leq m - 1 & \text{in } x \in \partial\Omega. \end{cases} \tag{9}$$

We define the energy functional of (9) as

$$E(t)=\frac{1}{2}\|u_t\|^2+\frac{1}{2}\left\|A^{1/2}u\right\|^2+\int_0^1\int_\Omega \frac{\xi(x)\,|z(x,\rho,t)|^{q(x)}}{q(x)}\mathrm{d}x\mathrm{d}\rho-b\int_\Omega \frac{|u|^{p(x)}}{p(x)}\mathrm{d}x, \tag{10}$$

for $t\geq 0$, where ξ is a continuous function satisfies

$$\tau\,|\mu_2|\,(q(x)-1)<\xi(x)<\tau\,(\mu_1 q(x)-|\mu_2|)\,,\, x\in\overline{\Omega}. \tag{11}$$

The following lemma gives that, under the condition $\mu_1>|\mu_2|$, $E(t)$ is decreasing.

Lemma 6 *Let* (u,z) *be a solution of (9), such that*

$$E'(t)\leq -C_0\int_\Omega \left(|u_t|^{q(x)}+|z(x,1,t)|^{q(x)}\right)dx\leq 0, \tag{12}$$

for some $C_0>0$.

Proof Multiplying the first eq. in (9) by u_t, integrating over Ω, then multiplying the second eq. of (9) by $\frac{1}{\tau}\xi(x)\,|z|^{q(x)-2}\,z$ and integrating over $\Omega\times(0,1)$, summing up, we get

$$\begin{aligned}&\frac{\mathrm{d}}{\mathrm{d}t}\left[\frac{1}{2}\|u_t\|^2+\frac{1}{2}\left\|A^{1/2}u\right\|^2+\int_0^1\int_\Omega \frac{\xi(x)\,|z(x,\rho,t)|^{q(x)}}{q(x)}\mathrm{d}x\mathrm{d}\rho-b\int_\Omega \frac{|u|^{p(x)}}{p(x)}\mathrm{d}x\right]\\ &=-\mu_1\int_\Omega |u_t|^{q(x)}\,\mathrm{d}x-\frac{1}{\tau}\int_\Omega\int_0^1 \xi(x)\,|z(x,\rho,t)|^{q(x)-2}\,zz_\rho(x,\rho,t)\,\mathrm{d}\rho\mathrm{d}x\\ &\quad-\mu_2\int_\Omega u_t z(x,1,t)\,|z(x,1,t)|^{q(x)-2}\,\mathrm{d}x.\end{aligned} \tag{13}$$

Now, we estimate the last two terms of the right-hand side of (13) as,

$$\begin{aligned}&-\frac{1}{\tau}\int_\Omega\int_0^1 \xi(x)\,|z(x,\rho,t)|^{q(x)-2}\,zz_\rho(x,\rho,t)\,d\rho dx\\ &=-\frac{1}{\tau}\int_\Omega\int_0^1 \frac{\partial}{\partial\rho}\left(\frac{\xi(x)\,|z(x,\rho,t)|^{q(x)}}{q(x)}\right)d\rho dx\\ &=\frac{1}{\tau}\int_\Omega \frac{\xi(x)}{q(x)}\left(|z(x,0,t)|^{q(x)}-|z(x,1,t)|^{q(x)}\right)dx\\ &=\int_\Omega \frac{\xi(x)}{\tau q(x)}\,|u_t|^{q(x)}\,\mathrm{d}x-\int_\Omega \frac{\xi(x)}{\tau q(x)}\,|z(x,1,t)|^{q(x)}.\end{aligned}$$

Using the Young's inequality, $r=\frac{q(x)}{q(x)-1}$ and $r'=q(x)$ for the last term to obtain

$$|u_t|\,|z(x,1,t)|^{q(x)-1}\leq\frac{1}{q(x)}\,|u_t|^{q(x)}+\frac{q(x)-1}{q(x)}\,|z(x,1,t)|^{q(x)}.$$

Consequently, we deduce that

$$-\mu_2 \int_\Omega u_t z \, |z(x,1,t)|^{q(x)-2} \, dx$$
$$\leq |\mu_2| \left(\int_\Omega \frac{1}{q(x)} |u_t(t)|^{q(x)} \, dx + \int_\Omega \frac{q(x)-1}{q(x)} |z(x,1,t)|^{q(x)} \, dx \right).$$

So,

$$\frac{dE(t)}{dt} \leq -\int_\Omega \left[\mu_1 - \left(\frac{\xi(x)}{\tau q(x)} + \frac{|\mu_2|}{q(x)} \right) \right] |u_t(t)|^{q(x)} \, dx$$
$$- \int_\Omega \left(\frac{\xi(x)}{\tau q(x)} - \frac{|\mu_2|(q(x)-1)}{q(x)} \right) |z(x,1,t)|^{q(x)} \, dx.$$

As a result, for all $x \in \overline{\Omega}$, the relation (11) satisfies,

$$f_1(x) = \mu_1 - \left(\frac{\xi(x)}{\tau q(x)} + \frac{|\mu_2|}{q(x)} \right) > 0,$$

$$f_2(x) = \frac{\xi(x)}{\tau q(x)} - \frac{|\mu_2|(q(x)-1)}{q(x)} > 0.$$

Since $q(x)$, and hence $\xi(x)$, is bounded, we infer that $f_1(x)$ and $f_2(x)$ are also bounded. So, if we define

$$C_0(x) = \min\{f_1(x), f_2(x)\} > 0 \text{ for any } x \in \overline{\Omega},$$

and take $C_0(x) = \inf_{\overline{\Omega}} C_0(x)$, then $C_0(x) \geq C_0 > 0$. Hence,

$$E'(t) \leq -C_0 \left[\int_\Omega |u_t(t)|^{q(x)} \, dx + \int_\Omega |z(x,1,t)|^{q(x)} \, dx \right] \leq 0.$$

Setting

$$H(t) = -E(t), \tag{14}$$

hence,

$$H'(t) = -E'(t) \geq 0,$$

$$0 < H(0) \leq H(t) \leq b \int_\Omega \frac{|u|^{p(x)}}{p(x)} dx \leq \frac{b}{p^-} \varrho(u), \tag{15}$$

where

$$\varrho(u) = \varrho_{p(\cdot)}(u) = \int_\Omega |u|^{p(x)} \, dx.$$

Similar to the work of Messaoudi et al. [17], we give the following technical lemma without proof.

Lemma 7 *Suppose that condition (2) holds, such that*

$$\varrho^{s/p^-}(u) \le C\left(\left\|A^{1/2}u\right\|^2 + \varrho(u)\right), \tag{16}$$

where $C > 1$ *depending on* Ω *only. Then, we have following inequalities: (i)*

$$\|u\|_{p^-}^s \le C\left(\left\|A^{1/2}u\right\|^2 + \|u(t)\|_{p^-}^{p^-}\right), \tag{17}$$

(ii)

$$\varrho^{s/p^-}(u) \le C\left(|H(t)| + \|u_t\|^2 + \varrho(u) + \int_0^1 \int_\Omega \frac{\xi(x)\,|z(x,\rho,t)|^{q(x)}}{q(x)} dx d\rho\right), \tag{18}$$

(iii)

$$\|u\|_{p^-}^s \le C\left(|H(t)| + \|u_t\|^2 + \|u\|_{p^-}^{p^-} + \int_0^1 \int_\Omega \frac{\xi(x)\,|z(x,\rho,t)|^{q(x)}}{q(x)} dx d\rho\right), \tag{19}$$

for any $u \in H_0^m(\Omega)$ *and* $2 \le s \le p^-$. *Let* (u, z) *be a solution of (9), then iv)*

$$\varrho(u) \ge C\,\|u\|_{p^-}^{p^-}, \tag{20}$$

(v)

$$\int_\Omega |u|^{q(x)}\, dx \le C\left(\varrho^{q^-/p^-}(u) + \varrho^{q^+/p^-}(u)\right). \tag{21}$$

The nonexistence result is given by the theorem as follows:

Theorem 1 *Let conditions (2) and (8) be provided and assume that* $E(0) < 0$. *Then, the solution (9) blows up in finite time* T^*, *and*

$$T^* \le \frac{1-\alpha}{\Psi\alpha\,[L(0)]^{\alpha/(1-\alpha)}},$$

where $L(t)$ *and* α *are given in (22) and (23), respectively.*

Proof Define

$$L(t) = H^{1-\alpha}(t) + \varepsilon \int_\Omega u u_t \mathrm{d}x, \tag{22}$$

where ε small to be chosen later and

$$0 \le \alpha \le \min\left\{\frac{p^- - 2}{2p^-}, \frac{p^- - q^+}{p^-(q^+ - 1)}\right\}. \tag{23}$$

Differentiation $L(t)$ with respect to t, and taking into consideration the first equation in (9), we have

$$L'(t) = (1-\alpha) H^{-\alpha}(t) H'(t) + \varepsilon \left\| u_t^2 \right\| - \varepsilon \left\| A^{1/2} u \right\|^2 + \varepsilon b \int_\Omega |u|^{p(x)} \mathrm{d}x - \varepsilon \mu_1 \int_\Omega u u_t(x,t) |u_t(x,t)|^{q(x)-2} dx - \varepsilon \mu_2 \int_\Omega u z(x,1,t) |z(x,1,t)|^{q(x)-2} dx.$$

By using the definition of the $H(t)$ and for $0 < a < 1$, such that

$$\begin{aligned} L'(t) \geq{} & C_0 (1-\alpha) H^{-\alpha}(t) \left[\int_\Omega |u_t|^{q(x)} \mathrm{d}x + \int_\Omega |z(x,1,t)|^{q(x)} \mathrm{d}x \right] \\ & + \varepsilon \left((1-a) p^- H(t) + \frac{(1-a) p^-}{2} \|u_t\|^2 + \frac{(1-a) p^-}{2} \left\| A^{1/2} u \right\|^2 \right) \\ & + \varepsilon (1-a) p^- \int_0^1 \int_\Omega \frac{\xi(x) |z(x,\rho,t)|^{q(x)}}{q(x)} \mathrm{d}x \mathrm{d}\rho \\ & + \varepsilon \left\| u_t^2 \right\| - \varepsilon \left\| A^{1/2} u \right\|^2 + \varepsilon a b \int_\Omega |u|^{p(x)} \mathrm{d}x \\ & - \varepsilon \mu_1 \int_\Omega u u_t(x,t) |u_t(x,t)|^{q(x)-2} \mathrm{d}x \\ & - \varepsilon \mu_2 \int_\Omega u z(x,1,t) |z(x,1,t)|^{q(x)-2} \mathrm{d}x. \end{aligned}$$

Hence,

$$\begin{aligned} L'(t) \geq{} & C_0 (1-\alpha) H^{-\alpha}(t) \left[\int_\Omega |u_t|^{q(x)} \mathrm{d}x + \int_\Omega |z(x,1,t)|^{q(x)} \mathrm{d}x \right] \\ & + \varepsilon (1-a) p^- H(t) + \varepsilon \frac{(1-a) p^- + 2}{2} \|u_t\|^2 + \varepsilon \frac{(1-a) p^- - 2}{2} \left\| A^{1/2} u \right\|^2 \\ & + \varepsilon (1-a) p^- \int_0^1 \int_\Omega \frac{\xi(x) |z(x,\rho,t)|^{q(x)}}{q(x)} dx d\rho + \varepsilon a b \varrho(u) \\ & - \varepsilon \mu_1 \int_\Omega u u_t(x,t) |u_t(x,t)|^{q(x)-2} dx \\ & - \varepsilon \mu_2 \int_\Omega u z(x,1,t) |z(x,1,t)|^{q(x)-2} dx. \end{aligned} \tag{24}$$

Utilizing Young's inequality, we get

$$\int_{\Omega} |u_t|^{q(x)-1} |u| \, dx \leq \frac{1}{q^-} \int_{\Omega} \delta^{q(x)} |u|^{q(x)} \, dx + \frac{q^+ - 1}{q^+} \int_{\Omega} \delta^{-\frac{q(x)}{q(x)-1}} |u_t|^{q(x)} \, \mathrm{d}x \tag{25}$$

and

$$\int_{\Omega} |z(x,1,t)|^{q(x)-1} |u| \, dx \leq \frac{1}{q^+} \int_{\Omega} \delta^{q(x)} |u|^{q(x)} \, \mathrm{d}x + \frac{q^+ - 1}{q^+} \int_{\Omega} \delta^{-\frac{q(x)}{q(x)-1}} |z(x,1,t)|^{q(x)} \, \mathrm{d}x. \tag{26}$$

Similar to [17], estimates (25) and (26) remain valid if δ is time-dependent. Let us choose δ such that

$$\delta^{-\frac{q(x)}{q(x)-1}} = kH^{-\alpha}(t),$$

where $k \geq 1$ is specified later, we get

$$\int_{\Omega} \delta^{-\frac{q(x)}{q(x)-1}} |u_t|^{q(x)} \, dx = kH^{-\alpha}(t) \int_{\Omega} |u_t|^{q(x)} \, \mathrm{d}x, \tag{27}$$

$$\int_{\Omega} \delta^{-\frac{q(x)}{q(x)-1}} |z(x,1,t)|^{q(x)} \, \mathrm{d}x = kH^{-\alpha}(t) |z(x,1,t)|^{q(x)} \, dx \tag{28}$$

and

$$\int_{\Omega} \delta^{q(x)} |u|^{q(x)} \, \mathrm{d}x = \int_{\Omega} k^{1-q(x)} H^{\alpha(q(x)-1)}(t) |u|^{q(x)} \, \mathrm{d}x \leq \int_{\Omega} k^{1-q^-} H^{\alpha(q^+-1)}(t) \int_{\Omega} |u|^{q(x)} \, \mathrm{d}x. \tag{29}$$

By using (20) and (21), we obtain

$$H^{\alpha(q^+-1)}(t) \int_{\Omega} |u|^{q(x)} \, \mathrm{d}x \leq C \left[(\varrho(u))^{q^-/p^- + \alpha(q^+-1)} + (\varrho(u))^{q^+/p^- + \alpha(q^+-1)} \right]. \tag{30}$$

From (23), we deduce that

$$s = q^- + \alpha p^- \left(q^+ - 1\right) \leq p^- \text{ and } s = q^+ + \alpha p^- \left(q^+ - 1\right) \leq p^-.$$

Then, by using lemma 7, yields

$$H^{\alpha(q^+-1)}(t) \int_{\Omega} |u|^{q(x)} \, dx \leq C \left(\left\| A^{1/2} u \right\|^2 + \varrho(u) \right). \tag{31}$$

Combining (25)-(31), we get

$$\begin{aligned}
L'(t) \geq\ & (1-\alpha) H^{-\alpha}(t)\left[C_0 - \varepsilon\left(\frac{q^+-1}{q^+}\right)ck\right]\int_\Omega |u_t|^{q(x)}\,dx \\
& + (1-\alpha) H^{-\alpha}(t)\left[C_0 - \varepsilon\left(\frac{q^+-1}{q^+}\right)ck\right]\int_\Omega |z(x,1,t)|^{q(x)}\,dx \\
& + \varepsilon\left(\frac{(1-a)p^- - 2}{2} - \frac{C}{q^- k^{1-q^-}}\right)\left\|A^{1/2}u\right\|^2 \\
& + \varepsilon(1-a)p^- H(t) + \varepsilon\frac{(1-a)p^- + 2}{2}\|u_t\|^2 \\
& + \varepsilon\left(ab - \frac{C}{q^- k^{1-q^-}}\right)\varrho(u) \\
& + \varepsilon(1-a)p^- \int_0^1\int_\Omega \frac{\xi(x)\,|z(x,\rho,t)|^{q(x)}}{q(x)}\mathrm{d}x\mathrm{d}\rho. \qquad (32)
\end{aligned}$$

Let us choose a small enough such that

$$\frac{(1-a)p^- + 2}{2} > 0$$

and k large enough so that

$$\frac{(1-a)p^- - 2}{2} - \frac{C}{q^- k^{1-q^-}} > 0 \text{ and } ab - \frac{C}{q^- k^{1-q^-}} > 0.$$

Once k and a are fixed, picking ε small enough such that

$$C_0 - \varepsilon\left(\frac{q^+-1}{q^+}\right)ck > 0$$

and

$$L(0) = H^{1-\alpha}(0) + \varepsilon\int_\Omega u_0 u_1 \mathrm{d}x > 0.$$

Consequently, (32) yields

$$L'(t) \geq \varepsilon\eta\left[H(t) + \|u_t\|^2 + \left\|A^{1/2}u\right\|^2 + \varrho(u) + \int_0^1\int_\Omega \frac{\xi(x)\,|z(x,\rho,t)|^{q(x)}}{q(x)}\mathrm{d}x\mathrm{d}\rho\right] \qquad (33)$$

for a constant $\eta > 0$. Hence, we get

$$L(t) \geq L(0) > 0, \forall t \geq 0.$$

Now, for some constants $\sigma, \Gamma > 0$, we denote

$$L'(t) \geq \Gamma L^{\sigma}(t).$$

Utilizing the Hölder inequality, we get

$$\left|\int_{\Omega} uu_t dx\right|^{1/(1-\alpha)} \leq C \|u\|_{p^-}^{1/(1-\alpha)} \|u_t\|_2^{1/(1-\alpha)},$$

and by using Young's inequality gives

$$\left|\int_{\Omega} uu_t \mathrm{d}x\right|^{1/(1-\alpha)} \leq C\left[\|u\|_{p^-}^{mu/(1-\alpha)} + \|u_t\|_2^{\Theta/(1-\alpha)}\right],$$

where $1/\mu + 1/\Theta = 1$. From (23), the choice of $\Theta = 2(1-\alpha)$ will make $\mu/(1-\alpha) = 2/(1-2\alpha) \leq p^-$. Hence,

$$\left|\int_{\Omega} uu_t \mathrm{d}x\right|^{1/(1-\alpha)} \leq C\left[\|u\|_{p^-}^{s} + \|u_t\|^2\right],$$

where $s = \mu/(1-\alpha)$. From (19), we have

$$\left|\int_{\Omega} uu_t \mathrm{d}x\right|^{1/(1-\alpha)} \leq C\left[|H(t)| + \|u_t\|^2 + \varrho(u) + \int_0^1 \int_{\Omega} \frac{\xi(x)|z(x,\rho,t)|^{q(x)}}{q(x)} dxd\rho\right]. \tag{34}$$

Hence, we get

$$\begin{aligned} L^{1/(1-\alpha)}(t) &= \left[H^{(1-\alpha)}(t) + \varepsilon \int_{\Omega} uu_t dx\right]^{1/(1-\alpha)} \\ &\leq 2^{\alpha/(1-\alpha)}\left[H(t) + \left|\int_{\Omega} uu_t \mathrm{d}x\right|^{1/(1-\alpha)}\right] \\ &\leq C\Big[H(t) + \|u_t\|^2 + \varrho(u) \\ &\quad + \int_0^1 \int_{\Omega} \frac{\xi(x)|z(x,\rho,t)|^{q(x)}}{q(x)} \mathrm{d}x\mathrm{d}\rho\Big]. \end{aligned} \tag{35}$$

So, for some $\Psi > 0$, from (33), we arrive

$$L'(t) \geq \Psi L^{1/(1-\alpha)}(t). \tag{36}$$

A simple integration of (36) over $(0, t)$ yields

$$L^{\alpha/(1-\alpha)}(t) \geq \frac{1}{L^{-\alpha/(1-\alpha)}(0) - \Psi\alpha t/(1-\alpha)},$$

which implies that the solution blows up in a finite time T^*,

$$T^* \leq \frac{1-\alpha}{\Psi\alpha\,[L(0)]^{\alpha/(1-\alpha)}}.$$

As a result, the proof is completed.

4 Conclusions

Recently, there has been published much work related to wave equations with time delay. To our best knowledge, there were no nonexistence results for the higher-order wave equation with delay and variable exponents. We have been obtained the nonexistence of solutions, under the sufficient conditions in a bounded domain.

Acknowledgements The authors are grateful to DUBAP (ZGEF.20.009) for research funds.

References

1. Adams RA (1975) Sobolev spaces. Pure Appl Math 65 (Academic Press, New York–London)
2. Abd-Elhameed WM, Machado JAT, Youssri YH (2021) Hypergeometric fractional derivatives formula of shifted Chebyshev polynomials: tau algorithm for a type of fractional delay differential equations. Int J Nonlinear Sci Numer Simul. https://doi.org/10.1515/ijnsns-2020-0124
3. Antontsev S (2011) Wave equation with $p(x, t)$-Laplacian and damping term: blow-up of solutions. C R Mecanique 339:751–755
4. Antontsev S (2011) Wave equation with $p(x, t)$-Laplacian and damping term: existence and blow-up. Differ Equ Appl 3:503–525
5. Antontsev S, Ferreira J, Pişkin E (2021) Existence and blow up of solutions for a strongly damped Petrovsky equation with variable-exponent nonlinearities. Electron J Differ Equ 6:1–18
6. Antontsev S, Ferreira J, Pişkin E, Yüksekkaya H, Shahrouzi M (2021) Blow up and asymptotic behavior of solutions for a p(x)-Laplacian equation with delay term and variable exponents. Electron J Differ Equ (84):1–20
7. Ball J (1977) Remarks on blow-up and nonexistence theorems for nonlinear evolution equations. Quart J Math 28:473–486
8. Chen Y, Levine S, Rao M (2006) Variable exponent, linear growth functionals in image restoration. SIAM J Appl Math 66:1383–1406
9. Diening L, Hasto P, Harjulehto P, Ruzicka MM (2011) Lebesgue and Sobolev spaces with variable exponents. Springer, Heidelberg
10. Georgiev V, Todorova G (1994) Existence of a solution of the wave equation with nonlinear damping and source terms. J Differ Equ 109:295–308

11. Ghegal S, Hamchi I, Messaoudi SA (2018) Global existence and stability of a nonlinear wave equation with variable-exponent nonlinearities. Appl Anal 1–11
12. Haraux A, Zuazua E (1988) Decay estimates for some semilinear damped hyperbolic problems. Arch Ration Mech Anal 150:191–206
13. Hafez RM, Youssri YH (2022) Shifted Gegenbauer-Gauss collocation method for solving fractional neutral functional-differential equations with proportional delays. Kragujev J Math 46(6)
14. He JH, Qie N, He CH (2021) Solitary waves travelling along an unsmooth boundary. Results Phys 24:104104
15. Levine H (1974) Some additional remarks on the nonexistence of global solutions to nonlinear wave equations. SIAM J Math Anal 5:138–146
16. Messaoudi SA, Talahmeh AA (2017) A blow-up result for a nonlinear wave equation with variable-exponent nonlinearities. Appl Anal 96:1509–1515
17. Messaoudi SA, Talahmeh AA, Al-Smail JH (2017) Nonlinear damped wave equation: existence and blow-up. Comput Math Appl 74:3024–3041
18. Kafini M, Messaoudi SA (2016) A blow-up result in a nonlinear wave equation with delay. Mediterr J Math 13:237–247
19. Messaoudi S (2001) Blow up in a nonlinearly damped wave equation. Math Nachr 231:1–7
20. Messaoudi SA, Kafini M (2019) On the decay and global nonexistence of solutions to a damped wave equation with variable-exponent nonlinearity and delay. Ann Pol Math 122(1). https://doi.org/10.4064/ap180524-31-10
21. Nicaise S, Pignotti C (2006) Stability and instability results of the wave equation with a delay term in the boundary or internal feedbacks. SIAM J Control Optim 45:1561–1585
22. Pişkin E (2019) Blow up of solutions for a class of nonlinear higher-order wave equation with variable exponents. Sigma J Eng Nat Sci 10(2):149–156
23. Pişkin E, Yüksekkaya H (2018) Non-existence of solutions for a Timoshenko equations with weak dissipation. Math Morav 22(2):1–9
24. Pişkin E, Yüksekkaya H (2018) Mathematical behavior of the solutions of a class of hyperbolic-type equation. J BAUN Inst Sci Technol 20(3):117–128
25. Pişkin E, Yüksekkaya H (2018) Blow up of solutions for a Timoshenko equation with damping terms. Middle East J Sci 4(2):70–80
26. Pişkin E, Yüksekkaya H (2020) Global attractors for the higher-order evolution equation. AMNS 5(1):195–210
27. Pişkin E, Yüksekkaya H (2020) Decay of solutions for a nonlinear Petrovsky equation with delay term and variable exponents. Aligarh Bull Math 39(2):63–78
28. Pişkin E, Yüksekkaya H (2021) Local existence and blow up of solutions for a logarithmic nonlinear viscoelastic wave equation with delay. Comput Methods Differ Equ 9(2):623–636
29. Pişkin E, Yüksekkaya H (2021) Blow-up of solutions for a logarithmic quasilinear hyperbolic equation with delay term. J Math Anal 12(1):56–64
30. Pişkin E, Yüksekkaya H (2021c) Blow up of solution for a viscoelastic wave equation with m-Laplacian and delay terms. Tbil Math J SI(7):21–32
31. Pişkin E, Yüksekkaya H (2021) Blow up of solutions for Petrovsky equation with delay term. J Nepal Math Soc 4(1):76–84
32. Pişkin E, Yüksekkaya H, Mezouar N (2021) Growth of solutions for a coupled Viscoelastic Kirchhoff system with distributed delay terms. Menemui Matematik (Discovering Mathematics) 43(1):26–38
33. Pişkin E, Yüksekkaya H (2021) Blow-up results for a viscoelastic plate equation with distributed delay. J Univ Math 4(2):128–139
34. Pişkin E, Yüksekkaya H (2021) Nonexistence of global solutions for a Kirchhoff-type viscoelastic equation with distributed delay. J Univ Math 4(2):271–282
35. Yüksekkaya H, Pişkin E, Boulaaras SM, Cherif BB, Zubair SA (2021) Existence, nonexistence, and stability of solutions for a delayed plate equation with the logarithmic source. Adv Math Phys 2021:1–11

36. Yüksekkaya H, Pişkin E, Boulaaras SM, Cherif BB (2021b) Existence, decay and blow-up of solutions for a higher-order Kirchhoff-type equation with delay term. J Funct Spaces 1-11. Article ID 4414545
37. Yüksekkaya H, Pişkin E (2021) Nonexistence of solutions for a logarithmic m-Laplacian type equation with delay term. Konuralp J Math 9(2):238–244
38. Pişkin E (2017) Sobolev spaces. Seckin Publishing (in Turkish)
39. Sahu PK, Ray SS (2017) A new Bernoulli wavelet method for accurate solutions of nonlinear fuzzy Hammerstein-Volterra delay integral equations. Fuzzy Sets Syst 309:131–144

Existence and Uniqueness of Mass Conserving Solutions to the Coagulation, Multi-fragmentation Equations with Compactly Supported Kernels

Arijit Das and Jitraj Saha

Abstract In this article, the existence result of a solution to continuous nonlinear, initial value problem is studied. In particular, we consider a special type of problem representing the time evolution of particle number density due to the coagulation, multi-fragmentation events among the particles present in a system. The existence theorem is proved with the kinetic kernels having compact support. The proof of the main theorem is based on the contraction mapping principle. Furthermore, the mass conservation property of the existed solution is also investigated.

Keywords Coagulation · Multi-fragmentation · Existence · Uniqueness · Mass conservation

2020 AMS Classification Primary 35Q70 · 45K05 · 45G05 · Secondary 47J35

1 Introduction

We consider the general nonlinear initial value problem representing particle coagulation and multi-fragmentation

$$\frac{\partial f(x,t)}{\partial t} = \frac{1}{2}\int_0^x K(x-y,y)f(x-y,t)f(y,t)\mathrm{d}y - f(x,t)\int_0^\infty K(x,y)f(y,t)\mathrm{d}y + \int_x^\infty b(x,y)S(y)f(y,t)\mathrm{d}y - S(x)f(x,t) \tag{1}$$

A. Das · J. Saha (✉)
Department of Mathematics, National Institute of Technology Tiruchirappalli, Tiruchirappalli, Tamil Nadu 620015, India
e-mail: jitraj@nitt.edu

S. S. Ray et al. (eds.), *Applied Analysis, Computation and Mathematical Modelling in Engineering*, Lecture Notes in Electrical Engineering 897,
https://doi.org/10.1007/978-981-19-1824-7_22

supported with the initial data,

$$f(x,0) = f_0(x)\,(\geq 0), \quad \text{for all } x > 0. \tag{2}$$

In Eq. (1), the function $f(x,t)$ denotes the distribution of particles of volume $x > 0$ at time $t \geq 0$. The coagulation kernel $K(x,y)$ represents the rate of coagulation (or aggregation) of particles with volume x and y to form larger agglomerates. The selection rate of particles of volume y for fragmenting into smaller particles is denoted by the function $S(y)$, and the distribution of daughter particles x formed due to the fragmentation of y is represented by the breakage function $b(x,y)$. Further, the first and the third terms of Eq. (1) denote the formation or *birth* of particles with volume x in the system. On the other hand, the second and the fourth terms denote the removal or *death* of particles having volume x from the system. In general, each of $K(x,y)$, $S(x)$ and $b(x,y)$ is nonnegative functions. Moreover, the coagulation kernel $K(x,y)$ is considered to be a symmetric function of x, y and the breakage function satisfies the relations

$$b(x,y) = 0 \quad \text{for all} \quad x \geq y, \quad \int_0^y b(x,y)\mathrm{d}x = \nu(y), \quad \text{and} \quad \int_0^y xb(x,y)\mathrm{d}x = y. \tag{3}$$

The first integral in (3) represents the total number of fragments $\nu(y)$ produced during the breakage event of particle volume y, and the second integral represents the volume conservative property of system during particle fragmentation. Owing to the multi-fragmentation events, two or more particle fragments are produced during fragmentation process and are considered to be bounded above by a positive, finite quantity $\nu(\geq 2)$.

The study of *coagulation-fragmentation* (CF) equations with different forms of kernels is of great interest in the fields of engineering sciences (e.g., chemical engineering, pharmaceutical, food processing, atmospheric engineering, etc. [1–4]). From theoretical point view, study on existence and uniqueness of solutions for the CF-equations is gathering high importance [5–8]. In most of the above mentioned articles, the authors have proved existence of solutions with a nonsingular, unbounded coagulation kernel. Very recently, [9–12] have studied the existence theory for the problems with singular coagulation kernels. In this regard, the articles of [10, 12] discuss the existence of weak solutions for the CF-equations; on the other hand, the articles [9, 11] discuss the existence of strong solutions for the same.

In the articles [9, 12] dealing with singular coagulation kernel, the authors have directly stated the existence of solution in compact domain. Therefore, the detailed derivation of the existence theory for compactly supported kernels is not available in the literature. The present study can be considered as a prequel of the articles [9, 10] proving the existence of solutions with singular kernels. In this context, the authors [13] proved the existence theory for coagulation with binary fragmentation equations, whereas in this article, we prove the same for coagulation equation with multiple fragmentation.

This article is constructed in the following order. In the next section, we define a functional space which is also forms a Banach space. Also, we state and prove the existence of strong solutions to the initial value problem (IVP) (1)–(2) along with the investigation of the mass conservation law for the existing solution. In Sect. 3, we draw the conclusion of our work and discuss the future scope of research.

2 Existence of Solutions

2.1 Functional Space

Referreing to [9, 10], we construct the following functional space.

For a fixed $T(>0)$, we consider the strip

$$\mathcal{S} := \{(x,t) : 0 < x < \infty, 0 \le t \le T\},$$

and define $\Upsilon_{r,\sigma}(T)$ to be the space of all continuous functions f with the bounded norm

$$\|f\|_{\Upsilon} := \sup_{0\le t\le T} \int_0^\infty \left(x^r + \frac{1}{x^{2\sigma}}\right) |f(x,t)|\, \mathrm{d}x, \quad r \ge 1,\ \sigma \ge 0. \tag{4}$$

Clearly $\Upsilon_{r,\sigma}(T)$ forms a Banach space. Furthermore, let $\Upsilon^+_{r,\sigma}(T)$ denotes the space of all nonnegative functions from $\Upsilon_{r,\sigma}(T)$.

The motivation behind using this norm is because the earlier authors [9, 10] have not proved the existence result for a compact supported kernel. Here, we have provided in details the proof of existence and uniqueness of strong solutions for the CF-equation (1), (2) under the assumptions that the coagulation and fragmentation kernels have compact support.

2.2 Existence of Solutions When Kernels Have Compact Support

Theorem 1 (Existence theorem for compactly supported kernels) *Let the functions $K(x,y)$, $b(x,y)$ and $S(x)$ be nonnegative and continuous for all $x, y \in (0,\infty)$ and the kernels $K(x,y)$ and $S(x)$ have compact support. If the initial data $f_0(x)$ belongs to $\Upsilon^+_{r,\sigma}(0) \cap \Upsilon^+_{0,0}(0)$ and the breakage function $b(x,y)$ satisfying the conditions*

$$\int_0^y x^{-2\sigma} b(x, y)\mathrm{d}x \le \bar{N} y^{-2\sigma} \quad \text{and} \quad 2 \le \nu(y) \le \nu \tag{5}$$

for some positive constant $\bar{N}$ and ν, *then the IVP* (1)–(2) *has at least one solution in* $\Upsilon^+_{r,\sigma}(T)$.

Proof We prove the theorem in the following steps;

- local existence of the solution, that is, there exists a $\tau > 0$ such that the IVP (1)–(2) has at least one solution $f_n \in \Upsilon^+_{r,\sigma}(\tau)$,
- nonnegativity of the local solution,
- global existence of the unique solution to the space $\Upsilon^+_{r,\sigma}(T)$.

Existence of local solution: Let there exists a fixed $R(> 0)$, such that for each $t \in [0, T]$ the kernels $K(x, y)$ and $S(x)$ have compact supports in the intervals $\left[\frac{1}{R}, R\right] \times \left[\frac{1}{R}, R\right]$ and $[0, R]$, respectively. In accordance with Eq. (1), the truncated problem can be written as

$$f(x,t) = f_0(x) + \int_0^t \left[\frac{1}{2} \int_0^x K(x-y, y) f(x-y, s) f(y, s)\mathrm{d}y \right.$$
$$\left. - f(x,s) \int_0^\infty K(x, y) f(y, s)\mathrm{d}y + \int_x^\infty b(x, y) S(y) f(y, s)\mathrm{d}y - S(x) f(x, s) \right] \mathrm{d}s. \tag{6}$$

Hence the solution to (1)–(2) for $x > 2R$ takes the value

$$f(x, t) = f_0(x). \tag{7}$$

The relation (7) gives an estimate of the solution function beyond the right hand side of the compact domain where, the *tails* of the solution $f(x, t)$ actually does not change with time and coincides with the *tails* of the initial data $f_0(x)$. We now proceed to prove the local existence of unique solution for $0 < x \le 2R$.

In this regard, we define the integral operator $\mathcal{G}$ as follows;

$$\mathcal{G}(f)(x, t) := \text{right hand side of equation (6)}.$$

Since K and S have compact supports and f_0 is a nonnegative continuous function therefore, the integral operator $\mathcal{G}$ is well defined on $\Upsilon_{r,\sigma}(\tau)$. We prove this result by using the contraction mapping principle. Let us first prove that for small $\tau > 0$, there exist a closed ball in $\Upsilon_{r,\sigma}(\tau)$ which is invariant relatively to the mapping $\mathcal{G}$. Let $L_0(> 0)$ be a constant such that

$$\|f\|_{\Upsilon}^{(\tau)} := \sup_{0\le t\le \tau} \int_0^{\infty} \left(x^r + \frac{1}{x^{2\sigma}}\right) |f(x,t)| \mathrm{d}x \le L_0. \tag{8}$$

Multiplying both sides of Eq. (6) with the weight $\left(x^r + \frac{1}{x^{2\sigma}}\right)$ and integrating with respect to x, we get

$$\|\mathcal{G}(f)\|_{\Upsilon}^{(\tau)} \le \|f_0\|_{\Upsilon}^{(\tau)} + \int_0^t \Bigg[\frac{1}{2}\int_0^{\infty}\left(x^r + \frac{1}{x^{2\sigma}}\right)\int_0^x K(x-y,y) f(x-y,s) f(y,s)\mathrm{d}y\mathrm{d}x$$
$$- \int_0^{\infty}\int_0^{\infty}\left(x^r + \frac{1}{x^{2\sigma}}\right) K(x,y) f(x,s) f(y,s)\mathrm{d}y\mathrm{d}x$$
$$+ \int_0^{\infty}\left(x^r + \frac{1}{x^{2\sigma}}\right)\left[\int_x^{\infty} b(x,y) S(y) f(y,s)\mathrm{d}y - S(x) f(x,s)\right]\mathrm{d}x\Bigg]\mathrm{d}s.$$

Since K and S have compact support, so their supremum exist. Let $\kappa_0 = \sup_{\frac{1}{R}\le x,y\le R} K(x,y)$ and $\zeta_0 = \sup_{0\le x\le R} S(x)$. Therefore, rearranging the terms in the above relation and using the inequalities

$$(x+y)^r \le 2^r\left(x^r + y^r\right), \quad (x+y)^{-2\sigma} \le x^{-2\sigma} + y^{-2\sigma}, \tag{9}$$

we can find the following estimate

$$\|\mathcal{G}(f)\|_{\Upsilon}^{(\tau)} \le \|f_0\|_{\Upsilon}^{(\tau)} + \int_0^t \Bigg[\frac{\kappa_0}{2}\int_0^{\infty}\int_0^{\infty}\left[2^r\left(x^r + y^r\right) + \frac{1}{x^{2\sigma}} + \frac{1}{y^{2\sigma}}\right] f(x,s) f(y,s)\mathrm{d}x\mathrm{d}y$$
$$+\zeta_0 \int_0^{\infty}\int_0^{y} y^r b(x,y) f(y,s)\mathrm{d}x\mathrm{d}y + \zeta_0 \int_0^{\infty}\int_0^{y} x^{-2\sigma} b(x,y) f(y,s)\mathrm{d}x\mathrm{d}y\Bigg]\mathrm{d}s$$

By using the hypothesis $(A3)$, we have

$$\|\mathcal{G}(f)\|_{\Upsilon}^{(\tau)} \le \|f_0\|_{\Upsilon}^{(\tau)} + \int_0^t \Bigg[2^r\kappa_0\int_0^{\infty}\int_0^{\infty}\left[\left(x^r + y^r\right) + \frac{1}{x^{2\sigma}} + \frac{1}{y^{2\sigma}}\right] f(x,s) f(y,s)\mathrm{d}x\mathrm{d}y$$
$$+\zeta_0\nu \int_0^{\infty} y^r f(y,s)\mathrm{d}y + \zeta_0 \bar{N} \int_0^{\infty} y^{-2\sigma} f(y,s)\mathrm{d}y\Bigg]\mathrm{d}s$$
$$\le \|f_0\|_{\Upsilon}^{(\tau)} + \int_0^t \Bigg[2^r\kappa_0\int_0^{\infty}\int_0^{\infty}\left[x^r + \frac{1}{x^{2\sigma}}\right]\left[y^r + \frac{1}{y^{2\sigma}}\right] f(x,s) f(y,s)\mathrm{d}x\mathrm{d}y$$

$$+\zeta_0\nu_0 \int_0^\infty (y^r + y^{-2\sigma}) f(y,s)\mathrm{d}y \Bigg] \mathrm{d}s$$

$$\leq \|f_0\|_\Upsilon^{(\tau)} + \int_0^\tau \left[2^r\kappa_0 \left(\|f\|_\Upsilon^{(\tau)} \right)^2 + \zeta_0\nu_0 \|f\|_\Upsilon^{(\tau)} \right] \mathrm{d}s.$$

$$\|\mathcal{G}(f)\|_\Upsilon^{(\tau)} \leq \|f_0\|_\Upsilon^{(\tau)} + \int_0^\tau \left[2^r\kappa_0 \left(\|f\|_\Upsilon^{(\tau)} \right)^2 + \zeta_0\nu_0 \|f\|_\Upsilon^{(\tau)} \right] \mathrm{d}s, \tag{10}$$

where $\nu_0 = \max\{\nu, \bar{N}\}$. Further let $\zeta_1 := \max\{\|f_0\|_\Upsilon^{(\tau)}, 2^r\kappa_0, \zeta_0\nu_0\}$ then the inequality (10) reduces to

$$\|\mathcal{G}(f)\|_\Upsilon^{(\tau)} \leq \zeta_1 \left(1 + \tau L_0 + \tau L_0^2\right).$$

Hence, $\|\mathcal{G}(f)\|_\Upsilon^{(\tau)} \leq L_0$, if $\zeta_1 \left(1 + \tau L_0 + \tau L_0^2\right) \leq L_0$. This inequality hold if $\tau < \frac{1}{\zeta_1}$ and

$$\frac{1 - \zeta_1\tau - \sqrt{(1-\zeta_1\tau)^2 - 4\zeta_1^2\tau}}{2\zeta_1\tau} \leq L_0 \leq \frac{1 - \zeta_1\tau + \sqrt{(1-\zeta_1\tau)^2 - 4\zeta_1^2\tau}}{2\zeta_1\tau}. \tag{11}$$

Now, we proceed to show that the mapping $\mathcal{G}$ is contracting. From (6), we have

$$\|\mathcal{G}(f) - \mathcal{G}(g)\|_\Upsilon^{(\tau)} \leq \int_0^t \Bigg[\frac{1}{2} \int_0^\infty \left(x^r + \frac{1}{x^{2\sigma}} \right) \int_0^x K(x-y,y) \left| A(x-y,y,s) \right| \mathrm{d}y\mathrm{d}x$$

$$+ \int_0^\infty \int_0^\infty \left(x^r + \frac{1}{x^{2\sigma}} \right) K(x,y) \left| A(x,y,s) \right| \mathrm{d}y\mathrm{d}x$$

$$+ \int_0^\infty \left(x^r + \frac{1}{x^{2\sigma}} \right) \int_x^\infty b(x,y) S(y) \left| f - g \right| (y,s)\mathrm{d}y\mathrm{d}x$$

$$+ \int_0^\infty \left(x^r + \frac{1}{x^{2\sigma}} \right) S(x) \left| f - g \right| (x,s)\mathrm{d}x \Bigg] \mathrm{d}s,$$

where $A(x,y,s) = f(x,s)f(y,s) - g(x,s)g(y,s)$. Now we perform the following mathematical operation, First change the order of the first and third integral. Then put $\bar{x} = x - y$, $\bar{y} = y$ and again replace $\bar{x}$ by x and $\bar{y}$ by y for the first integral to get

$$\|\mathcal{G}(f)-\mathcal{G}(g)\|_{\Upsilon}^{(\tau)} \leq \int_0^t \Bigg[\frac{\kappa_0}{2}\int_0^\infty\int_0^\infty \left((x+y)^r+\frac{1}{(x+y)^{2\sigma}}\right)|A(x,y,s)|\,\mathrm{d}y\mathrm{d}x$$
$$+\kappa_0\int_0^\infty\int_0^\infty\left(x^r+\frac{1}{x^{2\sigma}}\right)|A(x,y,s)|\,\mathrm{d}y\mathrm{d}x$$
$$+\int_0^\infty\int_x^\infty\left(x^r+\frac{1}{x^{2\sigma}}\right)b(x,y)S(y)\,|f-g|\,(y,s)\mathrm{d}y\mathrm{d}x$$
$$+\int_0^\infty S(x)\,|f-g|\,(x,s)\mathrm{d}x\Bigg]\mathrm{d}s$$

Further rearranging the terms, we get

$$\|\mathcal{G}(f)-\mathcal{G}(g)\|_{\Upsilon}^{(\tau)} \leq \int_0^t \Bigg[(2^r+1)\kappa_0\int_0^\infty\int_0^\infty \left(x^r+\frac{1}{x^{2\sigma}}\right)\left(y^r+\frac{1}{y^{2\sigma}}\right)|A(x,y,s)|\,\mathrm{d}y\mathrm{d}x$$
$$+\int_0^\infty\int_x^\infty\left(x^r+\frac{1}{x^{2\sigma}}\right)b(x,y)S(y)\,|f-g|\,(y,s)\mathrm{d}y\mathrm{d}x$$
$$+\int_0^\infty S(x)\,|f-g|\,(x,s)\mathrm{d}x\Bigg]\mathrm{d}s.$$

Now,

$$\int_0^\infty\int_0^\infty\left(x^r+\frac{1}{x^{2\sigma}}\right)\left(y^r+\frac{1}{y^{2\sigma}}\right)|A(x,y,s)|\,\mathrm{d}y\mathrm{d}x$$
$$=\int_0^\infty\int_0^\infty\left(x^r+\frac{1}{x^{2\sigma}}\right)\left(y^r+\frac{1}{y^{2\sigma}}\right)|f(x,s)f(y,s)-g(x,s)g(y,s)|\,\mathrm{d}y\mathrm{d}x$$
$$=\int_0^\infty\int_0^\infty\left(x^r+\frac{1}{x^{2\sigma}}\right)\left(y^r+\frac{1}{y^{2\sigma}}\right)|(f(x,s)$$
$$-g(x,s))f(y,s)-(g(y,s)-f(y,s))\,g(x,s)|\,\mathrm{d}y\mathrm{d}x$$
$$\leq\int_0^\infty\int_0^\infty\left(x^r+\frac{1}{x^{2\sigma}}\right)\left(y^r+\frac{1}{y^{2\sigma}}\right)\big[|f-g|\,(x,s)\,|f(y,s)|$$
$$+|f-g|\,(y,s)\,|g(x,s)|\big]\,\mathrm{d}y\mathrm{d}x$$

$$\leq \|f-g\|_{\Upsilon}^{(\tau)}\|f\|_{\Upsilon}^{(\tau)}+\|f-g\|_{\Upsilon}^{(\tau)}\|g\|_{\Upsilon}^{(\tau)}$$
$$=\|f-g\|_{\Upsilon}^{(\tau)}\left(\|f\|_{\Upsilon}^{(\tau)}+\|g\|_{\Upsilon}^{(\tau)}\right).$$

Therefore,

$$\|\mathcal{G}(f)-\mathcal{G}(g)\|_{\Upsilon}^{\tau} \leq \int_0^{\tau}\Big[(2^r+1)\kappa_0\|f-g\|_{\Upsilon}^{(\tau)}\left(\|f\|_{\Upsilon}^{(\tau)}+\|g\|_{\Upsilon}^{(\tau)}\right)$$
$$+\zeta_0\nu_0\|f-g\|_{\Upsilon}^{(\tau)}+\zeta_0\|f-g\|_{\Upsilon}^{(\tau)}\Big]\mathrm{d}s$$
$$\leq \int_0^{\tau}\Big[2\left(2^r+1\right)\kappa_0 L_0\|f-g\|_{\Upsilon}^{(\tau)}+\zeta_0\left(\nu_0+1\right)\|f-g\|_{\Upsilon}^{(\tau)}\Big]\mathrm{d}s.$$

Let $\zeta_2 := \max\{2\,(2^r+1)\,\kappa_0,\, \zeta_0\,(\nu_0+1)\}$ then the above inequality reduced to.

$$\|\mathcal{G}(f)-\mathcal{G}(g)\|_{\Upsilon}^{(\tau)} \leq \tau\zeta_2\,(L_0+1)\,\|f-g\|_{\Upsilon}^{(\tau)}. \tag{12}$$

So, the mapping $\mathcal{G}$ is contractive in $\Upsilon_{r,\sigma}^{+}(\tau)$ for $\tau < [\zeta_2\,(L_0+1)]^{-1}$. Using this result together with the inequalities (11), there exist an invariant ball of radius L_0 for sufficiently small $\tau > 0$. In that ball $\mathcal{G}$ is contractive. Consequently, that ball contains a fixed point of $\mathcal{G}$.
Nonnegativity: Case I: If $f_0 > 0$, then we assume (x_0, t_0) be the point such that

$$f\,(x_0,t_0)=0 \quad \text{and} \quad f(x,t)\neq 0 \quad \text{for all } 0<x<\max\{x_0,R\},\ 0\leq t<t_0. \tag{13}$$

Since, f is continuous, so this is possible only if $f(x,t) > 0$ for all $0 < x \leq \max\{x_0, R\}$, $0 \leq t < t_0$. Otherwise, (x_0, t_0) fails to satisfy (13) for which $f\,(x_0, t_0) = 0$. Again since, the solution is continuous and satisfy (6) it must be continuously differentiable with respect to t. Therefore,

$$\left.\frac{\partial f\,(x,t)}{\partial t}\right|_{(x_0,t_0)} = \frac{1}{2}\int_0^{x_0} K(x_0-y,y)f\,(x_0-y,t_0)f\,(y,t_0)\mathrm{d}y$$
$$+\int_{x_0}^{R} b(x_0,y)S(y)f\,(y,t_0)\mathrm{d}y. \tag{14}$$

- If $x_0 \leq R$, then $f\,(x,t) > 0$ for all $0 < x \leq R$ and $0 \leq t < t_0$. The positivity of right hand side of (14) implies $\left.\dfrac{\partial f\,(x,t)}{\partial t}\right|_{(x_0,t_0)} > 0$.

- If $x_0 > R$, then $\int_{x_0}^{R} b(x_0, y)S(y)f(y, t_0)\mathrm{d}y = -\int_{R}^{x_0} b(x_0, y)S(y)f(y, t_0)\mathrm{d}y = 0$ by the virtue of (3) and $S(x)$ being compactly supported over $[0, R]$. Therefore, $\left.\frac{\partial f(x,t)}{\partial t}\right|_{(x_0,t_0)} > 0$.

Moreover, positivity of the time derivative proves that there exist a point (x_0, t), with $t < t_0$ such that $f(x_0, t) < 0$. This contradicts the assumption that (x_0, t_0) is the point with the property (13). Hence, no such point (x_0, t_0) exists. Consequently, $f(x, t)$ is strictly positive provided the initial data is strictly positive.

Case II: Let f_0 is not strictly positive. Then we construct the sequence of positive function $\{f_0^n\}$ which satisfies the condition of Theorem 1 and converges to f_0 uniformly in $\Upsilon_{r,\sigma}(\tau)$. We already proved that the family of operators $\mathcal{G}_n : \Upsilon_{r,\sigma}(\tau) \to \Upsilon_{r,\sigma}(\tau)$ defined as

$$\begin{aligned}\mathcal{G}_n(f)(x,t) = f_0^n(x) + \int_0^t \Bigg[\frac{1}{2}\int_0^x K(x-y, y)f(x-y, s)f(y, s)\mathrm{d}y \\ - \int_0^\infty K(x, y)f(x, s)f(y, s)\mathrm{d}y \\ + \int_x^\infty b(x, y)S(y)f(y, s)\mathrm{d}y - S(x)f(x, s)\Bigg]\mathrm{d}s\end{aligned}$$

is a contraction mapping. Therefore as $n \to \infty$, we have

$$\sup_{\|f\|_\Upsilon^{(\tau)} \le L} \|\mathcal{G}_n(f) - \mathcal{G}(f)\|_\Upsilon^{(\tau)} \le \int_0^\infty \left(x^r + \frac{1}{x^{2\sigma}}\right)\left|f_0^n(x) - f_0(x)\right|\mathrm{d}x \to 0.$$

Since the mapping is contractive in $\Upsilon_{r,\sigma}(\tau)$, therefore

$$\begin{aligned}\|f^n - f\|_\Upsilon^{(\tau)} = \|\mathcal{G}_n(f^n) - \mathcal{G}(f)\|_\Upsilon^{(\tau)} &\le \|\mathcal{G}_n(f^n) - \mathcal{G}(f^n)\|_\Upsilon^{(\tau)} + \|\mathcal{G}(f^n) - \mathcal{G}(f)\|_\Upsilon^{(\tau)} \\ &\le \|\mathcal{G}_n(f^n) - \mathcal{G}(f^n)\|_\Upsilon^{(\tau)} + \bar{\xi}\|f^n - f\|_\Upsilon^{(\tau)}.\end{aligned}$$

Which implies

$$\left(1 - \bar{\xi}\right)\|f^n - f\|_\Upsilon^{(\tau)} = \|\mathcal{G}_n(f^n) - \mathcal{G}(f^n)\|_\Upsilon^{(\tau)} \to 0 \quad \text{whenever} \quad n \to \infty.$$

This proves that the solution f is nonnegative for a nonnegative initial data.

Global existence of unique solution: We first observe that the boundedness of moments

$$\mathcal{M}_k(t) = \int_0^\infty x^k f(x,t)\mathrm{d}x; \quad \text{where} \quad 0 \le k \le r \quad \text{and} \quad k = -2\sigma,$$

for compactly supported kernels. Simple calculations will lead us to the following results;

$$\mathcal{M}_1(t) \le \bar{m}_1, \quad \mathcal{M}_{-2\sigma}(t) \le \bar{m}_{-2\sigma}, \quad \mathcal{M}_0(t) \le \exp(\zeta_0 \nu T), \quad \mathcal{M}_2(t) \le \kappa_0 \bar{m}_1^2,$$

and so on. Here terms $\bar{m}_k, k = -2\sigma, 0, 1, \ldots, r$ are all constants. Moreover, it can be noted that for $k = 2, 3, \ldots, r$ the boundedness of $(k+1)$-th moment depends upon the boundedness of $k-$th moment. Thus, all the above results together implies

$$\|f\|_\Upsilon \le \bar{m}_r + \bar{m}_{-2\sigma}.$$

Therefore, the solution of IVP (1)–(2) is bounded in the norm $\|.\|_\Upsilon$. Taking into account nonnegativity of the local solution, we prolong it for all $0 \le t \le T$. Recalling Theorem 2.2 of [13], the global existence of unique solution belonging to $\Upsilon^+_{r,\sigma}(T)$ can easily be proved.

2.3 Conservation of Mass

For the mass conservation law, we multiplying with the weight x and integrating, Eq. (1) is written as

$$\begin{aligned}
\frac{\mathrm{d}\mathcal{M}_1(t)}{\mathrm{d}t} &= \frac{\mathrm{d}}{\mathrm{d}t}\int_0^\infty xf(x,t)\mathrm{d}x \\
&= \frac{1}{2}\int_0^\infty \int_0^x xK(x-y,y)f(x-y,t)f(y,t)\mathrm{d}y\mathrm{d}x \\
&+ \int_0^\infty \int_x^\infty xb(x,y)S(y)f(y,t)\mathrm{d}y\mathrm{d}x \\
&- \int_0^\infty \int_0^\infty xK(x,y)f(x,t)f(y,t)\mathrm{d}y\mathrm{d}x - \int_0^\infty xS(x)f(x,t)\mathrm{d}x
\end{aligned} \tag{15}$$

By using Fubini's theorem and the virtue that the kernels are compactly supported the right hand side of (15) reduced to zero. Therefore, the mass conservation law obeyed unconditionally.

3 Conclusion

In this work, a thorough mathematical investigation on the existence of strong solutions to the continuous CF-equation (1)–(2) is performed. Here, both the coagulation and fragmentation kernel includes a class of compactly supported functions. Initially, we have proved the local existence of nonnegative solutions for these compactly supported kernels. Next, we have proved the global existence of the unique solution to the space $\Upsilon^{+}_{r,\sigma}(T)$ of the IVP (1)–(2). The study is completed by examining the mass conservation law of the existing solution. This method can be applied to various kind of coagulation-fragmentation models like collision induced fragmentation equation, coagulation equation with collisional breakage equation, etc. The large time dynamics of the particulate system equipped with singular kinetic rates can be considered as a future scope research.

Acknowledgements The author AD thanks Ministry of Education (MoE), Govt. of India for their funding support during his Ph.D. program and JS thanks NITT for their support through seed grant (file No: NITT/R & C/SEED GRANT/19-20/P-13/MATHS/JS/E1) during this work.

References

1. Hauk T, Bonaccurso E, Roisman IV, Tropea C (2015) Ice crystal impact onto a dry solid wall. Particle fragmentation. Proc R Soc A 471:2181, 20150399. https://doi.org/10.1098/rspa.2015.0399
2. Hare C, Bonakdar T, Ghadiri M, Strong J (2018) Impact breakage of pharmaceutical tablets. Int J Pharm 536(1):370–376. https://doi.org/10.1016/j.ijpharm.2017.11.066
3. Wibowo C, Ng KM (2001) Product oriented process synthesis and development: creams and pastes. AIChE J 47(12):2746–2767. https://doi.org/10.1109/HPDC.2001.945188
4. Roisman IV, Tropea C (2015) Impact of a crushing ice particle onto a dry solid wall. Proc R Soc A Math Phys Eng Sci 471(2183), 20150525. https://doi.org/10.1098/rspa.2015.0525
5. Stewart IW, Meister E (1989) A global existence theorem for the general coagulation-fragmentation equation with unbounded kernels. Math Methods Appl Sci 11(5):627–648. https://doi.org/10.1002/mma.1670110505
6. Costa FP (1995) Existence and uniqueness of density conserving solutions to the coagulation-fragmentation equations with strong fragmentation. J Math Anal Appl 892–914
7. Dubovski PB, Stewart IW (1996) Existence, uniqueness and mass conservation for the coagulation-fragmentation equation. Math Methods Appl Sci 19(7):571–591. https://doi.org/10.1002/(SICI)1099-1476(19960510)19:7
8. Banasiak J, Lamb W, Langer M (2013) Strong fragmentation and coagulation with power-law rates. J Eng Math 82(1):199–215. https://doi.org/10.1007/s10665-012-9596-3
9. Saha J, Kumar J (2015) The singular coagulation equation with multiple fragmentation. Z Angew Math Phys 66(3):919–941. https://doi.org/10.1007/s00033-014-0452-3

10. Camejo CC, Warnecke G (2015) The singular kernel coagulation equation with multifragmentation. Math Methods Appl Sci 38(14):2953–2973
11. Ghosh D, Saha J, Kumar J (2020) Existence and uniqueness of steady-state solution to a singular coagulation-fragmentation equation. J Comput Appl Math 380. https://doi.org/10.1016/j.cam.2020.112992
12. Barik PK, Giri AK, Laurenot P (2020) Mass-conserving solutions to the Smoluchowski coagulation equation with singular kernel. Proc R Soc Edinb Sect A Math 150(4):1805–1825. https://doi.org/10.1017/prm.2018.158
13. Dubovski PB (1994) Mathematical theory of coagulation. National University

Approximation by Szász–Kantorovich-Type Operators Involving Boas–Buck-Type Polynomials

P. N. Agrawal and Sompal Singh

Abstract In the present article, we introduce a new sequence of Sz*ász*–Kantorovich-type operators—Buck-type polynomials which include Brenke-type polynomials, Sheffer polynomials and Appell polynomials. We estimate the error in the approximation by these operators in terms of the Lipschitz-type maximal function, Peetre's K-functional and Ditzian–Totik modulus of smoothness. We also study the order of convergence of these operators for unbounded functions in a weighted space by using the weighted modulus of continuity. Further, we study a quantitative Voronovoskaya-type theorem and Gr*ü*ss Voronovskaya-type theorem.

Keywords Boas–Buck-type polynomials · Lipschitz-type space · Peetre's K-functional · Ditzian–Totik modulus of smoothness · Weighted modulus of continuity

Mathematics Subject Classification MSC 41A36 · MSC 41A25 · MSC 26A15

1 Introduction

In recent years, there appears to have been a great deal of interest among researchers in linking the polynomials defined by means of generating functions with the theory of approximation. The research in this direction dates back to the paper by Jakimovski and Leviatan [19] who gave a generalization of Szász operators by means of Appell polynomials. Subsequently, Ismail [15] presented generalized Szász operators with the aid of Sheffer polynomials. Varma et al. [38] constructed a variant of Szász operators by involving Brenke-type polynomials. Later, Sucu et al. [34] gave an extension of Szász operators by using Boas–Buck-type polynomials which include Brenke-type polynomials, Sheffer polynomials and Appell polynomials, as follows:

P. N. Agrawal · S. Singh (✉)
Department of Mathematics, Indian Institute of Technology Roorkee, Roorkee 247667, India
e-mail: ssingh@ma.iitr.ac.in

S. S. Ray et al. (eds.), *Applied Analysis, Computation and Mathematical Modelling in Engineering*, Lecture Notes in Electrical Engineering 897,
https://doi.org/10.1007/978-981-19-1824-7_23

$$B_m(\hbar; z) = \frac{1}{D(1)B(mzE(1))} \sum_{k=0}^{\infty} p_k(mz)\hbar\left(\frac{k}{m}\right), \ z \geq 0, \ m \in \mathbb{N}, \tag{1}$$

where the generating function of the Boas–Buck-type polynomials is given by

$$D(s)B(zE(s)) = \sum_{k=0}^{\infty} p_k(z)s^k. \tag{2}$$

and $D(s)$, $B(s)$ and $E(s)$ are analytic functions such that

$$D(s) = \sum_{k=0}^{\infty} d_k s^k \ , \ (d_0 \neq 0)$$

$$B(s) = \sum_{j=0}^{\infty} b_j s^j, \ (b_j \neq 0, \ \ for\ any\ j)$$

$$E(s) = \sum_{k=1}^{\infty} e_k s^k \ (e_1 \neq 0),$$

under the following assumptions:

(i) $D(1) \neq 0, \ \frac{a_{j-i}b_i}{D(1)} \geq 0, \ $ for $0 \leq i \leq j, \ j = 0, 1, 2 \ldots$

(ii) $B : [0, \infty) \rightarrow (0, \infty)$,

(iii) The power series of generating function and analytic function B(s) converges for $|s| < R \ (R > 1)$,

(iv) $\lim_{x\rightarrow\infty}(E'(1))^j \frac{B^{(j)}(z)}{B(z)} = 1, \ $ for $j = 1, 2, 3, 4$.

Atakut and Büyükyazici [3] considered Kantorovich–Szász-type operators involving Brenke-type polynomials and studied the convergence properties of these operators. Garg et al. [12] discussed some more approximation properties of the operators defined in [3]. Büyükyazici et al. [6] defined Chlodowsky-type Jakimovski–Leviatan operators and studied the rate of convergence by means of weighted modulus of continuity. Sidharth et al. [33] considered Chlodowsky–Szász–Appell-type operators to approximate the functions of two variables. Neer et al. [26] extended the study to the Chlodowsky variant of Jakimovski–Leviatan–Durrmeyer-type operators. Özarslan and Duman [29] introduced modified Bernstein–Kantorovich-type operators by means of a parameter $\alpha > 0$ and studied two direct approximation theorems. Following the paper [30], Kumar [24] similarly defined generalized λ–Bernstein–

Kantorovich-type operators and investigated some approximation properties of these operators. Motivated by the ideas developed in these papers, for $\gamma > 0$, we construct a new kind of Szász–Kantorovich-type operators involving Boas–Buck-type polynomials as

$$L_m(\hbar; z) = \frac{1}{D(1)B(\alpha_m z E(1))} \sum_{k=0}^{\infty} p_k(\alpha_m z) \int_0^1 \hbar\left(\frac{k + s^{\gamma}}{\beta_m}\right) \mathrm{d}s, \tag{3}$$

where $\{\alpha_m\}$ and $\{\beta_m\}$ are strictly increasing sequences of positive real numbers such that $\frac{1}{\beta_m} \to 0$, as $m \to \infty$ and $\frac{\alpha_m}{\beta_m} = 1 + O\left(\frac{1}{\beta_m}\right)$, as $m \to \infty$. Further, the functions $D(s)$, $B(s)$ and $E(s)$ are as defined in (2) and satisfy the assumptions listed above from (i)–(iv).

In the present paper, we derive some convergence estimates for the operators given by (3) by using the well known Lipschitz-type maximal function, Ditzian–Totik modulus of smoothness and the Peetre's K-functional. Further, we also consider the weighted approximation and investigate the quantitative Voronovskaya and Grüss–Voronovskaya-type theorems. Note that, for $\alpha_m = m,\ \beta_m = m + 1$ and $\gamma = 1$, the operators (3) include the Szász–Kantorovich-type operators, based on Boas–Buck-type polynomials.

2 Preliminaries

Let us assume $\mathbb{R}_+ = [0, \infty)$.

Lemma 1 *[34]. From the generating function (2) of Boas–Buck-type polynomials, for $z \in \mathbb{R}_+$ we have*

$$(i)\quad \sum_{k=0}^{\infty} p_k(\alpha_m z) = D(1)B(\alpha_m z E(1)),$$

$$(ii)\quad \sum_{k=0}^{\infty} k p_k(\alpha_m z) = D'(1)B(\alpha_m z E(1)) + \alpha_m z D(1)E'(1)B'(\alpha_m z E(1)),$$

$$\begin{aligned}(iii)\quad \sum_{k=0}^{\infty} k^2 p_k(\alpha_m z) = {} & (D' + D'')(1)B(\alpha_m z E(1)) + \alpha_m z(2D'E' \\ & + D(E' + E''))(1)B'(\alpha_m z E(1)) \\ & + \alpha_m^2 z^2\left(D(E')^2\right)(1)B''(\alpha_m z E(1)),\end{aligned}$$

$$(iv)\ \sum_{k=0}^{\infty} k^3 p_k(\alpha_m z) = \Big(D' + 3D'' + D'''\Big)(1)B(\alpha_m zE(1)) + \alpha_m z\Big(6D'E' + DE'$$
$$+ 3D''E' + 3DE'' + DE''' + 3D'E''\Big)(1)B'(\alpha_m zE(1))$$
$$+ \alpha_m^2 z^2\Big(3D'(E')^2 + 3D(E')^2 + 3DE'E''\Big)(1)B''(\alpha_m zE(1))$$
$$+ \alpha_m^3 z^3 D(E')^3(1)B'''(\alpha_m zE(1)),$$
$$(v)\ \sum_{k=0}^{\infty} k^4 p_k(\alpha_m z) = \Big(D' + 7D'' + 6D''' + D''''\Big)(1)B(\alpha_m zE(1)) + \alpha_m z\Big(14D'E'$$
$$+ 18D''E' + 18D'E'' + 6D''E'' + 4D'''E' + 4D'E'''$$
$$+ DE' + 7DE'' + 6DE''' + DE''''\Big)(1)B'(\alpha_m zE(1))$$
$$+ \alpha_m^2 z^2\Big(18D'(E')^2 + 6D''(E')^2 + 9D'E'E'' + 3D'E'$$
$$+ 7D(E')^2 + 18AE'E'' + 3D(E'')^2 + 4DE'E'''\Big)(1)B''(\alpha_m zE(1))$$
$$+ \alpha_m^3 z^3\Big(4D'(E')^3 + 4D(E')^3 + 6D(E')^2E''\Big)(1)B'''(\alpha_m zE(1))$$
$$+ \alpha_m{}^4 z^4\Big(D(E')^4\Big)(1)B''''(\alpha_m zE(1)).$$

Consequently, by a simple calculation we obtain the raw moments as :

Lemma 2 *Let* $e_i(s) = s^i,\ i = 0, 1, 2, 3, \ldots$ *For the operator* $L_m(\hbar; z)$ *(3) and* $z \in \mathbb{R}_+$, *we have*

$$(i)\ L_m(1; z) = 1,$$
$$(ii)\ L_m(s; z) = \frac{1}{\beta_m}\left(\frac{1}{(\gamma+1)} + \frac{D'(1)}{D(1)} + \alpha_m z.E'(1)\frac{B'(\alpha_m zE(1))}{B(\alpha_m zE(1))}\right),$$
$$(iii)\ L_m(s^2; z) = \frac{1}{\beta_m^2}\left(\frac{1}{2\gamma+1}\right) + \frac{2}{(\gamma+1)\beta_m^2}\left\{\frac{D'(1)}{D(1)} + \alpha_m z.E'(1)\frac{B'(\alpha_m zE(1))}{B(\alpha_m zE(1))}\right\}$$
$$+ \frac{1}{\beta_m^2}\left\{\frac{(D''+D')(1)}{D(1)} + \alpha_m z\left(\frac{2(D'E')(1)}{D(1)} + (E'+E'')(1)\right)\right.$$
$$\left.\times \frac{B'(\alpha_m zE(1))}{B(\alpha_m zE(1))} + (\alpha_m)^2 z^2 (E'(1))^2 \frac{B''(\alpha_m zE(1))}{B(\alpha_m zE(1))}\right\},$$

$$
\begin{aligned}
(iv)\ L_m(s^3; z) = {} & \frac{1}{\beta_m^3}\left\{\frac{1}{3\gamma+1} + \frac{3}{2\gamma+1}\left(\frac{D'(1)}{D(1)} + \alpha_m z E'(1)\frac{B'(\alpha_m z E'(1))}{B(\alpha_m z E(1))}\right)\right. \\
& + \frac{3}{\gamma+1}\left(\frac{(D''+D')(1)}{D(1)} + \alpha_m z\left(\frac{(2D'E')(1)}{D(1)}\right.\right. \\
& \left.\left. + (E'+E'')(1)\right)\frac{B'(\alpha_m z E(1))}{B(\alpha_m z E(1))} + \alpha_m^2 z^2 \frac{(E'(1))^2 B''(\alpha_m z E(1))}{B(\alpha_m z E(1))}\right) \\
& + \frac{(D'+3D''+D''')(1)}{D(1)} + \alpha_m z\left(\frac{(6D'E')(1)}{D(1)} + \frac{(3D''E')(1)}{D(1)}\right. \\
& \left. + \frac{(3D'E'')(1)}{D(1)} + (E'+3E''+E''')(1)\right)\frac{B'(\alpha_m z E(1))}{B(\alpha_m z E(1))} \\
& + 3\alpha_m^2 z^2\left(\frac{(D'(E')^2)(1)}{D(1)} + (E'(1))^2 + (E'E'')(1)\right)\frac{B''(\alpha_m z E(1))}{B(\alpha_m z E(1))} \\
& \left. + \alpha_m^3 z^3 (E'(1))^3 \frac{B'''(\alpha_m z E(1))}{B(\alpha_m z E(1))}\right\},
\end{aligned}
$$

$$
\begin{aligned}
(v)\ L_m(s^4; z) = {} & \frac{1}{\beta_m^4}\left\{\frac{1}{4\gamma+1} + \frac{4}{3\gamma+1}\left(\frac{D'(1)}{D(1)} + \alpha_m z E'(1)\frac{B'(\alpha_m z E(1))}{B(\alpha_m z E(1))}\right)\right. \\
& + \left(\frac{6}{2\gamma+1}\right)\left[\frac{(D'+D'')(1)}{D(1)} + \alpha_m z\left(\frac{(2D'E')(1)}{D(1)}\right.\right. \\
& \left.\left. + (E'+E'')(1)\right)\frac{B'(\alpha_m z E(1))}{B(\alpha_m z E(1))} + \alpha_m^2 z^2 (E'(1))^2 \frac{B''(\alpha_m z E(1))}{B(\alpha_m z E(1))}\right] \\
& + \frac{4}{\gamma+1}\left[\frac{(D'+3D''+D''')(1)}{D(1)} + \alpha_m z\left(\frac{(6D'E')(1)}{D(1)} + \frac{(3D''E')(1)}{D(1)}\right.\right. \\
& \left. + \frac{(3D'E'')(1)}{D(1)} + (E'+3E''+E''')(1)\right)\frac{B'(\alpha_m z E(1))}{B(\alpha_m z E(1))} \\
& + 3\alpha_m^2 z^2\left(\frac{(D'(E')^2)(1)}{D(1)} + (E'(1))^2\right. \\
& \left.\left. + (E'E'')(1)\right)\frac{B''(\alpha_m z E(1))}{B(\alpha_m z E(1))} + \alpha_m^3 z^3 (E'(1))^3 \frac{B'''(\alpha_m z E(1))}{B(\alpha_m z E(1))}\right] \\
& + \frac{(D'+7D''+6D'''+D'''')(1)}{D(1)} \\
& + \alpha_m z\left(\frac{(14A'E' + 18D''E' + 18D'E'' + 6D''E'' + 4D'''E')(1)}{D(1)}\right. \\
& \left. + \frac{(4D'E''' + DE' + 7AE'' + 6AE''' + AE'''')(1)}{D(1)}\right)\frac{B'(\alpha_m z E(1))}{B(\alpha_m z E(1))} \\
& + \alpha_m^2 z^2\left(\frac{(18D'(E')^2 + 6D''(E')^2 + 12D'E'E'' + 7D(E')^2)(1)}{D(1)}\right. \\
& \left. + \frac{(18DE'E'' + 3D(E'')^2 + 4DE'E''')(1)}{D(1)}\right)\frac{B''(\alpha_m z E(1))}{B(\alpha_m z E(1))} \\
& + \alpha_m^3 z^3\left(\frac{(4D'(E')^3 + 6D(E')^3 + 6D(E')^2E'')(1)}{D(1)}\right)\frac{B'''(\alpha_m z E(1))}{B(\alpha_m z E(1))} \\
& \left. + \alpha_m^4 z^4 (E'(1))^4 \frac{B''''(\alpha_m z E(1))}{B(\alpha_m z E(1))}\right\}.
\end{aligned}
$$

Now, let $\mu_{m,n}(z)$ denote the n-th-order central moment for the operators $L_m(\hbar; z)$ defined by (3). Then

$$\mu_{m,n}(z) = L_m((s-z)^n; z), n \in \mathbb{N}^0,$$

where $\mathbb{N}^0 = \mathbb{N} \cup \{0\}$. From Lemma 2 , we easily derive the following result:

Lemma 3 *For the operator $L_m(\hbar; z)$ defined by (3) and for $z \in \mathbb{R}_+$, we have the following identities:*

$$(i)\ \mu_{m,0}(z) = 1,$$

$$(ii)\mu_{m,1}(z) = \frac{1}{\beta_m}\left(\frac{1}{\gamma+1} + \frac{D'(1)}{D(1)}\right) + \left(\frac{\alpha_m}{\beta_m}\frac{B'(\alpha_m zE(1))}{B(\alpha_m zE(1))}E'(1) - 1\right)z,$$

$$\begin{aligned}(iii)\mu_{m,2}(z) = &\left\{1 - 2\left(\frac{\alpha_m}{\beta_m}\right)\frac{B'(\alpha_m zE(1))}{B(\alpha_m zE(1))}E'(1) + \left(\frac{\alpha_m}{\beta_m}\right)^2 \frac{B''(\alpha_m zE(1))}{B(\alpha_m zE(1))}(E'(1))^2\right\}z^2 \\ &+ \left\{\frac{\alpha_m}{\beta_m^2}\left(\frac{(2D'E')(1)}{D(1)} + \frac{\gamma+3}{\gamma+1}E'(1) + E''(1)\right)\frac{B'(\alpha_m zE(1))}{B(\alpha_m zE(1))}\right. \\ &\left. - \frac{2}{\beta_m}\left(\frac{1}{\gamma+1} + \frac{D'(1)}{D(1)}\right)\right\}z + \frac{1}{\beta_m^2}\left\{\frac{(D'+D'')(1)}{D(1)} + \frac{1}{2\gamma+1} + \frac{2}{\gamma+1}\frac{D'(1)}{D(1)}\right\},\end{aligned}$$

$$\begin{aligned}(iv)\mu_{m,4}(z) = &\left\{\left(\frac{\alpha_m}{\beta_m}\right)^4 (E'(1))^4 \frac{B''''(\alpha_m zE(1))}{B(\alpha_m zE(1))} - 4\left(\frac{\alpha_m}{\beta_m}\right)^3 (E'(1))^3 \frac{B'''(\alpha_m zE(1))}{B(\alpha_m zE(1))}\right. \\ &\left. + 6\left(\frac{\alpha_m}{\beta_m}\right)^2 (E'(1))^2 \frac{B''(\alpha_m zE(1))}{B(\alpha_m zE(1))} - 4\left(\frac{\alpha_m}{\beta_m}\right)(E'(1))\frac{B'(\alpha_m zE(1))}{B(\alpha_m zE(1))} + 1\right\}z^4 \\ &+ \left\{\frac{\alpha_m^3}{\beta_m^4}\frac{B'''(\alpha_m zE(1))}{B(\alpha_m zE(1))}\left(\frac{4(E'(1))^3}{\gamma+1} + \frac{(4D'(E')^3 + 6D(E')^3 + 6D(E')^2E'')(1)}{D(1)}\right)\right. \\ &- 12\frac{\alpha_m^2}{\beta_m^3}\frac{B''(\alpha_m zE(1))}{B(\alpha_m zE(1))}\left(\frac{(E'(1))^2}{\gamma+1} + \frac{(D'E'^2)(1)}{D(1)} + ((E')^2 + E'E'')(1)\right) \\ &+ 6\frac{\alpha_m}{\beta_m^2}\frac{B'(\alpha_m zE(1))}{B(\alpha_m zE(1))}\left(\frac{2E'(1)}{\gamma+1} + \frac{(2D'E')(1)}{D(1)} + (E' + E'')(1)\right) \\ &\left. - \frac{4}{\beta_m}\left(\frac{1}{\gamma+1} + \frac{D'(1)}{D(1)}\right)\right\}z^3 + \left\{\frac{\alpha_m^2}{\beta_m^4}\frac{B''(\alpha_m zE(1))}{B(\alpha_m zE(1))}\left(\frac{6(E'(1))^2}{2\gamma+1}\right.\right. \\ &+ \frac{12}{\gamma+1}\left(\frac{(D'(E')^2)(1)}{D(1)} + (E'^2 + E'E'')(1)\right) \\ &+ \frac{1}{D(1)}\left(18D'(E')^2 + 6D''(E')^2 + 9D'E'E'' + 3D'E' + 7D(E')^2\right. \\ &\left.\left. + 18DE'E'' + 3D(E'')^2 + 4DE'E'''\right)(1)\right) \\ &- 4\frac{\alpha_m}{\beta_m^3}\left(\frac{3E'(1)}{2\gamma+1} + \frac{3}{\gamma+1}\left(\frac{(2D'E')(1)}{D(1)} + (E' + E'')(1)\right) + \frac{(6D'E')(1)}{D(1)}\right.\end{aligned}$$

$$+\frac{(3D''E')(1)}{D(1)}+E'(1)+(3E''+E''')(1)+\frac{(3D'E'')(1)}{D(1)}\bigg)\frac{B'(\alpha_m zE(1))}{B(\alpha_m zE(1))}$$

$$+\frac{6}{\beta_m^2}\left(\frac{1}{2\gamma+1}+\frac{2}{\gamma+1}\frac{D'(1)}{D(1)}+\frac{(D'+D'')(1)}{D(1)}\right)\Bigg\}z^2$$

$$+\frac{\alpha_m z}{\beta_m^4}\left\{\frac{4E'(1)}{3\gamma+1}+\frac{6}{2\gamma+1}\left(\frac{(2D'E')(1)}{D(1)}+E'(1)+E''(1)\right)\right.$$

$$+\frac{4}{\gamma+1}\left(\frac{(6D'E')(1)}{D(1)}+\frac{(3D''E')(1)}{D(1)}+\frac{(3D'E'')(1)}{D(1)}+(E'+3E''+E''')(1)\right)$$

$$+\frac{(14D'E'+18D''E'+18D'E''+6D''E'')(1)}{D(1)}$$

$$+\left.\frac{(4D'''E'+4D'E'''+DE'+7DE''+6DE'''+DE'''')(1)}{D(1)}\right\}\frac{B'(\alpha_m zE(1))}{B(\alpha_m zE(1))}$$

$$-\frac{4}{\beta_m^3}\left(\frac{1}{3\gamma+1}+\frac{3}{2\gamma+1}\frac{D'(1)}{D(1)}+\frac{3}{\gamma+1}\frac{(D'+D'')(1)}{D(1)}+\frac{(D'+3D''+D''')(1)}{D(1)}\right)$$

$$+\frac{1}{\beta_m^4}\left(\frac{1}{4\gamma+1}+\frac{4}{3\gamma+1}\frac{D'(1)}{D(1)}+\frac{6}{2\gamma+1}\frac{(D'+D'')(1)}{D(1)}\right.$$

$$\left.+\frac{4}{\gamma+1}\frac{(D'+3D''+D''')(1)}{D(1)}+\frac{(D'+7D''+6D'''+D'''')(1)}{D(1)}\right).$$

In our further consideration, let us make the following assumptions:

$$(a)\ \lim_{m\to\infty}\frac{B^{(j)}(\alpha_m zE(1))}{B(\alpha_m zE(1))}=\frac{1}{(E'(1))^j},\ for\ j=1,2,3,\ldots,$$

$$(b)\ \lim_{m\to\infty}\beta_m\left\{\frac{B'(\alpha_m zE(1))}{B(\alpha_m zE(1))}E'(1)-1\right\}=0,$$

$$(c)\ \lim_{m\to\infty}\beta_m\left\{1-\frac{2B'(\alpha_m zE(1))}{B(\alpha_m zE(1))}E'(1)+\frac{B''(\alpha_m zE(1))}{B(\alpha_m zE(1))}(E'(1))^2\right\}=0,$$

$$(d)\ \lim_{m\to\infty}\beta_m{}^2\left\{(E'(1))^4\frac{B''''(\alpha_m zE(1))}{B(\alpha_m zE(1))}-4(E'(1))^3\frac{B'''(\alpha_m zE(1))}{B(\alpha_m zE(1))}\right.$$

$$\left.+6(E'(1))^2\frac{B''(\alpha_m zE(1))}{B(\alpha_m zE(1))}-4E'(1)\frac{B'(\alpha_m zE(1))}{B(\alpha_m zE(1))}+1\right\}=0,$$

$$(e)\ \lim_{m\to\infty}\beta_m\left\{(E'(1))^4\frac{B''''(\alpha_m zE(1))}{B(\alpha_m zE(1))}-3(E'(1))^3\frac{B'''(\alpha_m zE(1))}{B(\alpha_m zE(1))}\right.$$

$$\left.+3(E'(1))^2\frac{B''(\alpha_m zE(1))}{B(\alpha_m zE(1))}-E'(1)\frac{B'(\alpha_m zE(1))}{B(\alpha_m zE(1))}\right\}=0,$$

$$(f)\ \lim_{m\to\infty}\beta_m\Bigg\{(E'(1))^3\frac{B'''(\alpha_m zE(1))}{B(\alpha_m zE(1))}\left(\frac{4}{\gamma+1}+\frac{4D'(1)}{D(1)}+\frac{6D(1)}{D(1)}+\frac{6E''(1)}{E'(1)}\right)$$
$$-12(E'(1))^2\frac{B''(\alpha_m zE(1))}{B(\alpha_m zE(1))}\left(\frac{1}{\gamma+1}+\frac{D'(1)}{D(1)}+1+\frac{E''(1)}{E'(1)}\right)$$
$$+6E'(1)\frac{B'(\alpha_m zE(1))}{B(\alpha_m zE(1))}\left(\frac{2}{\gamma+1}+\frac{2D'(1)}{D(1)}+1+\frac{E''(1)}{E'(1)}\right)$$
$$-4\left(\frac{1}{\gamma+1}+\frac{D'(1)}{D(1)}\right)\Bigg\}=0.$$

Remark 1 Using the above assumptions and Lemma 2, we obtain

$$(i)\ \lim_{m\to\infty}\beta_m L_m((s-z);z)=\frac{1}{\gamma+1}+\frac{D'(1)}{D(1)}+zO(1),$$
$$(ii)\ \lim_{m\to\infty}\beta_m L_m((s-z)^2;z)=\left\{1+\frac{E''(1)}{E'(1)}\right\}z,$$
$$(iii)\ \lim_{m\to\infty}{\beta_m}^2 L_m((s-z)^4;z)=3\left(1+\frac{E''(1)}{E'(1)}\right)^2 z^2.$$

Remark 2 Using the assumptions listed above from (a)–(f) and Lemma 2, for sufficiently large m and $z>0$, we have

$$\mu_{m,2}(z)\le\frac{C}{\beta_m}\phi^2(z),\ \ and\ \ \mu_{m,4}(z)\le\frac{C}{\beta_m^2}\phi^4(z),$$

where $\phi(z)=\sqrt{z}$ and C is a constant which depends on D and E.

3 Main Results

Let $\tilde{C_B}(\mathbb{R}_+)$ be the space of all real-valued bounded uniformly continuous functions $\hbar$ on $\mathbb{R}_+$, endowed with the norm

$$\|\hbar\|_\infty=\sup_{z\in\mathbb{R}_+}|\hbar(z)|.$$

The following result shows that the operators (3) are bounded linear operators on $\tilde{C_B}(\mathbb{R}_+)$.

Theorem 1 *For $\hbar\in\tilde{C_B}(\mathbb{R}_+)$ and each $z\in\mathbb{R}_+$, we have*

$$|L_m(\hbar;z)|\le\|\hbar\|_\infty.$$

Proof Applying the definition (3) and using Lemma 2, we have

$$\begin{aligned}|L_m(\hbar; z)| &\leq \left|\frac{1}{D(1)B(\alpha_m z E(1))}\sum_{k=0}^{\infty} p_k(\alpha_m z)\int_0^1 \hbar\left(\frac{k+s^\gamma}{\beta_m}\right)\mathrm{d}s\right| \\ &\leq \frac{1}{D(1)B(\alpha_m z E(1))}\sum_{k=0}^{\infty} p_k(\alpha_m z)\int_0^1 \left|\hbar\left(\frac{k+s^\gamma}{\beta_m}\right)\right|\mathrm{d}s \\ &\leq \|\hbar\|_\infty L_m(1; z) \\ &\leq \|\hbar\|_\infty,\end{aligned}$$

which completes the proof.

3.1 *Local Approximation*

Theorem 2 *Let $\hbar \in \tilde{C_B}(\mathbb{R}_+)$. Then for every $z \in \mathbb{R}_+$, the operators $L_m(.; z)$ given by (3), verify the following inequality:*

$$|L_m(\hbar; z) - \hbar(z)| \leq 2\omega(\hbar; (\mu_{m,2}(z))^{\frac{1}{2}}).$$

Proof Applying the property of modulus of continuity and Lemma 2, for any $\delta > 0$, we get

$$\begin{aligned}|L_m(\hbar; z) - \hbar(z)| &\leq \frac{1}{D(1)B(\alpha_m z E(1))}\sum_{k=0}^{\infty} p_k(\alpha_m z)\int_0^1 \left|\hbar\left(\frac{k+s^\gamma}{\beta_m}\right) - \hbar(z)\right|\mathrm{d}s \\ &\leq \omega(\hbar, \delta)\left\{1 + \frac{1}{\delta}\frac{1}{D(1)B(\alpha_m z E(1))}\sum_{k=0}^{\infty} p_k(\alpha_m z)\int_0^1 \left|\frac{k+s^\gamma}{\beta_m} - z\right|\mathrm{d}s\right\}.\end{aligned}$$

Now, using the Cauchy–Schwarz inequality and Lemma 2, we obtain

$$\begin{aligned}|L_m(\hbar; z) - \hbar(z)| &\leq \omega(\hbar, \delta)\left\{1 + \frac{1}{\delta}\left(\frac{1}{D(1)B(\alpha_m z E(1))}\sum_{k=0}^{\infty} p_k(\alpha_m z)\int_0^1 \left(\frac{k+s^\gamma}{\beta_m} - z\right)^2\mathrm{d}s\right)^{\frac{1}{2}}\right\} \\ &\leq \omega(\hbar, \delta)\left\{1 + \frac{1}{\delta}\left(\mu_{m,2}(z)\right)^{\frac{1}{2}}\right\}.\end{aligned}$$

On choosing $\delta := (\mu_{m,2}(z))^{\frac{1}{2}}$, we reach the desired result.

Next, we analyze the approximation of functions in a Lipschitz-type space introduced by Szász [35]. The Lipschitz-type space [35] is defined as :

$$Lip_M^*(\rho) := \left\{ \hbar \in C(\mathbb{R}_+) : |\hbar(s) - \hbar(z)| \leq M_\hbar \frac{|s-z|^\rho}{(s+z)^{\frac{\rho}{2}}};\ s \geq 0,\ z \in (0, \infty) \right\} \tag{4}$$

for some $M_\hbar > 0$ and $\rho \in (0, 1]$.

Theorem 3 *If* $\hbar \in Lip_M^*(\rho)$ *and* $\rho \in (0, 1]$, *then* $\forall\, z > 0$,

$$|L_m(\hbar(s); z) - \hbar(z)| \leq M_\hbar \left(\frac{\mu_{m,2}(z)}{z} \right)^{\frac{\rho}{2}}.$$

Proof Since $\hbar \in Lip_M^*(\rho)$, using the Definition (4)

$$|\hbar(s) - \hbar(z)| \leq M_\hbar \frac{|s-z|^\rho}{(s+z)^{\frac{\rho}{2}}}.$$

Applying the operator $L_m(.; z)$ on both sides of the above inequality, we get

$$|L_m(\hbar; z) - \hbar(z)| \leq M_\hbar L_m \left(\frac{|s-z|^\rho}{(s+z)^{\frac{\rho}{2}}}; z \right).$$

Now, using Hölder inequality with $p = \frac{2}{\rho},\ \ q = \frac{2}{(2-\rho)}$ and using *Lemma* 2

$$|L_m(\hbar; z) - \hbar(z)| \leq M_\hbar \left(L_m \left(\frac{(s-z)^2}{(s+z)}; z \right) \right)^{\frac{\rho}{2}}.$$

Since $(s + z) \geq z,\ \forall\, s \geq 0$ *and* $z > 0$, we have

$$|L_m(\hbar; z) - \hbar(z)| \leq \frac{M_\hbar}{z^{\frac{\rho}{2}}} (L_m(s-z)^2; z)^{\frac{\rho}{2}},$$

which yields the required result.

Lipschitz-type maximal function of order $\rho \in (0, 1]$: The Lipschitz-type maximal function of order ρ introduced by Lenze [25] is defined as:

$$\tilde{\omega}_\rho(\hbar; z) = \sup_{s \neq z \text{ and } z, s \in \mathbb{R}_+} \frac{|\hbar(s) - \hbar(z)|}{|s-z|^\rho}, \tag{5}$$

where $\rho \in (0, 1]$. For contributions of researchers on approximation of functions in this space, we refer to [20–22, 27].

In our next result, we study a local direct estimate for the operators (3), using the Lipschitz-type maximal function of order ρ.

Theorem 4 *Let* $\hbar \in \tilde{C}_B(\mathbb{R}_+)$ *and* $\rho \in (0, 1]$, *then* $\forall\, z \in [0, \infty)$

$$|L_m(\hbar; z) - \hbar(z)| \leq \tilde{\omega}_\rho(\hbar; z)(\mu_{m,2}(z))^{\frac{\rho}{2}}.$$

Proof In view of the equation (5),

$$|\hbar(s) - \hbar(z)| \leq \tilde{\omega}_\rho(\hbar; z)|s - z|^\rho.$$

Applying the operator $L_m(.; z)$ on both sides of the above inequality, and then using Lemma 2 and Hölder inequality with $p = \frac{2}{\rho},\ q = \frac{2}{(2-\rho)}$, we get

$$\begin{aligned}|L_m(\hbar; z) - \hbar(z)| &\leq \tilde{\omega}_\rho(\hbar; z)Ł_m(|s - z|^\rho; z)\\ &\leq \tilde{\omega}_\rho(\hbar; z)(Ł_m(s - z)^2; z)^{\frac{\rho}{2}}(Ł_m(1^2; z))^{\frac{(2-\rho)}{2}}\\ &= \tilde{\omega}_\rho(\hbar; z)(\mu_{m,2}(z))^{\frac{\rho}{2}}.\end{aligned}$$

This completes the proof.

Now we shall study a result which provides an error in terms of first- and second-order modulus of continuity via Peetre's K-functional. For this purpose, we need the following definitions:

Peetre's K-functional: For $\hbar \in \tilde{C}_B(\mathbb{R}_+)$, the Peetre's K-functional is given by

$$K(\hbar; \delta) = \inf_{g \in \tilde{C}_B^2(\mathbb{R}_+)} \left\{\|\hbar - g\|_\infty + \delta\|g\|_{\tilde{C}_B^2(\mathbb{R}_+)}, \delta > 0\right\},$$

where $\tilde{C}_B^2(\mathbb{R}_+) = \left\{g \in \tilde{C}_B(\mathbb{R}_+) : g', g'' \in \tilde{C}_B(\mathbb{R}_+)\right\}$ with the norm

$$\|g\|_{\tilde{C}_B^2} = \|g\|_\infty + \|g'\|_\infty + \|g''\|_\infty.$$

It is well known that $\forall\, \delta > 0$, there holds

$$K(\hbar; \delta) \leq M\,(\omega_2(\hbar; \sqrt{\delta}) + min\{1, \delta\}\|\hbar\|_\infty), \tag{6}$$

where M is a positive constant that does not depend on δ and f, and $\tilde{\omega}_2(\hbar; \delta)$ is the second-order modulus of continuity of $\hbar \in \tilde{C}_B(\mathbb{R}_+)$, which is defined as follows:

$$\omega_2(\hbar; \delta) = \sup_{0<|h|\leq\delta}\ \sup_{z, z\pm h \in \mathbb{R}_+} |\hbar(z + h) - 2\hbar(z) + \hbar(z - h)|. \tag{7}$$

For $\hbar \in \tilde{C}_B(\mathbb{R}_+)$, the first order modulus of continuity of f is defined as

$$\omega(\hbar; \delta) = \sup_{|s-z|\leq\delta} |\hbar(s) - \hbar(z)|. \tag{8}$$

For similar studies by other researchers, the interested reader may refer to (cf. [13, 16–18, 23] etc.).

Theorem 5 *For $\hbar \in \tilde{C}_B(\mathbb{R}_+)$, then*

$$|L_m(\hbar; z) - \hbar(z)| \leq 4K(\hbar; \delta_m(z)) + \omega(\hbar; |\mu_{m,1}(z)|),$$

where

$$\delta_m(z) = \frac{\mu_{m,2}(z) + \mu_{m,1}^2(z)}{8}.$$

Consequently,

$$|L_m(\hbar; z) - \hbar(z)| \leq M\left(\omega_2\left(\hbar; \sqrt{\delta_m(z)}\right) + min\{1, \delta_m(z)\}\|\hbar\|_\infty\right) + \omega(\hbar; |\mu_{m,1}(z)|).$$

Proof For $\hbar \in \tilde{C}_B(\mathbb{R}_+)$, we define an auxiliary operator $L_m^\star(\hbar; z)$ as follows:

$$L_m^\star(\hbar; z) = L_m(\hbar; z) - \hbar\left(\frac{1}{\beta_m}\left(\frac{1}{\gamma+1} + \frac{D'(1)}{D(1)}\right) + \frac{\alpha_m}{\beta_m} zE'(1)\frac{B'(\alpha_m zE(1))}{B(\alpha_m zE(1))}\right) + \hbar(z).$$

From Lemma 2, it follows that $L_m^\star(1; z) = 1$, $L_m^\star(s; z) = z$, and therefore

$$L_m^\star((s-z); z) = 0. \tag{9}$$

Further, in view of Lemma 2, we obtain

$$|L_m^\star(\hbar; z)| \leq |L_m(\hbar; z)| + \left|\hbar\left(\frac{1}{\beta_m}\left(\frac{1}{\gamma+1} + \frac{D'(1)}{D(1)}\right) + \frac{\alpha_m}{\beta_m} zE'(1)\frac{B'(\alpha_m zE(1))}{B(\alpha_m zE(1))}\right)\right| + |\hbar(z)|$$

$$\leq 3\|\hbar\|_\infty. \tag{10}$$

For any $g \in \tilde{C}_B^2(\mathbb{R}_+)$, using Taylor's theorem we may write

$$g(s) = g(z) + (s-z)g'(z) + \int_z^s (s-v)g''(v)\mathrm{d}v.$$

Now operating by $L_m^{\star}(.;z)$ on the above equation and using (9), we get

$$
\begin{aligned}
L_m^{\star}(g;z)-g(z) &= L_m^{\star}\bigg(\int_z^s (s-v)g''(v)\mathrm{d}v;z\bigg)\\
&= L_m\bigg(\int_z^s (s-v)g''(v)\mathrm{d}v;z\bigg)\\
&- \int_z^{\frac{1}{\beta_m}\left(\frac{1}{(\gamma+1)}+\frac{D'(1)}{D(1)}+\alpha_m zE'(1)\frac{B'(\alpha_m zE(1))}{B(\alpha_m zE(1))}\right)} \bigg\{\frac{1}{\beta_m}\bigg(\frac{1}{(\gamma+1)}+\frac{D'(1)}{D(1)}\\
&+ \alpha_m zE'(1)\frac{B'(\alpha_m zE(1))}{B(\alpha_m zE(1))}\bigg)-v\bigg\}g''(v)\mathrm{d}v.
\end{aligned}
$$

Hence,

$$
\begin{aligned}
|L_m^{\star}(\hbar;z)-\hbar(z)| &\le \bigg|L_m\bigg(\int_z^s (s-v)g''(v)\mathrm{d}v;z\bigg)\bigg|\\
&+ \bigg|\int_z^{\frac{1}{\beta_m}\left(\frac{1}{(\gamma+1)}+\frac{D'(1)}{D(1)}+\alpha_m zE'(1)\frac{B'(\alpha_m zE(1))}{B(\alpha_m zE(1))}\right)} \bigg\{\frac{1}{\beta_m}\bigg(\frac{1}{(\gamma+1)}+\frac{D'(1)}{D(1)}\\
&+ \alpha_m zE'(1)\frac{B'(\alpha_m zE(1))}{B(\alpha_m zE(1))}\bigg)-v\bigg\}\times g''(v)\mathrm{d}v\bigg|\\
&\le L_m\bigg(\bigg|\int_z^s |s-v||g''(v)|\mathrm{d}v\bigg|;z\bigg)\\
&+ \bigg|\bigg(\int_z^{\frac{1}{\beta_m}\left(\frac{1}{(\gamma+1)}+\frac{D'(1)}{D(1)}+\alpha_m zE'(1).\frac{B'(\alpha_m zE(1))}{B(\alpha_m zE(1))}\right)} \bigg|\frac{1}{\beta_m}\bigg(\frac{1}{(\gamma+1)}+\frac{D'(1)}{D(1)}\\
&+ \alpha_m zE'(1)\frac{B'(\alpha_m zE(1))}{B(\alpha_m zE(1))}\bigg)-v\bigg||g''(v)|\mathrm{d}v\bigg)\bigg|
\end{aligned}
$$

$$
\begin{aligned}
&\leq \|g''\|_{\infty}\left\{L_m\left(\left|\int_z^s |s-v|\mathrm{d}v\right|;z\right)\right. \\
&+\left|\left(\int_z^{\frac{1}{\beta_m}\left(\frac{1}{(\gamma+1)}+\frac{D'(1)}{D(1)}+\alpha_m zE'(1).\frac{B'(\alpha_m zE(1))}{B(\alpha_m zE(1))}\right)}\left|\frac{1}{\beta_m}\left(\frac{1}{(\gamma+1)}+\frac{D'(1)}{D(1)}\right.\right.\right.\right. \\
&\left.\left.\left.\left.+\alpha_m zE'(1)\frac{B'(\alpha_m zE(1))}{B(\alpha_m zE(1))}\right)-v\right|\mathrm{d}v\right)\right|\Bigg\}. \\
&\leq \frac{\|g\|_{\tilde{C}_B^2}}{2}\left\{L_m((s-z)^2;z)+\left\{\frac{1}{\beta_m}\left(\frac{1}{(\gamma+1)}+\frac{D'(1)}{D(1)}\right.\right.\right. \\
&\left.\left.\left.+\alpha_m zE'(1).\frac{B'(\alpha_m zE(1))}{B(\alpha_m zE(1))}\right)-z\right\}^2\right\} \\
&\leq \frac{\|g\|_{\tilde{C}_B^2}}{2}\left\{L_m((s-z)^2;z)+(L_m(s-z);z)^2\right\}.
\end{aligned}
$$

$$
\leq \frac{\|g\|_{\tilde{C}_B^2}}{2}\{\mu_{m,2}(z)+\mu_{m,1}^2(z)\}. \tag{11}
$$

Thus, for $\hbar \in \tilde{C}_B(\mathbb{R}_+)$ and any $g \in \tilde{C}_B^2(\mathbb{R}_+)$, we have

$$
\begin{aligned}
|L_m(\hbar;z)-\hbar(z)| &= \left|L_m^{\star}(\hbar;z)-\hbar(z)+\hbar\left(\frac{1}{\beta_m}\left(\frac{1}{(\gamma+1)}+\frac{D'(1)}{D(1)}\right.\right.\right. \\
&\left.\left.\left.+\alpha_m zE'(1).\frac{B'(\alpha_m zE(1))}{B(\alpha_m zE(1))}\right)\right)-\hbar(z)\right| \\
&\leq |L_m^{\star}(\hbar-g;z)|+|L_m^{\star}(g;z)-g(z)|+|g(z)-\hbar(z)| \\
&+\left|f\left(\frac{1}{\beta_m}\left(\frac{1}{(\gamma+1)}+\frac{D'(1)}{D(1)}+\alpha_m zE'(1)\frac{B'(\alpha_m zE(1))}{B(\alpha_m zE(1))}\right)\right)-\hbar(z)\right|.
\end{aligned}
$$

Hence in view of (8), (10) and (11), we get

$$
\begin{aligned}
|L_m(\hbar;z)-\hbar(z)| &\leq 4\|\hbar-g\|_{\infty}+\|g\|_{\tilde{C}_B^2}\frac{\{\mu_{m,2}(z)+\mu_{m,1}^2(z)\}}{2}+\omega(\hbar;|\mu_{m,1}(z)|) \\
&\leq 4\{\|\hbar-g\|_{\infty}+\delta_m(z)\|g\|_{\tilde{C}_B^2}\}+\omega(\hbar;|\mu_{m,1}(z)|),
\end{aligned}
$$

Now taking the greatest lower bound on the right-hand side over all $g \in \tilde{C}_B^2(\mathbb{R}_+)$, we obtain

$$|L_m(\hbar; z) - \hbar(z)| \leq 4K(\hbar; \delta_m(z)) + \omega(\hbar; |\mu_{m,1}(z)|),$$

which proves the first assertion of the theorem. Finally, using the relation between Peetre's K-functional and the second-order modulus of continuity given by (6), we arrive at the second assertion of the theorem. This completes the proof.

Our next theorem determines the rate of convergence of the operators (3) for bounded uniformly continuously differentiable functions in $\mathbb{R}_+$.

Theorem 6 *Let $f, \hbar' \in \tilde{C}_B[0, \infty)$, then $\forall\, z \geq 0$,*

$$|L_m(\hbar; z) - \hbar(z)| \leq \left\{|\hbar'(z)| + \frac{3}{2}\omega\left(\hbar'; \sqrt{\mu_{m,2}(z)}\right)\right\}\sqrt{\mu_{m,2}(z)}.$$

Proof Since $f, \hbar' \in \tilde{C}_B^1[0, \infty)$, we may write

$$\hbar(s) - \hbar(z) = (s - z)\hbar'(z) + \int_z^s (\hbar'(u) - \hbar'(z))du.$$

Now, by using the well-known property of usual modulus of continuity, for $\delta > 0$ and $\hbar' \in \tilde{C}_B^1[0, \infty)$, we have

$$|\hbar'(v) - \hbar'(z)| \leq \omega(\hbar', \delta)\left(\frac{|v - z|}{\delta} + 1\right).$$

Hence,

$$\left|\int_z^s (\hbar'(v) - \hbar'(z))\mathrm{d}v\right| \leq \omega(\hbar', \delta)\left(\frac{(s - z)^2}{2\delta} + |s - z|\right).$$

Thus,

$$|L_m(\hbar; z) - \hbar(z)| \leq |\hbar'(z)|L_m(|s - z|; z) + \omega(\hbar', \delta)\left(\frac{1}{2\delta}L_m((s - z)^2; z)\right.$$
$$\left. + L_m(|s - z|; z)\right).$$

Now applying the Cauchy–Schwarz inequality and Lemma 2, we get

$$\begin{aligned}|L_m(\hbar; z) - \hbar(z)| &\leq (|\hbar'(z)| + \omega(\hbar', \delta))\sqrt{L_m((s-z)^2; z)}\sqrt{L_m(1; z)} \\ &\quad + \frac{1}{2\delta}\omega(\hbar', \delta)\Big(L_m((s-z)^2; z)\Big) \\ &= (|\hbar'(z)| + \omega(\hbar', \delta))\sqrt{\mu_{m,2}(z)} + \frac{1}{2\delta}\omega(\hbar', \delta)\Big(\mu_{m,2}(z)\Big).\end{aligned}$$

Now choosing $\delta = \sqrt{\mu_{m,2}(z)}$, we reach the desired result.

Next, our goal is to determine the order of approximation for our operators (3) by means of the Ditzian–Totik modulus of smoothness. Let $\phi(z) = \sqrt{z}$ and for $\hbar \in \tilde{C}_B(\mathbb{R}_+)$, the Ditzian–Totik modulus of smoothness for the first order is given by

$$\omega_\phi(\hbar; \delta) = \sup_{0 \leq f \leq \delta}\left\{\left|\hbar\Big(z + \frac{f\phi(z)}{2}\Big) - \hbar\Big(z - \frac{f\phi(z)}{2}\Big)\right|; z \pm \frac{f\phi(z)}{2} \in \mathbb{R}_+\right\},$$

and an appropriate Peetre's K-functional is defined as

$$K_\phi(\hbar; \delta) = \inf_{g \in \tilde{C}_B(\mathbb{R}_+)}\{\|\hbar - g\|_\infty + \delta\|\phi g'\|_\infty\},\ \delta > 0,$$

where $AC_{loc}(\mathbb{R}_+)$ denotes the set of all locally absolutely continuous functions on every compact interval $[a, b] \subseteq \mathbb{R}_+$. From [9], it is known that $\exists$ a positive constant $M > 0$ such that

$$M^{-1}\omega_\phi(\hbar; \delta) \leq K_\phi(\hbar; \delta) \leq M\,\omega_\phi(\hbar; \delta). \tag{12}$$

Theorem 7 *For any $\hbar \in \tilde{C}_B(\mathbb{R}_+)$ and $z \in (0, \infty)$,*

$$|L_m(\hbar; z) - \hbar(z)| \leq M\,\omega_\phi\left(\hbar; \sqrt{\frac{\mu_{m,2}(z)}{z}}\right).$$

Proof Using Taylor's theorem, for any $g \in \tilde{C}_B^2(\mathbb{R}_+)$, we get

$$g(s) = g(z) + \int_z^s g'(v)\mathrm{d}v$$

$$= g(z) + \int_z^s \frac{g'(v)\phi(v)}{\phi(v)}\mathrm{d}v.$$

Hence,

$$|g(s) - g(z) \leq \|\phi g'\|_\infty \left| \int_z^s \frac{1}{\phi(v)}\mathrm{d}v \right|$$

$$\leq \|\phi g'\|_\infty 2|\sqrt{s} - \sqrt{z}|$$

$$\leq \|\phi g'\|_\infty \frac{|s-z|}{\sqrt{s}+\sqrt{z}},$$

$$\leq 2\|\phi g'\|_\infty \frac{|s-z|}{\sqrt{z}}$$

$$= 2\|\phi g'\|_\infty \frac{|s-z|}{\phi(z)}.$$

Consequently for any $g \in \tilde{C}_B^2(\mathbb{R}_+)$, using Lemma 3, we get

$$|L_m(\hbar; z) - \hbar(z)| \leq |L_m(\hbar - g; z)| + |L_m(g; z) - g(z)| + |g(z) - \hbar(z)|$$
$$\leq 2\|\hbar - g\|_\infty + \frac{2\|\phi g'\|_\infty}{\phi(z)} L_m(|s-z|; z).$$

Now applying Cauchy–Schwarz inequality, we are led to

$$|L_m(\hbar; z) - \hbar(z)| \leq 2\|\hbar - g\|_\infty + \frac{2\|\phi g'\|_\infty}{\phi(z)} L_m((s-z)^2; z)^{\frac{1}{2}}$$
$$\leq 2\|\hbar - g\|_\infty + \frac{2\|\phi g'\|_\infty}{\phi(z)} \sqrt{\mu_{m,2}(z)}.$$

Now taking the greatest lower bound on the right-hand side over all $g \in \tilde{C}_B^2(\mathbb{R}_+)$, we reach

$$|L_m(\hbar; z) - \hbar(z)| \leq 2K_\phi\left(\hbar; \sqrt{\frac{\mu_{m,2}(z)}{z}}\right),$$

which yields us the desired result with the help of relation (12) between Ditzian–Totik modulus of smoothness and Peetre's K-functional.

In the next theorem, we discuss the approximation of unbounded functions by the operators (3).

3.2 Weighted Approximation

Let us consider a weighted space defined as

$$B_2(\mathbb{R}_+) := \left\{\hbar : (\mathbb{R}_+) \to R \middle| |\hbar(z)| \leq M_\hbar(1+z^2)\right\}, \tag{13}$$

where $M_\hbar > 0$, is a positive constant which depends only on $\hbar$. Clearly, $B_2(\mathbb{R}_+)$ is normed linear space with the norm given by

$$\|\hbar\|_2 = \sup_{s\in\mathbb{R}_+} \frac{|\hbar(s)|}{1+s^2}. \tag{14}$$

Further, let us consider the spaces

$$C_2(\mathbb{R}_+) := \{\hbar \in B_2(\mathbb{R}_+) \ : \ \hbar \textit{ is a continuous function}\}$$

and

$$C_2^\star(\mathbb{R}_+) = \{\hbar \in C_2(\mathbb{R}_+) : \lim_{z\to\infty} \frac{\hbar(z)}{1+z^2} \textit{ is finite}\}.$$

Significant contributions have been made by several mathematicians in this direction (cf. [2, 9, 20] etc.).

Theorem 8 *Let $\hbar \in C_2^\star(\mathbb{R}_+)$, and $\omega_a(\hbar;\delta)$ be its modulus of continuity on $[0, a+1] \subset [0,\infty)$, where $a > 0$ then we have*

$$|L_m(\hbar; z) - \hbar(z)| \leq 4M_\hbar(1+a^2)\mu_{m,2}(z) + 2\omega_{a+1}\left(\hbar; \sqrt{\mu_{m,2}(z)}\right)$$

where $\mu_{m,2}(z)$ is the second-order central moment.

Proof From [14], for any $z \in [0, a+1]$ and $t \in [0,\infty)$, we have

$$|\hbar(s) - \hbar(z)| \leq 4M_\hbar(1+a^2)(s-z)^2 + \left(1 + \frac{|s-z|}{\delta}\right)\omega_{a+1}(\hbar;\delta), \ \ \delta > 0.$$

Thus, using Cauchy–Schwarz inequality, we have

$$\begin{aligned}|L_m(\hbar; z) - \hbar(z)| &\leq 4M_\hbar(1+a^2)Ł_m((s-z)^2; z) + \omega_{a+1}(\hbar;\delta)\left(1 + \frac{\sqrt{L_m(s-z)^2}}{\delta}\right)\\ &\leq 4M_\hbar(1+a^2)\mu_{m,2}(z) + \omega_{a+1}(\hbar;\delta)\left(1 + \frac{\sqrt{\mu_{m,2}(z)}}{\delta}\right).\end{aligned}$$

On choosing $\delta = \sqrt{\mu_{m,2}(z)}$, we get the desired result.

Theorem 9 *For each $\hbar \in C_2^{\star}(\mathbb{R}_+)$ and $\mu > 0$, we have*

$$\lim_{n\to\infty} \frac{|L_m(\hbar; z) - \hbar(z)|}{(1+z^2)^{1+\mu}} = 0.$$

Proof For any fixed point $z_0 \in \mathbb{R}_+$, we have

$$\begin{aligned}\sup_{z\in[0,\infty)} \frac{|L_m(\hbar; z) - \hbar(z)|}{(1+z^2)^{1+\mu}} &\leq \sup_{z\leq z_0} \frac{|L_m(\hbar; z) - \hbar(z)|}{(1+z^2)^{1+\mu}} + \sup_{z>z_0} \frac{|L_m(\hbar; z) - \hbar(z)|}{(1+z^2)^{1+\mu}} \\ &\leq \sup_{z\leq z_0}\{|L_m(\hbar; z) - \hbar(z)|\} + \sup_{z>z_0} \frac{|L_m(\hbar; z)| + |\hbar(z)|}{(1+z^2)^{1+\mu}}.\end{aligned}$$

Using the fact that $|\hbar(s)| \leq \|\hbar\|_2(1+s^2)$, we have

$$\begin{aligned}\sup_{z\in[0,\infty)} \frac{|L_m(\hbar; z) - \hbar(z)|}{(1+z^2)^{1+\mu}} &\leq \|L_m(\hbar; z) - \hbar(z)\|_{C[0,z_0]} + \|\hbar\|_2 \sup_{z>z_0} \frac{|L_m((1+s^2); z)|}{(1+z^2)^{1+\mu}} \\ &\quad + \sup_{z>z_0} \frac{|\hbar(z)|}{(1+z^2)^{1+\mu}} \\ &= J_1 + J_2 + J_3. \end{aligned} \tag{15}$$

In view of Theorem 8, the sequence $\{L_m(\hbar; z)\}$ uniformly converges to the function $\hbar(z)$ on every closed interval $[0, a]$, as $m \to \infty$, so for a given $\epsilon > 0$, $\exists\, N_1 \in \mathbb{N}$ such that

$$J_1 = \|L_m(\hbar; z) - \hbar(z)\|_{C[0,z_0]} < \frac{\epsilon}{3}, \forall\, m \geq N_1. \tag{16}$$

Now, we shall find an estimate of

$$sup_{z>z_0} \frac{|L_m((1+s^2); z)|}{(1+z^2)^{1+\mu}}.$$

By using Lemma 2, we can find a positive integer $N_2 \in \mathbb{N}$ such that

$$|L_m((1+s^2); z) - (1+z^2)| < \frac{\epsilon}{3\|\hbar\|_2}, \forall\, m \geq N_2,$$

or,

$$|L_m((1+s^2); z)| < (1+z^2) + \frac{\epsilon}{3\|\hbar\|_2}, \forall\, m \geq N_2.$$

Hence,

$$J_2 = \|\hbar\|_2 \sup_{z>z_0} \frac{|L_m((1+s^2);z)|}{(1+z^2)^{1+\mu}} < \|\hbar\|_2 \sup_{z>z_0} \frac{1}{(1+z^2)^{1+\mu}} \left((1+z^2) + \frac{\epsilon}{3\|\hbar\|_2}\right), \forall\, m \geq N_2,$$
$$< \|\hbar\|_2 \sup_{z>z_0} \frac{1}{(1+z^2)^{\mu}} + \frac{\epsilon}{3}, \forall\, m \geq N_2.$$

Thus,

$$J_2 < \frac{\|\hbar\|_2}{(1+z_0^2)^{\mu}} + \frac{\epsilon}{3}, \forall\, m \geq N_2. \tag{17}$$

From (14), we get

$$J_3 = \sup_{z>z_0} \frac{|\hbar(z)|}{(1+z^2)^{1+\mu}} \leq \frac{\|\hbar\|_2}{(1+z_0^2)^{\mu}}. \tag{18}$$

Let $N_0 = max\{N_1, N_2\}$. Combining (16)–(18), we obtain

$$J_1 + J_2 + J_3 < \frac{2\|\hbar\|_2}{(1+z_0^2)^{\mu}} + \frac{2\epsilon}{3}, \forall\, m \geq N_0. \tag{19}$$

Now, choosing z_0 to be so large that

$$\frac{2\|\hbar\|_2}{(1+z_0^2)^{\mu}} < \frac{\epsilon}{3},$$

the proof is completed by combining (15) and (19).

3.3 Quantitative Voronovskaja-Type Theorem for the Operators L_m

In this subsection, we present a quantitative Voronovskaja-type theorem for the operators (3).

Theorem 10 *If $\hbar \in \tilde{C}_B^2(R_+)$, then for each $z > 0$, there holds the following inequality:*

$$\left|\beta_m\left\{L_m(\hbar;z) - \hbar(z) - \mu_{m,1}(z)\frac{\hbar'(z)}{1!} - \frac{1}{2}\hbar''(z)\mu_{m,2}(z)\right\}\right| \leq M'\phi^2(z)\omega_\phi(\hbar''; {\beta_m}^{-1/2}),$$

where M' is some positive constant.

Proof For $\hbar \in \tilde{C}_B^2(R_+)$ and $z > 0$, using Taylor's series formula, we have

$$\hbar(v) - \hbar(z) = (v - z)\hbar'(z) + \int_z^v (v - u)\hbar''(u)du.$$

Hence,

$$\hbar(v) - \hbar(z) - (v - z)\hbar'(z) - \frac{(v - z)^2}{2}\hbar''(z) = \int_z^v (v - u)(\hbar''(u) - \hbar''(z))du.$$

Now applying the linear positive operator $L_m(.; z)$ on both sides of the above equation, we have

$$\left| L_m(\hbar; z) - \hbar(z) - \mu_{m,1}(z)\hbar'(z) - \mu_{m,2}(z)\frac{\hbar''(z)}{2} \right| \\ \leq L_m\left(\left| \int_z^v |v - u|(\hbar''(u) - \hbar''(z))du \right|; z \right). \tag{20}$$

For any $g \in \tilde{C}_B^2(\mathbb{R}_+)$ Finta ([10], p. 337) estimated the right-hand quantity of Eq. (20) as follows:

$$\left| \int_z^v |v - u|(\hbar''(u) - \hbar''(z))du \right| \leq 2\|\hbar'' - g\|(v - z)^2 \\ + 2\|\phi g'\|\phi^{-1}(z)|v - z|^3. \tag{21}$$

Now, combining (20) and (21)

$$\left| L_m(\hbar; z) - \hbar(z) - \mu_{m,1}\hbar'(z) - \frac{\hbar''(z)}{2}\mu_{m,2}(z) \right| \leq 2\|\hbar'' - g\|L_m((v - z)^2; z) \\ + 2\|\phi g'\|\phi^{-1}(z)L_m(|v - z|^3; z),$$

then by using the Cauchy–Schwarz inequality and Remark 1, we obtain

$$\left|L_m(\hbar; z) - \hbar(z) - \mu_{m,1}\hbar'(z) - \frac{\hbar''(z)}{2}\mu_{m,2}(z)\right| \leq 2\|\hbar'' - g\|\mu_{m,2}(z)$$
$$+ (2\|\phi g'\|\phi^{-1}(z)\{L_m((v-z)^2; z)\}^{\frac{1}{2}})$$
$$\times (\{L_m((v-z)^4; z)\}^{\frac{1}{2}})$$
$$\leq 2\|\hbar'' - g\|\mu_{m,2}(z)$$
$$+ 2\|\phi g'\|\phi^{-1}(z)\{\mu_{m,2}(z)\}^{\frac{1}{2}}\{\mu_{m,4}(z)\}^{\frac{1}{2}}$$
$$\leq 2\|\hbar'' - g\|\frac{C}{\beta_m}\phi^2(z) + \frac{2C}{\beta_m^{\frac{3}{2}}}\|\phi g'\|\phi^2(z)$$
$$\leq \frac{2C}{\beta_m}\phi^2(z)\left(\|\hbar'' - g\| + \frac{\|\phi g'\|}{\sqrt{\beta_m}}\right).$$

Now, taking the greatest lower bound on the right-hand side over all $g \in \tilde{C}_B^2(\mathbb{R}_+)$, we obtain

$$\left|\beta_m\left\{L_m(\hbar; z) - \hbar(z) - \mu_{m,1}(z)\frac{\hbar'(z)}{1} - \mu_{m,2}(z)\frac{\hbar''(z)}{2}\right\}\right| \leq M'\phi^2(z)K_\phi(\hbar''; \beta_m{}^{-1/2})$$

.

where $M' = 2C$. Now, applying the relation (12) we reach our desired result.

3.4 Grüss–Voronovskaya-type Theorem for the Operators L_m

First of all, Gal and Gonska [11] introduced the Grüss–Voronovskaya-type theorem by means of the Grüss inequality which yields an estimate of the difference of integral of the two functions with the product of the integral of the two functions. Subsequently, several researchers worked in this direction (see, for instance, [1, 8, 37] etc.). inspired by these studies, we derive the Grüss–Voronovskaya-type theorem for the operators (3).

Theorem 11 *If $\hbar, g, \hbar', g', \hbar'', g'' \in \tilde{C}_B^2(\mathbb{R}_+)$, then for each $z \in \mathbb{R}_+$, the operators (3) satisfy the following equality:*

$$\lim_{m\to\infty} \beta_m\left[L_m(\hbar g; z) - L_m(\hbar; z)L_m(g; z)\right] = \hbar'(z)g'(z)\left(1 + \frac{E''(1)}{E'(1)}\right)x.$$

Proof We have the double derivative of the product of two functions $\hbar(z)$ and g(z) as

$$(\hbar g)''(z) = \hbar''(z)g(z) + 2\hbar'(z)g'(z) + g''(z)\hbar(z).$$

By making an appropriate arrangement, we get

$$\begin{aligned}\beta_m\{L_m((\hbar g); z) - L_m(\hbar; z)L_m(g; z)\} = \beta_m\Bigg\{ & L_m((\hbar g); z) - \hbar(z)g(z) \\ & - (\hbar g)'\frac{\mu_{m,1}(z)}{1!} - \frac{(\hbar g(z))''}{2!}\mu_{m,2}(z) \\ & - g(z)\Bigg[L_m(\hbar; z) - \hbar(z) \\ & - \frac{\hbar'(z)}{1!}\mu_{m,1}(z) - \frac{\hbar''(z)}{2!}\mu_{m,2}(z)\Bigg] \\ & - L_m(\hbar; z)\Bigg[L_m(g; z) - g(z) \\ & - \frac{g'(z)}{1!}\mu_{m,1}(z) - \frac{g''(z)}{2!}\mu_{m,2}(z)\Bigg] \\ & + 2\frac{\hbar'(z)g'(z)}{2!}\mu_{m,2}(z) \\ & + \frac{g''(z)}{2!}\mu_{m,2}(z)\Bigg(\hbar(z) - L_m(\hbar; z)\Bigg) \\ & + \frac{g'(z)}{1!}\mu_{m,1}(z)\Bigg[\hbar(z) - L_m(\hbar; z)\Bigg]\Bigg\}.\end{aligned}$$

Using Korovkin-type theorem, for each fixed point $z \in \mathbb{R}_+$, $L_m(\hbar; z) \to \hbar(z)$, as $m \to \infty$, and for $\hbar''(z) \in \tilde{C}_B(\mathbb{R}_+)$, $z \in \mathbb{R}_+$ by the quantitative Voronovskaja-type Theorem 10, we have

$$\lim_{m\to\infty} \beta_m\left(L_m(\hbar; z) - \hbar(z) - \frac{\hbar'(z)}{1!}\mu_{m,1}(z) - \frac{\hbar''(z)}{2!}\mu_{m,2}(z)\right) = 0.$$

Hence, applying Remark 1, we obtain the desired result.

4 Conclusion

In the past decade, there has been a great deal of interest to connect the approximation theory with the orthogonal polynomials. The present paper is an attempt in this direction. We propose a linking of the sequence of Szász–Kantorovich-type operators with the Boas–Buck-type polynomials which include several well-known polynomials such as Brenke-type polynomials, Sheffer polynomials and Appell polynomials and examine the convergence behavior of these operators by means of the approximation tools, e.g., modulus of continuity, Lipschitz-type maximal function and Peetre's K- functional. The quantitative Voronoskaya and Grüss–Voronovskaya-type theorems are also discussed. The proposed operator includes the well-known operator defined by *Öksüzer* et al. [28] for the case $\gamma = 1,\ \alpha_m = \beta_m = m$ and presents an extension of the operators studied by Atakut and B*üyü*kyazici [3], Yilik et al. [39] and Rao et al. [31] to the space of L_p- integrable functions. The results obtained in the paper are, therefore, applicable to a wide variety of operators involving such polynomials and will be of sufficient interest for the researchers working in the area of approximation theory and orthogonal polynomials. Further, our operator could be used to investigate the approximation of Lebesgue integrable functions on $\mathbb{R}_+$.

References

1. Acar T (2016) Quantitative q-Voronovskaja and q-Grüss-Voronovskaja-type results for q-Szász operators. Georgian Math J 23:459–468
2. Aral A, Acar T (2013) Weighted approximation by new Bernstein-Chlodowsky-Gadjiev operators. Filomat 27(2):371–380
3. Atakut C, B*üyü*kyazici (2016) Approximation by Kantorovich-Sz*á*sz type operators based on Brenke type polynomials. Numer Funct Anal Optim 37(12):1488–1502
4. Bojanic R, Cheng F (1989) Rate of convergence Bernstein polynomials for functions with derivatives of bounded variation. J Math Anal Appl 141(1):136–151
5. Bojanic R, Khan MK (1991) Rate of convergence of some operators of functions with derivatives of bounded variation. Atti Sem Mat Fis Univ Modena 39(2):495–512
6. B*üyü*kyazici İ, Tanberkan H, Serenbay SK, Atakut C (2014) Approximation by chlodowsky type Jakimovski-Leviatan operators. J Comput Appl Math 259(Part A):153–163
7. Cheng F (1983) On the rate of convergence of Bernstein polynomials of functions of bounded variation. J Approx Theory 39(3):259–274
8. Deniz E (2016) Quantitative estimates for Jain-Kantorovich operators. Commun Fac Sci Univ Ank Ser A1 Math Stat 65:121–132
9. Ditzian Z, Totik V (1987) Moduli of smoothness. Springer Series in Computational Mathematics, vol 9. Springer, New York
10. Finta Z (2011) Remark on Voronovskaja theorem for q-Bernstein operators. Stud Univ Babeş-Bolyai Math 56:335–339
11. Gal SG, Gonska H (2015) Grüss and Grüss-Voronovskaya-type estimates for some Bernstein -type polynomials of reals and complex variables. Jean J Approx 7:97–122
12. Garg T, Agrawal PN, Araci S (2017) Rate of approximation by Kantorovich-Szász type operators based on Brenke type polynomials. J Inequal Appl 2017:156
13. Goyal M, Gupta V, Agrawal PN (2015) Quantitative convergence results for a family of hybrid operators. Appl Math Comput 271:893–904
14. Ibikli E, Gadjieva EA (1995) The order of approximation of some unbounded functions by the sequence of linear positive operators. Turkish J Math 19(3):331–337
15. Ismail M (1974) On a generalization of Szász operators, Mathematica, 16(390(2)):259–267
16. İspir N (2001) On modified Baskakov operators on weighted spaces. Turkish J Math 25:355–365
17. İspir N (2007) Rate of convergence of generalized rational type Baskakov operators. Math Comput Modeling 46(5–6):625–631
18. İspir N, Atakut Ç (2002) Approximation by modified Sz*á*sz-Mirakjan operators on weighted spaces. Proc Indian Acad Sci Math Sci 112(4):571–578
19. Jakimovski A, Leviatan D (1969) Generalized Szász operators for the approximation in the infinite interval. Mathematica (Cluj) 11(34):97–103
20. Kajla A (2017) Direct estimates of certain Mihesan-Durrmeyer type operators. Adv Oper Theory 2(2):162–178

21. Kajla A, Acu AM, Agrawal PN (2017) Baskakov-Szász-type operators based on inverse Polya-Eggenberger distribution. Ann Funct Anal 8(1):106–123
22. Kajla A, Agrawal PN (2015) Approximation properties of Szàsz type operators based on Charlier polynomials. Turk J Math 39(6):990–1003
23. Kajla A, Agrawal PN (2015) Szász Durrmeyer types operators based on Charlier polynomials. Appl Math Comput 268:1001–1014
24. Kumar A (2020) Approximation properties of Generalized λ- bernstein- Kantorovich type operators. Rend Circ Mat Palermo II Ser. https://doi.org/10.1007/s12215-020-00509-2
25. Lenze B (1988) On lipschitz-type maximal functions and their smoothness spaces. Nederl Akad Wetensch Indag Math 50(1):53–63
26. Neer T, Acu AM, Agrawal PN (2019) Degree of approximation by Chlodowsky variant of Jakimovski-Leviatan-Durrmeyer type operators. RACSAM 113:3445–3459
27. Neer T, Agrawal PN, Araci S (2017) Stancu-Durrmeyer type operators based on q-integers. Appl Math Inf Sci 11(3):767–775
28. Öksuzer Ö, Karsli H, Taşdelen F (2016) Approximation by a Kantorovich variant of Szász operators based on Brenke-type polynomials. Mediterr J Math 13(5). https://doi.org/10.1007/s00009-016-0688-6
29. Özarslan MA, Duman O (2010) Local approximation behavior of modified SMK operators. Miskolc Math Notes 11(1):87–99
30. Özarslan MA, Duman O (2016) Smoothness properties of modified Bernstein- Kantorovich operators. Numer Funct Anal Optim 37(1):92–105
31. Rao N, Wafi A, Deepmala (2017) Approximation by Szász type operators including Sheffer polynomials. J Math Appl 40:135–148
32. Shaw SY, Liaw WC, Lin YL (1993) Rates for approximation of functions in BV[a, b] and DBV[a, b] by positive linear operators. Chinese J Math 21:171–193
33. Sidharth M, Acu AM, Agrawal PN (2017) Chlodowsky-Szasz-Appell type operators for functions of two variables. Ann Funct Anal 8(4):446–459
34. Sucu S, Icöz G, Varma S (2012) On some extensions of Szász operators including Boas-Buck-type polynomials. Abst Appl Anal 2012, 15pp. Article ID 680340
35. Szász O (1950) Generalization of S. Bernstein's polynomials to the infinite interval. J Res Nat Bur Standards 45:239–245
36. Taşdelen F, Aktaş R, Altin A (2012) A Kantorovich type of Szász operators including Brenke-type polynomials. Abstr Appl Anal 13pp. Art. ID 867203
37. Ulusoy G, Acar, T (2015) q-Voronovskaya type theorems for q-Baskakov operators. Math Methods Appl Sci 3391–3401
38. Varma S, Sucu S, İçóz G (2012) Generalization of Szász operators involving Brenke type polynomials. Comput Math Appl 64(2):121–127
39. Yilik ÖÖ, Garg T, Agrawal PN (2020) Convergence rate of Szász operators involving Boas?Buck-type polynomials. Proc Natl Acad Sci, India, Sect A Phys Sci. https://doi.org/10.1007/s40010-020-00663-3

Temperature Distribution in Living Tissue with Two-Dimensional Parabolic Bioheat Model Using Radial Basis Function

Rohit Verma and Sushil Kumar

Abstract This study concerns the numerical modeling and simulation of heat distribution inside the skin tissue for cancer treatment with external exponential heating. Here, we consider the two-dimensional Pennes bioheat model for thermal therapy based on Fourier's law of heat conduction. We approximate the temporal variable using finite difference approximation and the spatial variable using the radial basis functions (RBF)-based collocation method to solve the considered model. The effect of different parameters on thermal diffusion in skin tissue has also been studied.

Keywords Pennes bioheat model · Radial basis function · Finite difference method · Exponential heat flux · Collocation method

MSC Classification 65M70 · 35Q92 · 65M06 · 62P10 · 65D25

1 Introduction

Heat transfer in skin tissues is a complex phenomenon in nature. The heat transfer inside the living tissue is the most critical mathematical modeling problem with real applications starting from the Pennes bioheat model (PBHM). The heat exchange between the tissue and environment mainly involves the combination of thermal conduction in tissue, convection, metabolism, blood perfusion, and external heating source. There are several mathematical models available in the literature [1, 2]. Still, among these models, the Pennes model [3] is the most effective and exciting model

Rohit Verma and Sushil Kumar: These authors contributed equally to this work.

R. Verma (✉) · S. Kumar
Department of Mathematics and Humanities, S.V. National Institute of Technology Surat, Surat, Gujarat 395007, India
e-mail: rohitverma260194@gmail.com

S. Kumar
e-mail: sushilk@amhd.svnit.ac.in

S. S. Ray et al. (eds.), *Applied Analysis, Computation and Mathematical Modelling in Engineering*, Lecture Notes in Electrical Engineering 897,
https://doi.org/10.1007/978-981-19-1824-7_24

to analyze the heat transfer effect in living biological tissue. The model is based on Fourier's law

$$q(P,t) = -K\nabla T(P,t), \tag{1.1}$$

where $T(P,t)$ and $q(P,t)$ are the temperature and heat flux, respectively at position $P(x, y)$ and time t. Nowadays, the study on bioheat transfer models is getting more attention, which is beneficial for thermal treatments [4]. Ample literature is available to study the one-dimensional model, but the literature for the multidimensional domain is limited. Ahmadikia et al. [5] obtained the analytical solution for heat transfer inside the tissue using the parabolic and hyperbolic model with heat flux conditions, namely transient and constant heat flux. Also, the analytical study on PBHM with spatial transient heating inside the tissue or skin surface was given by Deng and Liu [6]. Analytic solution of two-dimensional bioheat equation was obtained by Azevedo et al. [7] with convective boundary condition. Biswas et al. [8] obtained the analytical solution of two-dimensional bioheat model with constant and sinusoidal conditions using the orthogonal eigenfunction expansion method (OEEM). Dutta et al. [9] developed an analytical solution for a two-dimensional hyperbolic model for temperature response in single-layer skin tissue using the hybrid form of separation of variables and finite integral transform method. Ezzat et al. [10] did the analytical study of a two-dimensional Pennes bioheat model with temperature-dependent thermal conductivity and rheological properties using Fourier series expansion-based numerical method.

Several techniques are available in the literature for the numerical solution of bioheat models. Some of these methods are very popular, like the finite volume method (FVM), finite element method (FEM), and finite difference method (FDM), and collocation method, etc. In these methods, the discretization techniques for the bioheat model's numerical solution are broadly and successfully implemented. These methods have the requirement of mesh generation, which is easy to implement for one-dimensional problems.

Ozen et al. [11] considered the thermal wave model and solved it using the finite difference method (FDM). Karaa et al. [12] gave the numerical solution of three-dimensional classical Pennes bioheat transfer equation with the different spatial heating conditions. Al-Humedi et al. [13] obtained the numerical solution for the one-dimensional unsteady state bioheat equation using shifted Legendre polynomial. Fahmy [14] studied the two-dimensional model based on the non-Fourier bioheat transfer model and Biot's theory by the boundary element method (BEM) in the complex shapes of soft tissue. Hosseininia et al. [15] did the numerical study of a two-dimensional variable-order fractional model of dual-phase lag using the Legendre wavelet method. Roohi et al. [16] proposed the fractional-order Legendre functions (F-OLFs) and Galerkin method to analyze the temperature distribution inside the tissue using space-time fractional-order bioheat equation. Stroher et al. [17] did the numerical study of two-dimensional Pennes bioheat model using a multigrid method with non-stationary and steady-state cases for skin health and with melanoma.

For multidimensional problems, mesh generation becomes a typical task due to complex geometry. In recent times, the meshless method has acquired a lot of attention due to overcoming the difficulty of mesh generation [18]. In 1990, Kansa [19, 20] developed a meshfree method and applied it for the solution of the partial differential equation. Dehghan and Shokri [21] proposed a numerical scheme for the two-dimensional sine-Gordon equation using radial basis function (RBF) approximation. The numerical method of telegraph equations with constant coefficients using FDM for the time derivative and RBFs for space derivative was given by Dehghan and Shokri [22]. Jiang et al. [23] gave a numerical solution for the time-dependent partial differential equations using the meshfree approach. Jamil et al. [24] used the meshfree RBF collocation approach to analyze heat transfer's effect inside the skin tissue. Simulation of the two-dimensional human eye model for transient bioheat transfer by HFS-FEM approach coupled with the RBF approximation was given by Zhang et al. [25]. Recently, Verma and Kumar [26] did a numerical study on heat transfer in the skin with two types of boundary conditions: constant heat flux and sinusoidal heat flux using the radial basis function (RBF) approximation.

A numerical analysis of the two-dimensional Pennes bioheat transfer model with exponential heat flux condition is presented in the present study. We use the Crank–Nicolson (C-N) finite difference approximation and Gaussian radial basis function, respectively, to approximate time and space derivatives. The applied C-N scheme is unconditionally stable for the diffusion problem and provides the freedom for the step size in the time direction. The Gaussian RBF is a purely meshless approach, so the mesh generation in space direction is not required. This proposed method is easy to implement in multidimensional as well as in complex geometry. We analyze the influence of thermophysical parameters ω, K, and η involved in the model on heat transfer in tissue.

The remainder of this paper is organized as follows: Sect. 2 contains a brief explanation of the models' governing equation, including initial and boundary conditions. Section 3 shows the numerical scheme for the proposed mathematical model using the RBF and Crank–Nicolson (C-N) FDM scheme for the spatial and temporal variables, respectively. Section 4 is devoted to the validation of the present numerical method with the analytical method. Section 5 analyzes the results and effect of parameters involved in exponential heat flux condition on the skin surface. Finally, Sect. 6 includes the conclusion that summarizes the results; and at last, references cited in the study are listed.

2 Mathematical Model

Figure 1 represents a schematic of a two-dimensional rectangular domain of the skin tissue with dimension $L \times W$, where $L = 0.03\,\text{m}$ is length of the tissue along x-direction and $W = 0.03\,\text{m}$ is taken as the skin thickness along y-direction. As shown, the left wall ($x = 0$) of the rectangular domain has been maintained at normal body temperature, and adiabatic conditions are used at other sides of the skin tissue. In this study, the z-directional heat flow has been neglected.

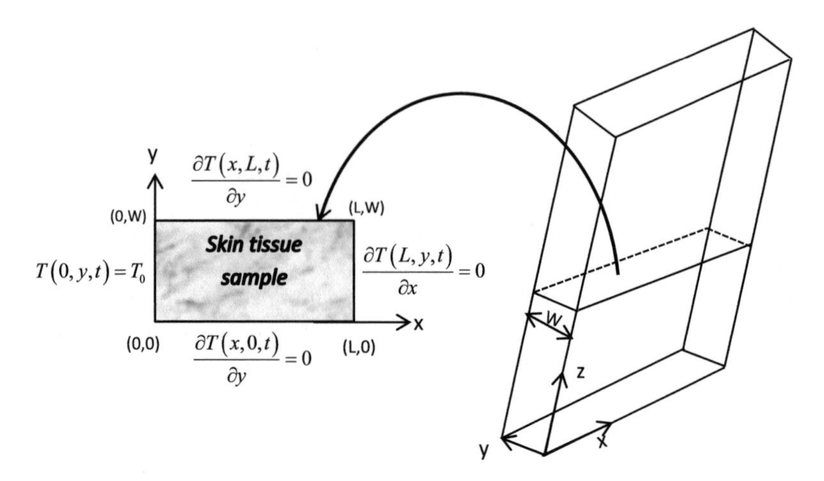

Fig. 1 A schematic diagram of 2D domain

2.1 Governing Equation

The governing equation for the heat transfer in biological tissue is given as [3]

$$\rho c \frac{\partial T}{\partial t} = K \left(\frac{\partial^2 T}{\partial x^2} + \frac{\partial^2 T}{\partial y^2} \right) + \rho_b c_b w_b (T_a - T) + Q_{\text{met}} + Q_{\text{ext}}, \tag{2.1}$$

where ρ, T, c, x and t symbolize the density, temperature, specific heat, distance, and time, respectively. The subscript b is used for the blood; and T_a, w_b, Q_{ext}, and Q_{met} are notations for arterial blood temperature, blood perfusion rate, external heat sources, and metabolic heat generation, respectively.

2.2 Initial Condition

Initial condition is given as

$$T(x, y, 0) = T_0. \tag{2.2}$$

2.3 Boundary Conditions

Boundary conditions are given as

$$T\,(0,\,y,\,t) = T_0,\ \ \frac{\partial T(L,y,t)}{\partial x} = 0,\qquad \frac{\partial T(x,0,t)}{\partial y} = 0,\ \ \frac{\partial T(x,W,t)}{\partial y} = 0. \tag{2.3}$$

3 Numerical Scheme

The domain $[0,\,L] \times [0,\,W] \times (0,\,t_f)$, t_f being the simulation time, is divided into equally spaced nodes $x_j = x_0 + j\Delta x,\ y_j = y_0 + j\Delta y,\ j = 1, 2, 3, \ldots, N$ and $t_n = n\Delta t, r = 1, 2, 3, \ldots, K$, where $\Delta x,\ \Delta y$, and Δt are step size in space $(x,\,y)$ and time t, respectively.

Rewriting Eq. (2.1) as

$$F_1\frac{\partial T}{\partial t} = K\left(\frac{\partial^2 T}{\partial x^2} + \frac{\partial^2 T}{\partial y^2}\right) - F_2T + F_3, \tag{3.1}$$

where $F_1 = \rho c$, $F_2 = \rho_b c_b w_b$ and $F_3 = \rho_b c_b w_b T_a + Q_{\text{ext}} + Q_{\text{met}}$. The approximation of the time derivative in Eq. (3.1), using the Crank–Nicolson (C-N) approximation at node $(x,\,y,\,t_n)$ gives

$$\begin{aligned}&\left(F_1 + \frac{1}{2}F_2(\Delta t)\right)T^{n+1}(x,\,y) - \frac{1}{2}K(\Delta t)\left(T_{xx}^{n+1}(x) + T_{yy}^{n+1}(y)\right)\\ &= \left(F_1 - \frac{1}{2}F_2\,(\Delta t)\right)T^n(x,\,y) + \frac{1}{2}K\,(\Delta t)\left(T_{xx}^n(x) + T_{yy}^n(x)\right) + (\Delta t)F_3.\end{aligned} \tag{3.2}$$

Equation (3.2) can be written as

$$\begin{aligned}G_1T^{n+1}(x,\,y) - G_2\left(T_{xx}^{n+1}(x) + T_{yy}^{n+1}(y)\right) = G_3T^n(x,\,y) + G_4\left(T_{xx}^n(x) + T_{yy}^n(y)\right)\\ + (\Delta t)F_3,\end{aligned} \tag{3.3}$$

where

$$\begin{aligned}&T^n(x,\,y) = T(x,\,y,\,t_n),\quad G_1 = F_1 + \frac{1}{2}F_2(\Delta t),\quad G_2 = \frac{1}{2}K(\Delta t),\\ &G_3 = F_1 - \frac{1}{2}F_2(\Delta t),\quad G_4 = \frac{1}{2}K(\Delta t).\end{aligned}$$

Assuming that there are total $(N \times N)$ interpolation points, So $T^n(x, y)$ is approximated using RBFs as

$$T^n(x, y) = \sum_{i=1}^{(N \times N)} \lambda_i^n \phi(r_i) + \sum_{k=1}^{M} \mu_k^n p_k(x, y) \quad x, y \in R^d, \tag{3.4}$$

where $\lambda_i^n = \lambda_i(t_n)$, $r_i = \sqrt{(x - x_i)^2 + (y - y_i)^2}$ and $\mu_k^n = \mu_k(t_n)$. For getting extra M condition, following constraints [27, 28] are imposed

$$\sum_{i=1}^{(N \times N)} \lambda_i^{n+1} p_k(x_i, y_i) = 0, \quad k = 1, 2, \ldots, M. \tag{3.5}$$

Approximating (3.3) using (3.4), (3.5) and collocating it at internal points (x_j, y_j), $j = 2, \ldots, N - 1$ we get the following linear system of equations

$$(G_1\mathbf{S} - G_2\mathbf{T})[\Lambda]^{n+1} = (G_3\mathbf{S} + G_4\mathbf{T})[\Lambda]^n + (\Delta t)F_4, \tag{3.6}$$

where

$$\mathbf{S} = \begin{pmatrix} \phi_{11} & \cdots & \phi_{(1)(NN)} & p_{11} & \cdots & p_{(1)(M)} \\ \vdots & \ddots & \vdots & \vdots & \ddots & \vdots \\ \phi_{(N-2)(N-2)(1)} & \cdots & \phi_{(N-2)(N-2)(NN)} & p_{(N-2)(N-2)(1)} & \cdots & p_{(N-2)(N-2)(M)} \\ p_{11} & \cdots & p_{(NN)(1)} & 0 & \cdots & 0 \\ \vdots & \ddots & \vdots & \vdots & \ddots & \vdots \\ p_{1M} & \cdots & p_{NM} & 0 & \cdots & 0 \end{pmatrix},$$

$$[\Lambda]^n = \left[\lambda_1^n, \lambda_2^n, \ldots \lambda_{NN}^n, \mu_1^n, \mu_2^n, \ldots \mu_M^n\right]',$$

$\phi_{ji} = \phi(r_{ji})$, $r_{ji} = \sqrt{(x_j - x_i)^2 + (y_j - y_i)^2}$, $p_{ji} = p_i(x_j, y_j)$, and $\mathbf{T} = [(s_{ij})_{xx} + (s_{ij})_{yy}]$.

To incorporate the boundary conditions at surface $x = 0$, $x = L$ and $y = 0$, $y = W$, Eq. (2.3) is approximated using Eq. (3.4), which gives a linear system of $(4N - 4)$ equations in $(M + (N \times N))$ unknowns $[\Lambda]$ at time level t_n, $n = 2, 3, \ldots, t_{K-2}$.

Equation (3.6) and boundary conditions form a linear system of $((N \times N) + M)$ equations in $(M + (N \times N))$ unknowns $[\Lambda]$ at time level t_n, $n = 2, 3, \ldots, t_{K-2}$. The value of $[\Lambda]^0$ is obtained using initial condition (Eq. 2.2).

4 Computer Code Validation

To validate the proposed scheme and computer code, we compare the numerical solution of the parabolic model considering $Q_{\text{ext}} = 0$ and $Q_{\text{met}} = 0$ with the analytical solution given by Shih et al. [29] for the parameters given in Sect. 5. The temperature profile for constant heat flux ($q(0, t) = q_0$) with respect to time is plotted in Fig. 2. The temperature profile shown in Fig. 2 is favorable enough for the reliability of the proposed numerical method and developed code.

5 Results and Discussion

The parameter values used are listed in Table 1 [29, 30]. The heat absorption within the tissue, heating by laser, ultrasound, or microwave, can truly be approximated by Beer's law as,

$$Q_{\text{ext}} = \eta P_0 (t) \exp\left[-\eta (L - x)\right]. \tag{5.1}$$

where $\eta = 200$, is the scattering coefficient and $P_0 (t) = 2.4 \times 10^4 \times \sin(\omega t)$, is the time-depending heating power on skin surface, where $0 \leqslant x \leqslant L, 0 \leqslant y \leqslant W$.

Here, we study the heat transfer on skin tissue for thermal therapy with the Pennes bioheat transfer model. Figure 3 shows the temperature profile with respect to distance at time $t = 0$ s, $t = 62.4$ s, $t = 96.8$ s, $t = 185.6$ s, $t = 313.6$ s, and $t = 400$ s. This study shows the impact of exponential heat flux conditions. We see that in Fig. 3, the maximum temperature fluctuates due to exponential heat flux. But the temperature fluctuation is not more than 66.0596°C. We analyze that, initially, the temperature increases with time, and then it decreases slowly and again increases with time. This process repeats with time.

Figure 4 shows the distribution of the temperature with respect to time at skin tissue positions (0.0126 m, 0.0126 m), (0.0142 m, 0.0142 m) and (0.0158 m, 0.0158 m). In the Fig. 4, we see that the maximum temperature is 39.72°C, 40.38°C and 41.17°C

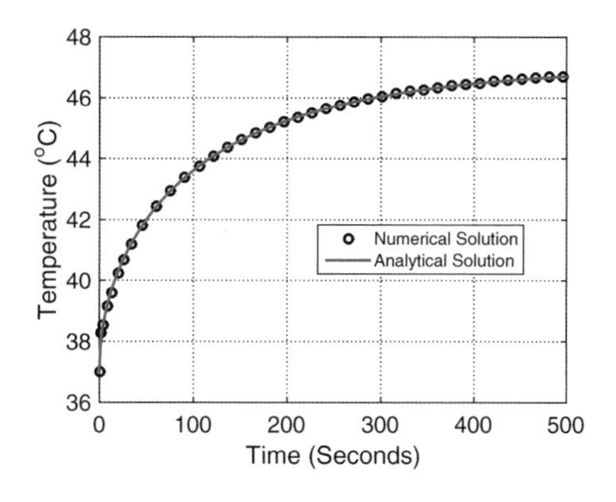

Fig. 2 Temperature profile for present numerical method and analytical method [29]

Table 1 Thermophysical parameters [29, 30]

Parameters	Values and units
Arterial blood temperature (T_a)	37°C
Length of the tissue along x-axis (L)	0.03 m
Length of the tissue along y-axis (W)	0.03 m
Density of tissue (ρ)	1050 kg/m^3
Density of blood (ρ_b)	1050 kg/m^3
Blood perfusion (W_b)	0.5 kg/m^3 s
Thermal conductivity of tissue (K)	0.5 W/(m °C)
Specific heat of the blood (c_b)	3770 J/(kg °C)
Specific heat of the tissue (c)	4180 J/(kg °C)
Metabolic heat generation (Q_{met})	420 W/m^3
Heating frequency (ω)	0.05/s

at skin tissue position (0.0126 m, 0.0126 m), (0.0142 m, 0.0142 m), and (0.0158 m, 0.0158 m), respectively. We observe that the temperature increases with the skin tissue depth.

In Fig. 5, we show the effects of the scattering coefficient (η) on temperature distribution in skin tissue. We see that maximum temperature in tissue is 39.72°C, 40.36°C, and 41.15°C for $\eta = 200$, $\eta = 150$, and $\eta = 100$, respectively at skin tissue position (0.0126 m, 0.0126 m). We observe that as the value of η decreases, the temperature increases.

Figure 6 shows the temperature distribution with respect to time at skin tissue position (0.0126 m, 0.0126 m) for heating frequency (ω). Here, we consider the different values of heating frequency $\omega = 0.05$, $\omega = 0.04$, $\omega = 0.03$, and $\omega = 0.02$ with fixed value $\eta = 200$. In Fig. 6, we see that amplitude of wave increases as the value of heating frequency decreases. Also, we observe in Fig. 6 that the maximum temperature is 39.72°C, 40.32°C, 41.19°C, and 43.71°C for $\omega = 0.05$, $\omega = 0.04$, $\omega = 0.03$, and $\omega = 0.02$, respectively, i.e., the wave attains a higher temperature on decreasing the heating frequency ω.

Thermal conductivity is an important thermophysical property and plays a major role in the study of thermal therapy. To show the effect of thermal conductivity, temperature with time at skin tissue position (0.0095 m, 0.0095 m) is plotted in Fig. 7 for $K = 0.5$, $K = 0.4$, $K = 0.3$, and $K = 0.2$ with $\omega = 0.05$ and $\eta = 200$. Figure 7 shows that the thermal wave gradually increases as time proceeds. Also, the temperature is increasing with an increase in thermal conductivity. The highest temperature (38.66°C) is attained by the thermal wave for $K = 0.5$.; i.e., the temperature is directly proportional to the thermal conductivity.

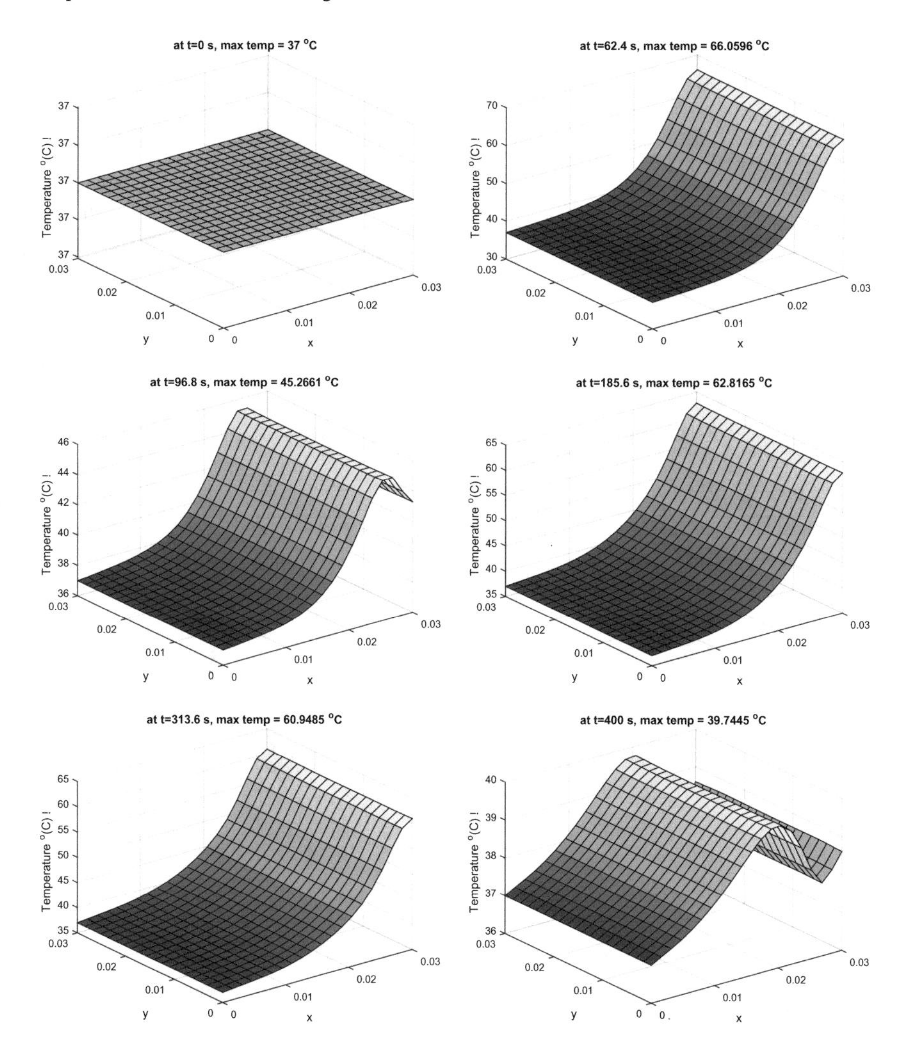

Fig. 3 Temperature profile with exponential heat flux at different time

6 Conclusion

The numerical modeling and simulation of two-dimensional bioheat models for heat distribution in skin tissues with exponential heat flux conditions were presented in the study. The mathematical model's numerical solution is obtained using Crank–Nicolson finite difference approximation and radial basis function approximation for time and space, respectively. Due to exponential heating (Eq. 5.1), the temperature profile of Pennes' bioheat model behaves like a sinusoidal wave. The effects of thermophysical properties of the skin on the temperature profile in the tissue were

Fig. 4 Temperature profile at different points (x, y)

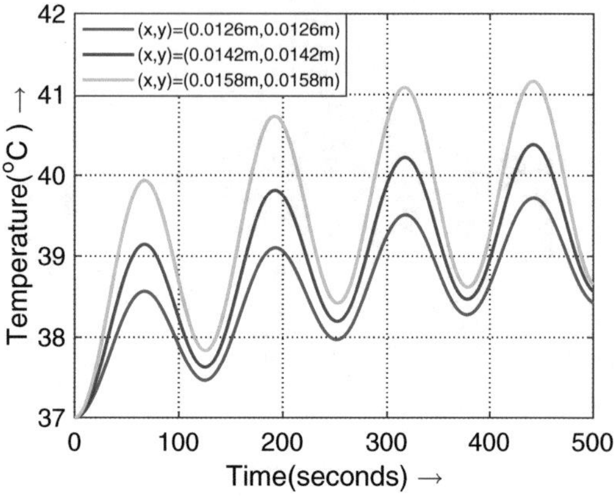

(x,y)=(0.0126m,0.0126m)
(x,y)=(0.0142m,0.0142m)
(x,y)=(0.0158m,0.0158m)
Temperature(°C) →
Time(seconds) →
42
41
40
39
38
37
0
100
200
300
400
500

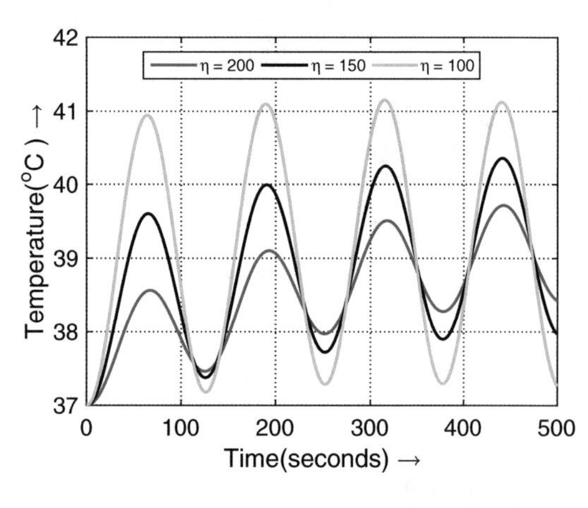

Fig. 5 Effect of η on temperature distribution at points (0.0126, 0.0126)

η = 200
η = 150
η = 100
Temperature(°C) →
Time(seconds) →
42
41
40
39
38
37
0
100
200
300
400
500

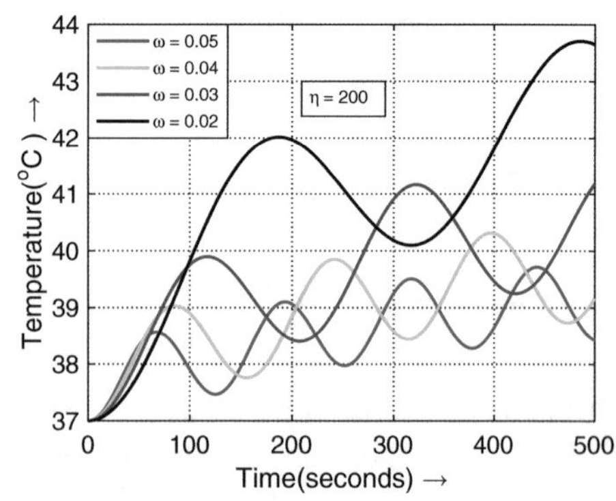

Fig. 6 Effect of ω with exponential heat flux at points (0.0126, 0.0126)

ω = 0.05
ω = 0.04
ω = 0.03
ω = 0.02
η = 200
Temperature(°C) →
Time(seconds) →
44
43
42
41
40
39
38
37
0
100
200
300
400
500

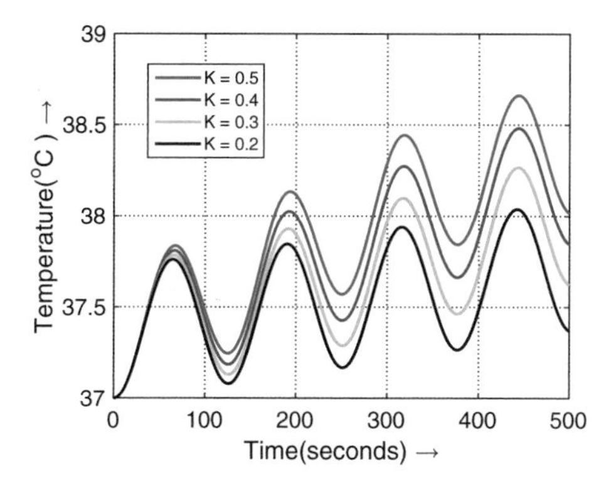

Fig. 7 Effect of thermal conductivity on temperature at points (0.0095, 0.0095)

also discussed. It was observed that as the value of heating frequency (ω) decreases, the temperature increases. Also, the temperature is observed to be inversely proportional to the η. The present study may be effective in predicting the bioheat model's temperature behavior and tissue damage.

Acknowledgements Authors are thankful to Science and Engineering Research Board, India, for providing the financial support for this study via project no. ECR/2017/003174.

References

1. Waterman FM, Tupchong L, Nerlinger RE, Matthews J (1991) Blood flow in human tumors during local hyperthermia. Int J Radiat Oncol Biol Phys 20(6):1255–1262
2. Xu F, Lu TJ (2011) Introduction to skin biothermomechanics and thermal pain. Springer Science Press, Beijing
3. Pennes HH (1948) Analysis of tissue and arterial blood temperatures in the resting human forearm. J Appl Physiol 1(2):93–122
4. Khafaji M, Zamani M, Golizadeh M, Bavi O (2019) Inorganic nanomaterials for chemo/photothermal therapy: a promising horizon on effective cancer treatment. Biophys Rev 11(3):335–352
5. Ahmadikia H, Fazlali R, Moradi A (2012) Analytical solution of the parabolic and hyperbolic heat transfer equations with constant and transient heat flux conditions on skin tissue. Int Commun Heat Mass Transfer 39(1):121–130
6. Deng Z-S, Liu J (2002) Analytical study on bioheat transfer problems with spatial or transient heating on skin surface or inside biological bodies. ASME J Biomech Eng 124(6):638–649
7. Azevedo M, Guedes R, Neto FS (2006) Analytical solution to the two dimensional transient bioheat equation with convective boundary conditions. In: Proceedings of the 11th Brazilian congress of thermal sciences and engineering. ABCM, Curitiba, PR, Brazil
8. Biswas P, Singh S, Srivastava A (2020) A unique technique for analytical solution of two dimensional dual phase lag bio-heat transfer problem with generalized time-dependent boundary conditions. Int J Therm Sci 147:106139
9. Dutta J, Kundu B (2020) Two-dimensional hybrid analytical approach for the investigation of thermal aspects in human tissue undergoing regional hyperthermia therapy. Proc Inst Mech Eng Part C J Mech Eng Sci 234(20):3951–3966

10. Ezzat MA (2021) Analytical study of two-dimensional thermo-mechanical responses of viscoelastic skin tissue with temperature-dependent thermal conductivity and rheological properties. Mech Based Design Struct Mach 1–18
11. Ozen S, Helhel S, Cerezci O (2008) Heat analysis of biological tissue exposed to microwave by using thermal wave model of bio-heat transfer (TWMBT). Burns J Int Soc Burn Injuries 34(1):45–49
12. Karaa S, Zhang J, Yang F (2005) A numerical study of a three dimensional bioheat transfer problem with different spatial heating. Math Comput Simul 68(4):375–388
13. Al-Humedi HO, Al-Saadawi FA (2021) The numerical solution of bioheat equation based on shifted Legendre polynomial. Int J Nonlinear Anal Appl 12(2):1061–1070
14. Fahmy MA (2019) Boundary element modeling and simulation of biothermomechanical behavior in anisotropic laser-induced tissue hyperthermia. Eng Anal Bound Elem 101:156–164
15. Hosseininia M, Heydari MH, Roohi R, Avazzadeh Z (2019) A computational wavelet method for variable-order fractional model of dual phase lag bioheat equation. J Comput Phys 395:1–18
16. Roohi R, Heydari M, Aslami M, Mahmoudi M (2018) A comprehensive numerical study of spacetime fractional bioheat equation using fractional-order Legendre functions. Eur Phys J Plus 133(10):412
17. Stroher GR, Santiago CD (2020) Numerical two-dimensional thermal analysis of the human skin using the multigrid method. Acta Scientiarum Technol 42:e40992–e40992
18. Liu G (2009) Meshfree methods: moving beyond the finite element method, 2nd edn. CRC Press, New York
19. Kansa EJ (1990) Multiquadrics—a scattered data approximation scheme with applications to computational fluid-dynamics I surface approximations and partial derivative estimates. Comput Math Appl 19(8–9):127–145
20. Kansa EJ (1990) Multiquadrics—a scattered data approximation scheme with applications to computational fluid-dynamics II solutions to parabolic, hyperbolic and elliptic partial differential equations. Comput Math Appl 19(8–9):147–161
21. Dehghan M, Shokri A (2008) A numerical method for solution of the two-dimensional sine-Gordon equation using the radial basis functions. Math Comput Simul 79(3):700–715
22. Dehghan M, Shokri A (2008) A numerical method for solving the hyperbolic telegraph equation. Numer Methods Partial Differ Equ Int J 24(4):1080–1093
23. Jiang Z, Su L, Jiang T (2014) A meshfree method for numerical solution of nonhomogeneous time-dependent problems. In: Abstract and applied analysis, vol 2014. Hindawi, pp 1–11
24. Jamil M, Ng E (2013) Evaluation of meshless radial basis collocation method (RBCM) for heterogeneous conduction and simulation of temperature inside the biological tissues. Int J Therm Sci 68:42–52
25. Zhang ZW, Wang H, Qin QH (2014) Analysis of transient bioheat transfer in the human eye using hybrid finite element model. In: Applied mechanics and materials, vol 553. Trans Tech Publ, pp 356–361
26. Verma R, Kumar S (2020) Computational study on constant and sinusoidal heating of skin tissue using radial basis functions. Comput Biol Med 121:103808
27. Kansa E, Hon Y (2000) Circumventing the ill-conditioning problem with multiquadric radial basis functions: applications to elliptic partial differential equations. Comput Math Appl 39(7–8):123–137
28. Hardy RL (1990) Theory and applications of the multiquadric-biharmonic method 20 years of discovery 1968–1988. Comput Math Appl 19(8–9):163–208
29. Shih T-C, Yuan P, Lin W-L, Kou H-S (2007) Analytical analysis of the Pennes bioheat transfer equation with sinusoidal heat flux condition on skin surface. Med Eng Phys 29(9):946–953
30. Damor R, Kumar S, Shukla A (2013) Numerical solution of fractional bioheat equation with constant and sinusoidal heat flux condition on skin tissue. Am J Math Anal 1(2):20–24

Correction to: Convergence and Comparison Theorems for Three-Step Alternating Iteration Method for Rectangular Linear System

Smrutilekha Das, Debadutta Mohanty, and Chinmay Kumar Giri

Correction to: Chapter "Convergence and Comparison Theorems for Three-Step Alternating Iteration Method for Rectangular Linear System" in: S. S. Ray et al. (eds.), *Applied Analysis, Computation and Mathematical Modelling in Engineering*, Lecture Notes in Electrical Engineering 897, https://doi.org/10.1007/978-981-19-1824-7_10

In the original version of the book, the following correction has been incorporated: In Chapter 10, the second author's affiliation has been changed from "Department of Mathematics, National Institute of Science and Technology, Berhampur, Odisha 761008, India" to "Department of Mathematics, Seemanta Mahavidyalaya, Jharpokharia, Odisha, 757086, India".

The correction chapter and the book have been updated with the change.

The updated original version of this chapter can be found at
https://doi.org/10.1007/978-981-19-1824-7_10

S. S. Ray et al. (eds.), *Applied Analysis, Computation and Mathematical Modelling in Engineering*, Lecture Notes in Electrical Engineering 897,
https://doi.org/10.1007/978-981-19-1824-7_25

Printed in the United States
by Baker & Taylor Publisher Services